Environmental Contamination Remediation and Management

There are many global environmental issues that are directly related to varying levels of contamination from both inorganic and organic contaminants. These affect the quality of drinking water, food, soil, aquatic ecosystems, urban systems, agricultural systems and natural habitats. This has led to the development of assessment methods and remediation strategies to identify, reduce, remove or contain contaminant loadings from these systems using various natural or engineered technologies. In most cases, these strategies utilize interdisciplinary approaches that rely on chemistry, ecology, toxicology, hydrology, modeling and engineering.

This book series provides an outlet to summarize environmental contamination related topics that provide a path forward in understanding the current state and mitigation, both regionally and globally.

Topic areas may include, but are not limited to, Environmental Fate and Effects, Environmental Effects Monitoring, Water Re-use, Waste Management, Food Safety, Ecological Restoration, Remediation of Contaminated Sites, Analytical Methodology, and Climate Change.

Sesan Abiodun Aransiola
Babafemi Raphael Babaniyi
Naga Raju Maddela
Olugbenga Solomon Bello
Editors

White Pollution: Biodiversity and hazards in marine plastisphere

Springer

Editors
Sesan Abiodun Aransiola
Dep. of Microbiology
University of Abuja
Abuja, Nigeria

Babafemi Raphael Babaniyi
Bioresources Development Centre
Natl Biotechnology Development Agency
Ogbomoso, Nigeria

Naga Raju Maddela
Instituto de Investigación
Universidad Tecnica de Manabi
Portoviejo, Ecuador

Olugbenga Solomon Bello
Ladoke Akintola University
Ogbomoso, Nigeria

ISSN 2522-5847 ISSN 2522-5855 (electronic)
Environmental Contamination Remediation and Management
ISBN 978-3-031-95546-4 ISBN 978-3-031-95547-1 (eBook)
https://doi.org/10.1007/978-3-031-95547-1

This Springer imprint is published by the registered company Springer Nature Switzerland AG
The registered company address is: Gewerbestrasse 11, 6330 Cham, Switzerland

Foreword

Plastic debris has become a ubiquitous contaminant and the contamination is a persistent environmental global problem. This plastic debris is harmful to ocean life, soil indwellers, and the atmospheric environment. Also, plastic pollution affects human health. Although it is well known that the majority of the plastic trash detected comes from land-based sources, there is a need to enlighten the global audience about recent happening in the world of plastisphere pollution. The questions are: do microorganisms have a role to play in plastisphere formation? Does formation of plastisphere affect the microbial flora of the soil which helps the plants to grow? If yes, does its presence in soil affects plants and cause a global food crisis? How this menace impacts the aquatic and atmospheric lives?

White pollution, also known as plastic pollution, is posing a significant threat to biodiversity around the world. This type of pollution refers to the accumulation of discarded plastic products in the environment, particularly in marine and terrestrial ecosystems. It is a pressing environmental issue that requires urgent action to mitigate its impact on biodiversity.

White pollution poses a direct threat to various forms of life, including marine animals, birds, and terrestrial wildlife. Animals can become entangled in plastic debris, leading to injury or death. Additionally, ingestion of plastic can cause internal injuries and blockages in the digestive system, ultimately resulting in starvation and death. Furthermore, the presence of plastic pollution in natural habitats can alter ecosystems and disrupt the natural balance of biodiversity.

Detecting and monitoring white pollution is crucial for understanding its prevalence and impact on biodiversity. There are several methods used for detecting and assessing the extent of plastic pollution in the environment. These methods include visual surveys, remote sensing technologies, and the analysis of water and sediment samples.

Visual surveys involve physically inspecting and documenting the presence of plastic debris in the environment. This method is commonly used in coastal areas, beaches, and marine environments where plastic pollution is visible to the naked eye. Remote sensing technologies, such as satellite imagery and aerial surveys, can provide a larger-scale assessment of plastic pollution in marine and terrestrial habitats. These technologies allow for the identification and mapping of areas affected by plastic pollution, providing valuable data for conservation efforts.

Furthermore, the analysis of water and sediment samples can provide insights into the extent of plastic pollution in aquatic environments. By collecting and analyzing samples from water bodies and riverbeds, researchers can quantify the concentration of plastic particles and identify sources of pollution. This information is essential for developing targeted strategies to mitigate the impact of plastic pollution on biodiversity. White pollution poses a grave threat to biodiversity, and detecting its prevalence is essential for effective conservation efforts. This knowledge is crucial for developing and implementing strategies to reduce plastic pollution and safeguard the diverse forms of life that depend on healthy and unpolluted ecosystems

White Pollution: Biodiversity and hazards in marine plastisphere is a good collection of independent chapters which presents full insights into the biodiversity and hazards of plastic pollution in the marine environment. In an extensive form, this book focuses on an insight to prevalence and distribution of plastisphere in the marine environment, the effects of marine white pollution on higher animals, plants, and the food chain, characterization of biodegradable plastic-producing bacteria isolates from Atlantic Ocean, interconnectedness of plastisphere, marine snow, and blue carbon: ecological implications and policy perspectives, and biodiversity and hazards in marine plastisphere in polar regions. Other critically reviewed topics include microbial ecology of the plastisphere in soil plant uptake; microbial relationship of marine microplastics, plastisphere, and soil biodiversity; plastisphere and environmental sustainability; white pollution biodiversity and methods of detection; white pollution and air quality; health crises through white pollution exposure; characterization technologies of white pollution in atmosphere; policies and regulations of white pollution in the environment; and climate change and white pollution as a result of renewable and non-renewable energy consumption in growing economies.

I therefore have no doubt that the above focused areas provide adequate information on white pollution biodiversity and the hazard in marine plastisphere do fill the expected scientific knowledge gaps.

Vice Director of Beijing Laboratory of Smart Environmental Protection, Faculty of Information Technology
Beijing University of Technology,
Beijing, China
April 10, 2024

Jian Tang

Preface

The term white pollution describes the various types of large number of plastic products in the ecosystem worldwide. Because plastic materials and their by-products are difficult to degrade, they remain constantly present in the soils, water bodies, and in the atmosphere. Persistent plastic materials undergo fragmentation and dispersion through currents and wind, which also leads to their limited degradation. As a result, marine, soil, and atmospheric environments worldwide contain an abundance of plastic particles. In contrast to the effects of ingestion or entanglement observed in larger organisms, the interaction between microorganisms and plastics is distinctly different. Plastics serve as habitats and swiftly attract dense biofilms composed of microorganisms on their surfaces, known as the "plastisphere." This plastisphere represents a portion of the global ecosystem characterized by floating plastic debris and the associated microbes and other organisms.

Plastisphere may deplete phosphorus, stripping it away from other organisms. Additionally, the plastisphere could act as a reservoir for the transmission of human diseases since fish and shellfish consume minute plastic fragments harboring *Vibrio* and cholera-like bacteria, leading to foodborne illnesses. Due to the long persistence of plastic, microbes within the plastisphere can be transported over significant distances, presenting a potential source of invasive species. If these microbes are transferred across various ecosystems within the ocean, air, and soil, they could impact native microbial populations and the larger organisms that depend on them. Additionally, the plastisphere has the potential to modify plastic debris, which can alter its impact on an ecosystem, making it either more or less harmful. As a society, we acknowledge that plastic marine debris poses a predicament. However, we lack solutions to even the most basic inquiries, such as the quantity and nature of plastic present in the ocean, soils, air, and their origin. The microorganisms that constitute the plastisphere may influence the destiny and repercussions of plastic within marine ecosystems. Microscopic and molecular sequence data demonstrate that the plastisphere is composed of primary producers, heterotrophs, symbionts, and predators. The interactions between phytoplankton and bacteria are crucial for facilitating the ecological cycles of the Earth and shaping the structure of food webs in the oceans. This interconnection between autotrophic organisms and other microorganisms is also evident within the plastisphere.

Going by the available facts and the global trends of plastics production, this book titled *White Pollution: Biodiversity and hazards in marine plastisphere* publication was conceived to address the new emerging occurrence, distribution and degradation of marine white plastisphere, plastisphere in soil plant uptake and the role in food security and the sources, prevalence and effects of the atmospheric plastisphere to humans and other organisms.

Therefore, this book is divided into three parts with five chapters each. The division of the volume include Part I: An insight to prevalence and distribution of plastisphere in the marine environment; the effects of marine white pollution on higher animals, plants, and the food chain; characterization of biodegradable plastic-producing bacteria isolates from Atlantic Ocean; interconnectedness of plastisphere, marine snow, and blue carbon; ecological implications and policy perspectives; and biodiversity and hazards in marine plastisphere in polar regions. Part II: **Plastisphere in soil plant uptake and food crisis** has the following topics: microbial ecology of the plastisphere in soil plant uptake, microbial relationship of marine microplastics, plastisphere and soil biodiversity, plastisphere and environmental sustainability and white pollution biodiversity, and methods of detection. Part III: **Sources, prevalence, and effects of the atmospheric plastisphere** was designed to include white pollution and air quality, health crises through white pollution exposure, characterization technologies of white pollution in atmosphere, policies and regulations of white pollution in the environment, and climate change and white pollution as a result of renewable and non-renewable energy consumption in growing economies.

The chapters were contributed by 63 academicians/scientists/researchers from 15 different countries (UK, Nigeria, India, Ghana, USA, Malaysia, Italy, Algeria, Sweden, Turkey, Ecuador, Syrian Arab Republic, South Africa, Saudi Arabia, and China).

Abuja, Nigeria — Sesan Abiodun Aransiola
Ogbomoso, Nigeria — Babafemi Raphael Babaniyi
Portoviejo, Ecuador — Naga Raju Maddela
Ogbomoso, Nigeria — Olugbenga Solomon Bello

Acknowledgments

With great honor, we acknowledge the support of the chapter's contributors for their valuable contributions and timely responses to the success of this volume. All contributors are immensely appreciated for their eagerness and inordinate support for this volume to be ready in the scheduled time. We, therefore, appreciate their teamwork and partnership. We also thank anonymous reviewers for their constructive criticism, which has helped us in improving the quality of this book by inviting the experts to contribute additional chapters. We greatly acknowledge the Springer Editorial and Production team for their support; without their guidelines, this project would not have been finished in such a very short time. It is an honest honor and privilege to work with them all. Finally, yet importantly, we are very much thankful to the management and colleagues at University of Abuja (Nigeria), National Biotechnology Development Agency (Nigeria), Universidad Técnica de Manabí (Ecuador), and Ladoke Akintola University of Technology (Nigeria) for their unrestricted backing and for the establishment of treasured propositions at the time of book proposal and final book preparation.

University of Abuja
Abuja, Nigeria
Sesan Abiodun Aransiola

National Biotechnology Development Agency (NABDA)
Ogbomoso, Nigeria
Babafemi Raphael Babaniyi

Universidad Técnica de Manabí
Portoviejo, Ecuador
Naga Raju Maddela

Ladoke Akintola University of Technology
Ogbomoso, Nigeria
Olugbenga Solomon Bello

Contents

Part I Occurrence, Distribution and Degradation of Marine White Plastisphere

1 An Insight into the Prevalence and Distribution of Plastisphere in the Marine Environment 3
Samuel Oluwatobi Agboola, Temitope Chris Alagbada, Gladys Mercy Inetabor, Toluwanimi Rachael Arowolo, Kayode Adesina Adegoke, Mohd Azmier Ahmad, and Olugbenga Solomon Bello

2 The Effects of Marine White Pollution on Higher Animals, Plants, and the Food Chain 35
Innocent Ojeba Musa, Abdulhakeem Idris Abdulhakeem, Mustapha Abdulsalam, Miracle Uwa Livinus, Olaitan Lateefat Salam, Shehu-Alimi Elelu, Auwal Sagir Muhammad, Job Oloruntoba Samuel, and Adamu Mustapha

3 Characterization of Biodegradable Plastic-Producing Bacteria Isolates from Atlantic Ocean 53
Scott O. Fayemi, Bunmi Kotun, O. A. Fatunbi, O. Oda, and J. Ayedun

4 Interconnectedness of Plastisphere, Marine Snow and Blue Carbon: Ecological Implications and Policy Perspectives 77
Ojo Emmanuel Ige, Adesuyi Emmanuel Omole, and Babafemi Raphael Babaniyi

5 Biodiversity and Hazards in Marine Plastisphere in Polar Region . . 99
Miracle Uwa Livinus, Sunday Zeal Bala, Mustapha Abdulsalam, Innocent Ojeba Musa, Madinat Hassan, and Priscilla Kini

Part II Plastisphere in Soil Plant Uptake and Food Crisis

6 **Microbial Ecology of the Plastisphere in Soil Plant Uptake** 125
Gloria Ginikanwa Ezeagu, Morenike Fadayomi, Sani Sambo Datsugwai Mohammed, Augustine Innalegwu Daniel, Ashwil Klein, and Namla Djadjiti

7 **Microbial Relationship of Marine Microplastics** 145
Sarah Chioma Nwakanma, Sesan Abiodun Aransiola, Emmanuel Kuyet Adejokun, Ojonoma Loretta Okwute, Loai Aljerf, and Naga Raju Maddela

8 **Plastisphere and Soil Biodiversity** 165
Evelyn Nwadinkpa Fatokun, Kehinde Olubunmi Adediran, and Christy Echakachi Manyi-Loh

9 **Plastisphere and Environmental Sustainability** 185
Samuel Oluwatobi Agboola, Oluwakemi Christiana Alao, Gideon Oluranti Ojediran, Adedamola Titi Ojedokun, Oluwatosin Stephen Ajayi, Kayode Adesina Adegoke, Mohd Azmier Ahmad, and Olugbenga Solomon Bello

10 **White Pollution Biodiversity and Methods of Detection** 209
Emmanuel Kuyet Adejokun, Sylvester Chukwudi Ugoh, Sesan Abiodun Aransiola, Sarah Chioma Nwakanma, Nasser Belboukhari, Alissar AlJerf, and Naga Raju Maddela

Part III Sources, Prevalence and Effects of the Atmospheric Plastisphere

11 **White Pollution and Air Quality** 239
Gbolahan Olamide Isaac, Babafemi Raphael Babaniyi, Taiwo Oluwasola Abe, and Ademola Bisi-Omotosho

12 **Health Cries Through White Pollution Exposure** 261
Srinivasan Kameswaran and Bellamkonda Ramesh

13 **Characterization Technologies of White Pollution in the Atmosphere** 285
Mustapha Abdulsalam, Innocent Ojeba Musa, Miracle Uwa Livinus, Shehu-Alimi Elelu, Ganiyat Omotayo Ibrahim, Salami Olaitan Lateefat, and Auwal Sagir Muhammad

14 **Policies and Regulations of White Pollution in the Environment**.... 307
Gabriel Gbenga Babaniyi, Olusegun Opeyemi Oni, Ulelu Jessica Akor, and Oluwatosin Emmanuel Daramola

15 **Climate Change and White Pollution as a Result of Renewable and Non-renewable Energy Consumption in Growing Economies** .. 325
Hakan Acaroğlu, Atabak Najafi, and Mustafa Güllü

Editors and Contributors

About the Editors

Sesan Abiodun Aransiola is a Ph.D. holder in Environmental Microbiology in the Department of Microbiology, Federal University of Technology, Minna, Nigeria. He obtained his first degree (B.Tech.) and Master Degree (M.Tech.) from the same department in 2009 and 2014, respectively. He has demonstrated his research expertise in the production of vermicasts from vermicoposting (vermitechnology) of organic wastes to assist plants in the remediation of polluted soil with heavy metals. Currently, he is a Lecturer at the Department of Microbiology, University of Abuja, Nigeria. His area of interest is Environmental Microbiology with research area in phytoremediation, vermicomposting, biosorption, and bioremediation of soil-contaminated environment. He is an award-winning researcher and has over 70 publications including book chapters, research, and review articles of good international repute with a high impact factor. He has co-edited scientific books of global interest which include *Microbial Biotechnology for Bioenergy* (Elsevier); *Ecological Interplays in Microbial Enzymology, Prospects for Soil Regeneration and Its Impact on Environmental Protection* (Springer); *Microbial Biofilms—Applications and Control, Marine Greens: Environmental, Agricultural, Industrial and Biomedical Applications* (CRC Press) with many more in the process. Also, has worked to the rank of an Assistant Chief Scientific Officer at Bioresources

Development Centre, National Biotechnology Development Agency, Nigeria, where he was involved in the use of vinasse (a by-product of ethanol), as bio-fertilizer to reclaim polluted soil for agricultural purposes. Abiodun is a member of the Nigerian Society for Microbiology and the American Society for Microbiology, among others.

Babafemi Raphael Babaniyi is a graduate of Industrial Chemistry (B.Sc.) at Caritas University, Amorji-Nike, Emene, Enugu State, Nigeria, in 2008. Later, he obtained his Master degree (M.Tech.) in Environmental Chemistry at Federal University of Technology, Akure, Ondo State, Nigeria, in 2021. Currently, he is a Ph.D. candidate in the Department of Chemistry (Environmental Chemistry), Federal University of Technology, Akure, Ondo State, Nigeria. He had over 25 publications including book chapters and articles of international repute. Additionally, he is an Assistant Chief Scientific Officer at Bioresources Development Centre, National Biotechnology Development Agency, Nigeria. His area of interest is Environmental Chemistry with research areas in Environmental Chemistry, Green Chemistry, Phytoremediation, Chemical Biology, Environmental Science, Bioremediation of Soil-Contaminated Environment, Plant Protection and Animal Health, Environmental Impact Assessment, and Biodegradable Plastics.

Naga Raju Maddela received his M.Sc. (1996–1998) and Ph.D. (2012) in Microbiology from Sri Krishnadevaraya University, Anantapuramu, India. During his doctoral program in the area of Environmental Microbiology, he investigated the effects of industrial effluents/insecticides on soil microorganisms and their biological activities, and he is working as Faculty in Microbiology since 1998, teaching undergraduate and postgraduate students. He worked on Prometeo Investigator (fellowship received from SENESCYT) at Universidad Estatal Amazónica, Ecuador, during 2013–15, and received "Postdoctoral Fellowship" (2016–2018) from Sun Yat-sen University, China. He also received external funding from "China Postdoctoral Science Foundation" in 2017 and internal funding from

"Universidad Técnica de Manabí" in 2020. He participated in national/international conferences, and presented research data in China, Cuba, Ecuador, India, and Singapore. Currently, he is working as a Full Professor at the Facultad de Ciencias de la Salud, Universidad Técnica de Manabí, Portoviejo, Ecuador. He has been actively publishing scientific articles, books (authored and edited), and chapters since 2007. As of now, he has published 70 articles, 12 books, and 40 book chapters.

Olugbenga Solomon Bello is a Professor of Physical and Environmental Chemistry in the Department of Pure and Applied Chemistry, Ladoke Akintola University of Technology, Ogbomoso, Oyo State, Nigeria. He completed his PhD in Physical Chemistry at the University of Ibadan, Nigeria, in 2008. He was the recipient of the 2019 Nigerian Academy of Science Gold Medal Award in Physical Sciences. He received the African Union Third World Academy of Science (AU-TWAS) Young Scientists National Award in 2012 and the Prof. Kayode Adebowale Young Scientist Prize in Chemical Sciences organized by the Nigerian Young Academy in 2016. He benefited from the USM-TWAS Postdoctoral and Visiting Scholar Fellowships at the School of Chemical Engineering University of Science Malaysia. He won TWAS research grant (2012 and 2015) and Tertiary Education Trust Fund (TETFUND) Institutional Based Research (IBR) in 2016. He was nominated and recognized as one of the "175 Faces of Chemistry" worldwide by the Royal Society of Chemistry, UK. He has published over 150 articles in peer-reviewed national and international journals of high repute. He was listed among the 2% most cited Scientists in the world by a group of Stanford University researchers. His research interest is focused on adsorption of heavy metals, dyes, pharmaceuticals, and pesticides from wastewaters using agricultural waste materials as adsorbents in lieu of expensive commercial activated carbon to bioremediate the environment. His accomplishment, as well as the significant impact of his research findings, has contributed immensely to environmental protection and conservation. It has positively impacted stakeholders like environmentalists, water purification experts, and those in the medical disciplines. He has mentored several postgraduate students. He has delivered keynote and lead papers at national and international conferences. He has served as external examiner and assessor to many universities at national and international levels. He has served on various committees as chairman and member at the departmental, faculty, and university levels. He is the Managing Editor of *Science Focus* journal published by the Faculty of Pure and Applied Sciences, LAUTECH, Ogbomoso.

Contributors

Abdulhakeem Idris Abdulhakeem Department of Fisheries and Aquaculture, Federal University Dustin-Ma, Dustin-Ma, Kastina State, Nigeria

Mustapha Abdulsalam Department of Microbiology, School of Sciences and Information Technology, Skyline University Nigeria, Kano, Nigeria

Taiwo Oluwasola Abe Department of Chemistry, Federal University of Technology, Akure, Nigeria

Hakan Acaroğlu Department of Economics, Faculty of Economics and Administrative Sciences, Eskisehir Osmangazi University, Eskisehir, Turkey

Kehinde Olubunmi Adediran Department of Biological Sciences, Faculty of Pure and Applied Sciences, Lead City University, Ibadan, Oyo State, Nigeria

Kayode Adesina Adegoke LAUTECH SDG 6 (Clean Water and Sanitation Research Group), Ogbomoso, Oyo State, Nigeria

LAUTECH SDG 11 (Sustainable Cities and Communities Research Group), Ogbomoso, Oyo State, Nigeria

Department of Industrial Chemistry, First Technical University, Ibadan, Oyo State, Nigeria

Emmanuel Kuyet Adejokun Department of Microbiology, Faculty of Science, University of Abuja, Abuja, Nigeria

Samuel Oluwatobi Agboola LAUTECH SDG 6 (Clean Water and Sanitation Research Group), Ogbomoso, Oyo State, Nigeria

LAUTECH SDG 11 (Sustainable Cities and Communities Research Group), Ogbomoso, Oyo State, Nigeria

Department of Pure and Applied Chemistry, Ladoke Akintola University of Technology, Ogbomoso, Oyo State, Nigeria

Mohd Azmier Ahmad School of Chemical Engineering, Engineering Campus, Universiti Sains Malaysia, Penang, Malaysia

Oluwatosin Stephen Ajayi LAUTECH SDG 6 (Clean Water and Sanitation Research Group), Ogbomoso, Oyo State, Nigeria

LAUTECH SDG 11 (Sustainable Cities and Communities Research Group), Ogbomoso, Oyo State, Nigeria

Department of Pure and Applied Chemistry, Ladoke Akintola University of Technology, Ogbomoso, Oyo State, Nigeria

Ulelu Jessica Akor Department of Animal Production, University of Ilorin, Ilorin, Nigeria

Temitope Chris Alagbada LAUTECH SDG 6 (Clean Water and Sanitation Research Group), Ogbomoso, Oyo State, Nigeria

LAUTECH SDG 11 (Sustainable Cities and Communities Research Group), Ogbomoso, Oyo State, Nigeria

Department of Pure and Applied Chemistry, Ladoke Akintola University of Technology, Ogbomoso, Oyo State, Nigeria

Oluwakemi Christiana Alao LAUTECH SDG 6 (Clean Water and Sanitation Research Group), Ogbomoso, Oyo State, Nigeria

LAUTECH SDG 11 (Sustainable Cities and Communities Research Group), Ogbomoso, Oyo State, Nigeria

Department of Pure and Applied Chemistry, Ladoke Akintola University of Technology, Ogbomoso, Oyo State, Nigeria

Loai Aljerf Key Laboratory of Organic Industries, Department of Chemistry, Faculty of Sciences, Damascus University, Damascus, Syrian Arab Republic

Faculty of Pharmacy, Al-Sham Private University, Damascus, Syrian Arab Republic

Alissar AlJerf Haitham Abdel-Salam School, Al-Jama'at District, Jaramanah, Ministry of Education, Damascus, Syrian Arab Republic

Sesan Abiodun Aransiola Department of Microbiology, Faculty of Science, University of Abuja, Abuja, Nigeria

Bioresources Development Centre, National Biotechnology Research Development Agency, Ogbomoso, Nigeria

Toluwanimi Rachael Arowolo LAUTECH SDG 6 (Clean Water and Sanitation Research Group), Ogbomoso, Oyo State, Nigeria

LAUTECH SDG 11 (Sustainable Cities and Communities Research Group), Ogbomoso, Oyo State, Nigeria

Department of Pure and Applied Chemistry, Ladoke Akintola University of Technology, Ogbomoso, Oyo State, Nigeria

J. Ayedun Department of Biological Sciences and Industril Biotechnology, College of Pure and Applied Sciences, Imota, Ghana

Babafemi Raphael Babaniyi Bioresources Development Center, National Biotechnology Research and Development Agency, Abuja, Nigeria

Gabriel Gbenga Babaniyi Department of Agricultural Development and Management, Agricultural and Rural Management Training Institute (ARMTI), Ilorin, Nigeria

Sunday Zeal Bala Department of Chemical Sciences, Faculty of Science and Computing, Karl Kumm University, Vom, Plateau State, Nigeria

Nasser Belboukhari Bioactive Molecules and Chiral Separation Laboratory, University of Bechar, Bechar, Algeria

Olugbenga Solomon Bello LAUTECH SDG 6 (Clean Water and Sanitation Research Group), Ogbomoso, Oyo State, Nigeria

LAUTECH SDG 11 (Sustainable Cities and Communities Research Group), Ogbomoso, Oyo State, Nigeria

Department of Pure and Applied Chemistry, Ladoke Akintola University of Technology, Ogbomoso, Oyo State, Nigeria

Ademola Bisi-Omotosho Department of Sustainability Management and Innovation, University of Westminster, London, UK

Augustine Innalegwu Daniel Department of Biochemistry, School of Life Sciences, Federal University of Technology, Minna, Niger State, Nigeria

Plant Omics Laboratory, Department of Biotechnology, Faculty of Natural Sciences, University of the Western Cape, Bellville, South Africa

Oluwatosin Emmanuel Daramola Department of Chemistry and Biochemistry, Texas Tech University, Lubbock, TX, USA

Namla Djadjiti Department of Biological Sciences, Faculty of Science, King Abdulaziz University, Jeddah, Saudi Arabia

Shehu-Alimi Elelu Department of Chemistry, Howard University, Washington, DC, USA

Gloria Ginikanwa Ezeagu Department of Biology, Microbiology & Biotechnology, Faculty of Science, Nile University of Nigeria, Abuja, Nigeria

Morenike Fadayomi Department of Biology, Microbiology & Biotechnology, Faculty of Science, Nile University of Nigeria, Abuja, Nigeria

Evelyn Nwadinkpa Fatokun Department of Biological Science and Biotechnology, College of Pure and Applied Sciences, Caleb University, Imota, Lagos State, Nigeria

O. A. Fatunbi Department of Microbiology and Industrial Biotechnology, College of Pure and Applied Sciences, Caleb University, Imota, Lagos, Nigeria

Scott O. Fayemi Department of Microbiology and Industrial Biotechnology, College of Pure and Applied Sciences, Caleb University, Imota, Lagos, Nigeria

Mustafa Güllü Republic of Turkey Ministry of National Education, Ankara, Turkey

Madinat Hassan Foresight Institute of Research and Translation (FIRAT), Kigali, Rwanda

Ganiyat Omotayo Ibrahim Department of Chemistry, Nottingham Trent University, Nottingham, UK

Ojo Emmanuel Ige Department of Chemistry, Federal University of Technology, Akure, Nigeria

Gladys Mercy Inetabor LAUTECH SDG 6 (Clean Water and Sanitation Research Group), Ogbomoso, Oyo State, Nigeria

LAUTECH SDG 11 (Sustainable Cities and Communities Research Group), Ogbomoso, Oyo State, Nigeria

Department of Pure and Applied Chemistry, Ladoke Akintola University of Technology, Ogbomoso, Oyo State, Nigeria

Gbolahan Olamide Isaac Department of Chemistry, Federal University of Technology, Akure, Nigeria

Srinivasan Kameswaran Department of Botany, Vikrama Simhapuri University College, Kavali-524201, Andhra Pradesh, India

Priscilla Kini Department of Toxicology, University of Maryland Eastern Shore, Princess Anne, MD, USA

Ashwil Klein Department of Biology, Microbiology & Biotechnology, Faculty of Science, Nile University of Nigeria, Abuja, Nigeria

Bunmi Kotun Department of Microbiology and Industrial Biotechnology, College of Pure and Applied Sciences, Caleb University, Imota, Lagos, Nigeria

Salami Olaitan Lateefat Department of Chemistry and Molecular Biology, University of Gothenburg, Gothenburg, Sweden

Miracle Uwa Livinus Department of Biochemistry, School of Sciences and Information Technology, Skyline University Nigeria, Kano, Nigeria

Naga Raju Maddela Departamento de Ciencias Biológicas, Facultad la Ciencias de la Salud, Universidad Técnica de Manabí, Portoviejo, Manabí, Ecuador

Christy Echakachi Manyi-Loh Centre of Applied Food Sustainability and Biotechnology, Department of Life Sciences, Faculty of Health and Environmental Science, Central University Technology, Bloemfontein, Free State, South Africa

Sani Sambo Datsugwai Mohammed Department of Biology, Microbiology & Biotechnology, Faculty of Science, Nile University of Nigeria, Abuja, Nigeria

Auwal Sagir Muhammad Department of Computer Science, Xiamen University, Xiamen, China

School of Informatics, Xiamen University, Xiamen, China

Innocent Ojeba Musa Department of Microbiology, School of Sciences and Information Technology, Skyline University Nigeria, Kano, Nigeria

Adamu Mustapha Department of Microbiology, Federal University of Technology Minna, Minna, Nigeria

Atabak Najafi Department of Electrical and Electronics Engineering, İstanbul Esenyurt University, İstanbul, Turkey

Sarah Chioma Nwakanma Department of Microbiology, Faculty of Science, University of Abuja, Abuja, Nigeria

O. Oda Department of Microbiology and Industrial Biotechnology, College of Pure and Applied Sciences, Caleb University, Imota, Lagos, Nigeria

Gideon Oluranti Ojediran LAUTECH SDG 6 (Clean Water and Sanitation Research Group), Ogbomoso, Oyo State, Nigeria

LAUTECH SDG 11 (Sustainable Cities and Communities Research Group), Ogbomoso, Oyo State, Nigeria

Department of Chemistry, Faculty of Physical Sciences, Alex Ekwueme Federal University, Ndufu-Alike, Ebonyi State, Nigeria

Adedamola Titi Ojedokun LAUTECH SDG 6 (Clean Water and Sanitation Research Group), Ogbomoso, Oyo State, Nigeria

LAUTECH SDG 11 (Sustainable Cities and Communities Research Group), Ogbomoso, Oyo State, Nigeria

Department of Pure and Applied Chemistry, Ladoke Akintola University of Technology, Ogbomoso, Oyo State, Nigeria

Ojonoma Loretta Okwute Department of Microbiology, Faculty of Science, University of Abuja, Abuja, Nigeria

Adesuyi Emmanuel Omole Department of Marine Biology, University of Padova, Padova, Italy

Olusegun Opeyemi Oni Department of Agricultural Development and Management, Agricultural and Rural Management Training Institute (ARMTI), Ilorin, Nigeria

Bellamkonda Ramesh Biochemistry and Molecular Biology, University of Nebraska Medical Center, Omaha, NE, USA

Olaitan Lateefat Salam Department of Chemistry and Molecular Biology, University of Gothenburg, Gothenburg, Sweden

Job Oloruntoba Samuel Department of Microbiology, Federal University of Technology Minna, Minna, Nigeria

Sylvester Chukwudi Ugoh Department of Microbiology, Faculty of Science, University of Abuja, Abuja, Nigeria

Part I
Occurrence, Distribution and Degradation of Marine White Plastisphere

Chapter 1
An Insight into the Prevalence and Distribution of Plastisphere in the Marine Environment

Samuel Oluwatobi Agboola, Temitope Chris Alagbada, Gladys Mercy Inetabor, Toluwanimi Rachael Arowolo, Kayode Adesina Adegoke, Mohd Azmier Ahmad, and Olugbenga Solomon Bello

Abstract A distinct microbial ecosystem known as the plastisphere has emerged as a result of the widespread plastic pollution in the marine environment. It has negative effects on human health, ecosystem functioning, biodiversity, and marine ecosystems, making it a serious threat to these environments. It is made up of microbial communities that live and infect the surfaces of plastic waste in aquatic environments. Comprehending the full extent of the ecological impact of plastic pollution requires an understanding of the prevalence and distribution of the plastisphere. An overview of the state of our understanding of the plastisphere is given in this study, with particular attention to its formation, dynamics, ecological significance, difficulties, and future possibilities. The present study examines the factors that influ-

S. O. Agboola · T. C. Alagbada · G. M. Inetabor · T. R. Arowolo · O. S. Bello (✉)
LAUTECH SDG 6 (Clean Water and Sanitation Research Group), Ogbomoso, Oyo State, Nigeria

LAUTECH SDG 11 (Sustainable Cities and Communities Research Group), Ogbomoso, Oyo State, Nigeria

Department of Pure and Applied Chemistry, Ladoke Akintola University of Technology, Ogbomoso, Oyo State, Nigeria
e-mail: osbello@lautech.edu.ng

K. A. Adegoke
LAUTECH SDG 6 (Clean Water and Sanitation Research Group), Ogbomoso, Oyo State, Nigeria

LAUTECH SDG 11 (Sustainable Cities and Communities Research Group), Ogbomoso, Oyo State, Nigeria

Department of Industrial Chemistry, First Technical University, Ibadan, Ibadan, Oyo, Nigeria

M. A. Ahmad
School of Chemical Engineering, Engineering Campus, Universiti Sains Malaysia, Penang, Malaysia

S. A. Aransiola et al. (eds.), *White Pollution: Biodiversity and hazards in marine plastisphere*, Environmental Contamination Remediation and Management,
https://doi.org/10.1007/978-3-031-95547-1_1

ence the prevalence and distribution of the plastisphere, taking into account the type of plastic, environmental conditions, and geographic location, through a synthesis of existing literature. The plastisphere's ecological ramifications are also explored, including its function in the cycling of nutrients, biogeochemical changes, and the movement of contaminants and potentially dangerous microbes. Regarding the precise makeup and ecological ramifications of the plastisphere, there are still unknowns. Given this, the difficulties in researching the plastisphere are also emphasized, along with suggestions for future lines of inquiry to deepen our knowledge of this puzzling ecological phenomenon. This chapter aims to inform strategies for mitigating plastic pollution and preserving the integrity and health of marine ecosystems for future generations by illuminating the plastisphere.

Keywords Pollution · Plastics · Ecosystem · Marine · Nutrient

1.1 Introduction

Plastics are becoming more and more common in the marine environment, despite increased awareness and international efforts to reduce plastic waste. Amidst this ecological catastrophe, there has been a rise in interest in a lesser-known but noteworthy phenomenon called the "plastisphere" (Shim and Thompson 2015; Musa et al. 2024). A diverse community of microbes inhabits and interacts with plastic debris in the marine environment, creating a complex ecosystem (Stubbins et al. 2021; Auta et al. 2022). The complex microbial communities that settle and flourish on the surfaces of plastic waste in aquatic environments are referred to as "plastispheres." With the introduction of plastics into marine environments, these communities have emerged as a new ecological niche. To fully grasp the ecological impact of plastic pollution, one must have a thorough understanding of the prevalence and distribution of the plastisphere.

The plastisphere has been linked to several ecological processes, such as biogeochemical transformations, nutrient cycling, and the movement of potentially hazardous microorganisms and pollutants, despite its relatively recent discovery (Wright et al. 2020). Furthermore, interactions between microbes in the plastisphere may have a significant impact on marine biodiversity, ecosystem performance, and ultimately human health (Alomar et al. 2016). Numerous variables, such as the kind of plastic, the surrounding environment, and the location, affect the plastisphere's prevalence and distribution. Through ocean currents, plastic debris can travel great distances and disperse microbial communities throughout various marine habitats (Eriksson et al. 2013). Furthermore, the plastisphere's structure and composition are shaped by the physical and chemical characteristics of plastics, which in turn shapes the ecological dynamics of the region.

The plastisphere's distribution and prevalence in the marine environment are investigated in this study. The mechanisms guiding the plastisphere's formation and dynamics are clarified by a thorough examination of the body of existing literature.

Along with discussing potential ramifications for marine ecosystems, this study evaluates the ecological significance of plastisphere and describes research methodologies.

1.2 Overview of Plastisphere

1.2.1 Definition and Composition

The term "Platisphere" refers to an artificial biome that is proportional to the ecosystem's level of plastic waste; it was first used in 1972 when plastic was discovered in the ocean, and it is the most common type of man-made debris (Zettler et al. 2013). The term "Platisphere" was first used in 2013 by Zettler et al. of the Sea Education Association in Woods Hole, United States. From ecotoxicity and new connections in marine food webs to the fate of plastics in the water column, every interaction this synthetic material (plastic) has with its surroundings is influenced by the plastisphere, which is located at the point of intersection between the material itself and its surrounding microenvironment. It also acts as a carrier of dangerous germs.

When discharged into marine habitats, microplastics (MPs) are quickly colonized by microbial populations, which create "plastispheres," or biofilms, on the plastic's surface (including fungus, diatoms, and bacteria) that have a great attraction for plastic material and can stick to the MPs in spite of changing environmental conditions. This microbiological community, known as the Plastisphere, is an ecology of microorganisms that live on floating plastic waste. Plastic waste can travel over large distances—up to 1000 km—in less than 60 days, from coastal waters in the US East Coast to the North Atlantic SubTropical Gyre (Zettler et al. 2013). The Platisphere also describes the thin coating of microbes and metazoans that forms on any piece of plastic in an aquatic environment. These ecosystems have developed to survive in settings built of plastic by humans (Oberbeckmann et al. 2014).

Comparable to the built environment across multiple earthly biomes, the microbial community atop plastic debris, or plastisphere, has a unique biodiversity. The colonies of bacteria found inside plastispheres may contribute to the development of antibiotic resistance in aquatic environments as well as the acquisition and spread of diseases. The plastisphere's distinct composition is made up of symbionts, predators, heterotrophs, and primary producers. The moderation of earth's ecological cycles and the organization of oceanic food webs is largely dependent on interactions between bacteria and phytoplankton. Mayali (2018) notes that the plastisphere exhibits this same relationship between autotrophic species and other microbes. Studies on plastic waste have shown that diatoms are almost always present and can even take over the plastic surface (Kettner et al. 2019; Michels et al. 2018; Cheng et al. 2020). These autotrophic organisms can regulate the microbial community's composition and supply organic resources to the plastisphere (Mayali 2018). In the plastisphere, diatoms such as Sellaphora, Amphora, Mastogloia, Navicula, Nitzschia, Cyclotella, Pleurosigma, and Nitzschia have been widely discovered. The plastisphere was also found to include cyanobacteria, including Phormidium and Rivularia (Zettler et al. 2013; Bryant et al. 2016).

1.2.2 Formation and Development of Plastisphere

Microbial adhesion, extracellular polymeric material secretion, and microbial proliferation are the three processes that normally occur in the plastisphere. These processes are similar to those of other biofilms. According to De Tender et al. (2017a, b), usually, plastisphere bacteria can be discovered in the adjacent waters or in the sediments on the bottom of the sea. Early in microbial colonization, bacterial communities are selected based on physical properties of plastics, including surface charge, roughness, hydrophobicity, and crystallinity. The choice of succeeding communities inside the plastic circle may likewise be influenced by these pioneer bacteria (Tu et al. 2020).

The primary stratum of the first biofilm is formed by pioneer bacteria that initially settle on the plastic surface by reversible adhesion (Tu et al. 2020). When microplastics colonize, their surface loses some of their hydrophobicity, which can affect how the particles move vertically in the ocean and open up new biological niches for other microbes. (Cai et al. 2019; Fazey and Ryan 2016; Kooi et al. 2017). The pioneer bacterium released exopolysaccharides which can improve the plastic surface's adherence, facilitating the colonization of additional microorganisms (Anjana et al. 2020). Subsequently, extracellular polymeric substrates, adhesion proteins, pili, and other active substances produced by secondary bacteria can aid in permanent attachment by increasing the quantity of attachable sites on the plastic surface (Dussud et al. 2018).

Competition or the combined effects of many microorganisms eventually result in the formation of a mature biofilm, which enhances adherence to the plastic surface and allows for additional species recruitment or loss on the biofilm. The surface of MPs is colonized by microorganisms, but in a certain order. Microorganisms must first penetrate MPs' hydrophobic surface in order to colonize their surface (Gong et al. 2019). Also, microorganisms can also quickly colonize Microplastics' surface; however, it takes a considerable time for a stable biofilm to form (Quero and Luna 2017).

According to Quero and Luna (2017), the pioneer colonizers, or Gammaproteobacteria and Alphaproteobacteria, are the microorganisms that first settled on microplastics. According to reports (De Tender et al. 2017a, b; Dussud et al. 2018; Harrison et al. 2014), on most types of plastic polymers, gammaproteobacteria are the predominant microorganisms in the initial biofilms. Identiifed during the initial stages of early biofilm formation were Thalassobius, Neptuniibacter, Phodobacteriaceae, and Poseobacter (Zhang et al. 2022). Early in the biofilm's development, diatoms and cyanobacteria were also discovered (Amaral-Zettler et al. 2015; Dang and Lovell 2016). Due to their widespread distribution, flexibility, and capacity to make use of the organic substrates that early colonists released, Bacteroidetes—particularly Flavobacteriaceae—tend to proliferate throughout time (Quero and Luna 2017; Zhang et al. 2022). These microorganisms are known as secondary colonizers.

Primary biofilm typically engages a week or less to become noticeable, the secondary biofilm, however takes several months to build (De Tender et al. 2017a, b; Wright et al. 2020). Chen et al. demonstrated that during the first 30 days of polyethylene (PE) surface development, the main biofilm colonists were Microtrichaceae (Acidimicrobiia), Rhodobacteraceae (Alphaproteobacteria), and Flavobacteriaceae (Bacteroidia). (Tu et al. 2020). The predominant populations progressively changed throughout the mid-term (75 days) from Bacillaceae (Bacilli) and Moraxellaceae (Gammaproteobacteria) to Flavobacteriaceae and Erythrobacteriaceae.

Flavobacteriaceae (Bacteroidia) returned to being the predominant family of polyethylene (PE) colonizers in the subsequent phases of biofilm formation (at 135 days), and their abundance within Pirellulaceae (Planctomycetes), Microtrichaceae (Acidimicrobiia), Rhodobacteraceae (Alphaproteobacteria), was also markedly increased. There are differences throughout the tests concerning the time course of the different biofilm phases. The total amount of nutrients and bacteria in the marine milieu where the plastic is placed most likely controls the rate at which biofilms develop.

An increase in plastic invading the aquatic environment—where it can create new ecological niches for microbial populations and form a structure known as a plastisphere—is the result of high plastic manufacturing rates, low recycling rates, and inadequate waste management. The transformation of bacteria from motile to colonized state occurs during the development of the plastisphere. According to Solano et al. (2014), this transition enables a single cell to accomplish quorum sensing-synchronized gene expression for processes like adhesion, communication, substance transport, and chemotaxis as well as the formation of a tissue-like structure. In these conditions, fluid channels are formed to facilitate the distribution of nutrients across cells (Amaral-Zettler et al. 2020).

1.2.3 Factors Influencing the Growth of Plastisphere

1.2.3.1 Biogeography

Numerous types of organisms, including algae, fungi, protozoa, and archaea, can be found in the marine plastisphere (Bryant et al. 2016; Kirstein et al. 2018). These extra kingdoms exhibit location-dependent variations in abundance and taxonomic complexity, as suggested by the influence of biogeography on their distribution. Therefore, the initial colonization of plastic surfaces is primarily opportunistic, propelled by pre-existing microbial communities that are able to take advantage of this niche (Oberbeckmann et al. 2014). This process is further impacted by variables including local temperature, salinity, and co-pollutant concentrations.

With increasing geographic distance, differences in plastispheres become increasingly noticeable. The study conducted by Wu et al. (2020a, b) on microplastics (MPs) at several sample sites revealed that MPs had strong stress tolerance, which lessens the effect of regional variability. Temperature, salinity, oxygenation, and the diversity of plants and animals at each location are specific factors that affect microbial diversity and nutrient load (Coons et al. 2021). The conditioning coating on plastics, known as dissolved organic matter (DOM), is likely to differ from place to place and influence prokaryotic binding and metabolism. DOM might also include co-pollutants. Wide variations in salinity, temperature, pressure, availability of light and nutrients, and even pollution define the biogeochemical characteristics of the world's seas despite their interconnection (Coons et al. 2021). As a result, separate microbial species must acclimate by constitutively expressing traits that increase their likelihoods of survival in the many niches that the water column becomes separated into (Nguyen et al. 2021). This creates distinct communities at each site that are tailored to the biogeochemical forces that are present there. In view of this, only microbes that are appropriate for a given place can attach to plastic once it is introduced there (Coons et al. 2021). There exists an understanding that this first attachment is indiscriminate, this means that the members of the plastisphere will predominantly reflect the community at large (Wright et al. 2020; Amaral-Zettler 2022).

The SAR11 clade, which includes Alphaproteobacteria, usually has a dominant position in the pelagic microbial community in most coastal habitats. In these marine ecosystems, Prochlorococcus and Synechococcus (Cyanobacteria) are also common, as reported by Bolaños et al. (2021), as well as Flavobacteria (Bacteroidetes), Rhodobacteraceae (Alphaproteobacteria), and other prominent classes such as Bacteroidetes, Alphaproteobacteria, and Gammaproteobacteriaas reported by Frère et al. (2018), Kirstein et al. (2018), and Scales et al. Nevertheless, the plastisphere will only enhance biofilm-forming species that can tolerate the competition and toxicity of co-pollutants within it. On plastic, Flavobacteria and Rhodobacteraceae are more common than the SAR11 clade, Synechococcus, and Prochlorococcus, among many other SAR and OM1 clades. (Frère et al. 2018; Bryant et al. 2016). Instead, a combination of the other less-dominant types of bacteria present in those areas makes up the dominating bacteria in these biofilms (Kirstein et al. 2018; Frère et al. 2018; De Tender et al. 2015; Bryant et al. 2016).

The biogeographical location has the greatest impact on the plastisphere community's makeup, according to Coons et al. (2021). Plastisphere communities are highly impacted by their geographic position, according to other earlier research (Oberbeckmann et al. 2014). Geographical differences distinguish communities more: For instance, the most prevalent and dominant group in the Western Pacific is Rhodobacterales; in the North Sea, communities were dominated by Bacteroidetes and Cyanobacteria. In the tropical regions of the Northern Hemisphere in the summer, plastisphere ecosystems showed greater species richness (Amaral-Zettler et al. 2015).

1.2.3.2 Local Environmental Conditions

Biofilm colonies' constitution can be strongly influenced by the nutrients found in the localseawaters. Low nutrient levels have been shown to cause numerous bacterial species to adhere to the exterior of the base, whereas high nutrition concentrations prevent numerous bacteria from attaching, thus being harmful to the production of specialized biofilms (Du et al. 2022). High nutrient levels in seawater lead to relatively low nutrient levels supplied by plastics, which decreases the plastisphere's substrate specificity. An increase in nutrient availability also stimulates the creation of dominant and resultant (secondary) biofilms, leading to a microbial species confluence inside the plastisphere.

The growth of plastic-associated biofilms can be influenced by various environmental conditions such as pressure, temperature, salinity, acceleration of tidal current, along with the presence of oxygen and light (Oberbeckmann et al. 2014). The plastisphere populations of distinct coastal locations differ due to their variation throughout different sea regions (Wang et al. 2021; Harrison et al. 2018). Wu et al. (2020a, b) observed that MPs at various sample sites had comparable operational taxonomic units (OTUs), leading them to hypothesize that MPs' high stress tolerance lessens the impact of regional variation. Biofilm formation is influenced by plastic characteristics, ambient factors, and colonization time. In the meantime, the biofilm communities in turn affect how plastics behave in the environment, particularly with regard to transportation, the buildup of contaminants, and the breakdown and fragmentation of plastics.

Temperature, salinity, and other factors vary between freshwater and marine habitats, as well as between WWTP and uncontrolled freshwaters. For example, plastisphere assemblages in deep seas are anticipated to be subject to selecting factors different from those in shallow habitats due to the low temperatures (<5C), lack of light, and high pressure. One variation is illustrated in the state of the open expanse of seas, which frequently has nutrient-poor conditions, which is unlike the inland and tidal streams that get significant outflows of vital nutrients from the environment around them (Horton et al. 2017).

Apart from the intake of organic matter and upwelling, human activities such as agriculture discharge large amounts of nutrients (such as phosphate and nitrogen). Pollution degradation has been linked to a number of plastisphere members (Park and Kim 2019; Schwabl et al. 2019; Corradini et al. 2019; Boucher and Friot 2017). It is likely that a number of pollutants influence biofilm development and actions on polymers. It is true that a variety of contaminants, including heavy metals, can adhere to microplastics (Huang et al. 2021; Krueger et al. 2015; Zhu and Wang 2020).

1.2.3.3 Plastic Type

The plastisphere's composition in the natural environment is contingent upon factors such as temperature, seasonal, geographical, and type of plastic (Oberbeckmann et al. 2014). Microbial adhesion, extracellular polymeric material secretion, and microbial development, normally encompasses the plastisphere just like other biofilms (Amaral-Zettler et al. 2015). All plastics are made of hydrocarbon chains, some of which include functional groups like carbonyls, ketones, or esters, or aromatic benzene rings like those found in polystyrene and polyethylene terephthalate (PET) or methyl groups like those found in polyethylene, polyamide, PET, and polypropylene (Crawford and Quinn 2017a, b).

According to Crawford and Quinn (2017a, b), the chain structures of hydrocarbons within the polymeric frame diverge as a result of these groups' encouragement of polymer branching. This process produces aliphatic plastics, which are characterized by loose packing, like polyethylene and polypropylene, or crystalline plastics, which are characterized by tight packing, like polystyrene and polyvinyl chloride (PVC). The surfaces of substances (plastic) submerged in seawater quickly accumulate dissolved organic matter (DOM) and biological nanoparticles, including trace metals. This process occurs minutes after the material is submerged (Lorite et al. 2011).

The processes that attract marine microorganisms to the outermost layer of plastic substances are made easier by functional groups present in the ecocorona. Van der Waals force, hydrophilic interaction, and electrostatic interaction are some of these methods. Variations in surface roughness, buoyancy, hydrophobicity, plastic charge, and crystallinity are often mentioned as contributing to the diversity of plastispheres (Tobias-Hünefeldt et al. 2021; Kirstein et al. 2018).

According to Frère et al. (2018) and Miao et al. (2020), there are notable differences between the microbial populations attached to polystyrene and PVC and those attached to aliphatic polymers. Variabilities in microbial diversity were observed by Zhang et al. (2022) in all plastic types that were studied. Moreover, Wright et al.'s meta-analysis from 2021 revealed that, in contrast to other plastics, aliphatic plastics draw a particular variety of microbes.

The different functional groups of aromatics that these replacement polymers contain may be connected to these phenomena (Crawford and Quinn 2017a, b). Variances in chemical additives between different types of plastic also have an impact on the observed disparities in plastispheres (Miao et al. 2020; Frère et al. 2018; Bakir et al. 2014).

On the creation of microbial communities, the effects of location and other characteristics that are specific to a certain area, like salinity, seem to outweigh the influence of plastic type (Coons et al. 2021). Moreover, during biofilm formation, the surface chemistry of different polymers begins to converge, which may account for the biofilm variety overlap seen in the later phases of the growth and establishment of this ecosystem that exists on plastics in the ocean-plastisphere (Woodall et al. 2018; Kirstein et al. 2018). This attenuates the impact of plastic type in the subsequent stages of plastisphere innidiation (colonization), notwithstanding the fact that the early biofilm creation remains contingent on the substratum's chemical composition, which fluctuates based on the kind of plastic. (Zhang et al. 2022; Tobias-Hünefeldt et al. 2021; Kirstein et al. 2018).

1.2.3.4 Surface Chemistry and Structure

Plastics are subject to temperature changes, ocean hydrodynamics, and ultraviolet radiation (UV-R) in the water (Gewert et al. 2015). Reactive oxygen species (ROS) may be produced, chemical adjuvants and other additives that contain chemicals may also be released. There can also be an increase in the electrostatic charge and permeability of plastic (Wang et al. 2020). Biofilms, also referred to as "plastispheres" (Zettler et al. 2013), are developing and adhere to plastic surfaces with precision and speed, aiding in the abiotic weathering of plastics (Hossain et al. 2019).

Within several minutes, environmental microorganisms rapidly colonize tiny plastic particles (Microplastics) (Harrison et al. 2014). Probably a majority of identical mechanisms drive the formation/development of plastisphere colonies in freshwater environments as they do in marine ones. For instance, studies on the development of biofilm on different constructed substrates (Fish et al. 2016) correlates with data advancing the importance of the surface properties (e.g., roughness and hydrophobicity) throughout the initial colonization of plastic particles (Oberbeckmann et al. 2015).

UV (ultraviolet) rays and waves have the power to change the chemical makeup of the surface and framework of polymers. Among other things, this can happen as a result of pit and fracture formation, a drop in relative molar mass, and a surge in chemical oxidization. Plasmid biofilm's development and composition can also be influenced by chemical, physical, and biological variables as well. Biofilm development is facilitated by a small number of these characteristics that have only been partially studied in relation to microplastics (Gewert et al. 2015; Brandon et al. 2016). The surface characteristics and buoyancy of polymers have also been observed to be influenced by microbes that colonize plastic (Oberbeckmann et al. 2015; Mincer et al. 2019; Kowalski et al. 2016). Given the likelihood of microplastics being carried into the oceans by streams, rivers, and sewage treatment plants (WWTP), it can be predicted that factors that lead to initial colonization (e.g., roughness of the substrate and adhesion by founding colonials) are particularly important in freshwaters (Sadri and Thompson 2014; Jambek et al. 2015).

Although polymer occupies more time in marine habitats than in rivers and streams, the impacts of molecular generation and/or degradation on plastisphere consortium may be more apparent there (Gewert et al. 2015). Microplastics can build up in ecosystems like lakes, where they can be exposed to intense UV radiation and survive for decades—as is projected for marine habitats (Corcoran et al. 2015; Eerkes-Medrano et al. 2015; Free et al. 2014). There have been observed local-scale variations in the polymer types that make up plastisphere assemblages (Oberbeckmann et al. 2015; Oberbeckmann et al. 2014; Amaral-Zettler et al. 2015), but it is unclear if there are any broad variations in the predominant plastic types in freshwater and marine environments.

1.2.3.5 Biological Interaction

The structure of biofilm communities is shaped by a confluence of bottom-up and top-down selection forces in the environment (Tobias-Hünefeldt et al. 2021). By secreting toxins and extracellular polymeric substances together, bacteria and algae may reduce the amount of grazers and diseases that adversely affect the consortia of bacterium (Moresco et al. 2021; Buchan et al. 2014).

Metalloproteases specific to peptidoglycan are secreted by pseudoalteromonas sp. bacteria, which targets Gram-indicating bacterium (Tang et al. 2020). According to Jackson et al. (2018), Gram-positive Actinobacteria respond by producing antibiotic enzymes, most likely as a defense or competitive strategy against Gram-negative bacterial species. In early plastispheres, the quantity of Gram-positive bacteria may have been lower due to this extensive enzyme synthesis. Notwithstanding their widespread usage in bacteria (Ma et al. 2014; Anand et al. 2013), these enzymes are not directly shown to exist within the plastisphere (Oberbeckmann et al. 2021; Delacuvellerie et al. 2022a, b). Further evidence that this hostile connection would not be an impediment to the survival, reproduction or elimination of species in these marine plastispheres comes from the discovery that many of the most widespread microbes remain quiescent on polymer (Oberbeckmann et al. 2021; Delacuvellerie et al. 2022a, b).

Independent of their activity, the microbes secrete the ions that are negatively charged and adhesive proteins essentially; thereby allowing Bacteroidetes, Alphaproteobacteria, and Actinobacteria, to cling to plastic (Hossain et al. 2019; Mammo et al. 2020; Pelicic 2019; Dussud et al. 2018). Microbial hydrophilic characteristics attracts diatoms to the plastisphere promptly in the microbiological lifecycle. (Buchan et al. 2014; Delacuvellerie et al. 2019; Schlundt et al. 2020a, b). Every diatom has a corresponding phycosphere where bacteria congregate for mutual benefit, namely species of Flavobacteria and Rhodobacteraceaeconstituents (Buchan et al. 2014; You et al. 2021).

This mechanism may account for the Flavobacteria and Rhodobacteraceae bacterial groups' dominance and strong metabolic activity in the plastisphere (Delacuvellerie et al. 2022a, b, 2019). When the plastisphere population stabilizes, diatom abundance doesn't change (Zettler et al. 2013; Oberbeckmann et al. 2014; Kirstein et al. 2018; Cheng et al. 2021), which adds credence to the idea that the plastisphere biofilm functions as a welcoming community (Pollet et al. 2018).

According to a study of the preexisting interconnections between genera inside the plastisphere, it is conceivable to foresee that the existence and distribution of organisms in early phases of transition will impact those of it's predecessors (Schlundt et al. 2020a, b; Lorite et al. 2011; Datta et al. 2016). These predictions are consistent with patterns seen in other ecological communities. When examining selection forces inside recently established niches, this idea of inclusion over inhibition is commonly used.

1.2.3.6 Movement and Transport Within Habitats

According to research by Sadri and Thompson (2014) and Jambek et al. (2015), microplastics are thought to enter marine ecosystems through rivers, streams, and wastewater treatment facilities (WWTPs), playing a significant role in the initial colonization process. These microplastics act as attachment points for "pioneer colonizers" and add to the roughness of the surface. The mobility of suspended particles is influenced by different physical processes in freshwater and marine environments, making it an important element among other things (Eerkes-Medrano et al. 2015). Freshwater and marine habitats differ primarily in that water flows continuously downstream. The process of "spiraling" is a term used to describe sediment movement in rivers (Newbold et al. 1982; Tank et al. 2010). Resuspension, deposition, downstream transport, and bed load transfer are the constituent parts of a single spiral. The retention-export ratio, turnover time, and deposition length and velocity are used to quantify this well-developed approach to modeling particle mobility (Webster et al. 1999).

Every spiral step is anticipated to affect the composition and activity of plastic-associated biofilms since it coincides with changes in the surrounding environment (Oberbeckmann et al. 2014). By identifying the dimensional and geometrical span ranges that determine how polymeric fragments move through riverine ecological systems, studies on the whirling dimensions of plastic particles can serve to shed light on possible modifications which the communities of microorganisms affected may experience during the period of multiple downriver swirls.

Another distinctive feature of rivers is flooding, which transfers particles within the oceanic and riverside elements of the riverbed (Gregory et al. 1991; Jones and Smock 1991). According to McCormick and Hoellein (2016), flooding causes more plastic to become stranded in debris dams and transports plastic from the riparian zone into aquatic environments. In marine ecosystems, tidal motions and surges from storms are similar mechanisms that leave behind polymer on benthic or ruined areas (Eerkes-Medrano et al. 2015).

The implications of migration from aquatic to terrestrial settings on plastic-associated biofilms may likely affect the plastisphere ecosystems. The water and particle residence times for every lake differ based on the water quantity and waves, and most lakes possess a minimum of one headwaters entrance and one downward outflow. Research on the plastisphere communities in lakes are insufficient and limited, although studies into the subject area is projected to benefit from a budgeting strategy that evaluates microplastic influx, outflow, and retention rates. These metrics will establish the residence durations of microplastics, which will likely impact the relationships between microbes and plastics in a variety of habitats, such as the benthic, littoral, and epilimnion zones.

The distribution of microplastics in lakes is probably also influenced by wind and wave activity (Eerkes-Medrano et al. 2015). Plasmid assemblages may experience a range of taxonomic and physiological changes as a result of the migration of microplastics from freshwater to marine settings, albeit the precise effects of this process are unknown (Osborn and Stojkovic 2014; Mincer et al. 2019).

For instance, it was discovered that Pseudomonas aeruginosa inhibited the formation of biofilms and decreased the rate at which this strain degraded benzoate when exposed to salt stress (0.5 M NaCl). Though current studies have looked at the temporal and spatial arrangement of colony/coalitions of maritime microbes on plastic, it is still unclear how the composition and structure of plastisphere environments for freshwater habitats fluctuate spatially and seasonally (Oberbeckmann et al. 2016; Oberbeckmann et al. 2014; Amaral-Zettler et al. 2015). On the basis of 6-week on site subjection of PET (polyethylene terephthalate) plastic containers/bottles, examination shows geographically or regionally influenced as well as periodic shifts in the chemical structure and arrangement of plastisphere ecosystems in the North Sea (Oberbeckmann et al. 2014; Oberbeckmann et al. 2016).

1.2.3.7 Method of Attachment

According to Scherwass et al., in order to create a kind of bacteria biofilm that protects against threats of predators, scarcity of food, and external stresses, bacteria adhere to surfaces like plastic. The selectivity of the early plastisphere may have been influenced by the microbes' capacity to bind to plastic, as a restricted range of microbial species exhibit morphologies that are favorable for substrate binding. For example, the plastisphere frequently contains Hyphomonas (Oberbeckmann et al. 2014; Zettler et al. 2013). This may be because they rarely express prosthecae, which are involved in substrate binding (Zettler et al. 2013).

According to Zettler et al. (2013), Bryant et al. (2016), Delacuvellerie et al. (2022a, b), Quorum sensing is a function of some Rhodobacteraceae, which are the extremely widespread Alphaproteobacteria found within the plastisphere. This function can also reinforce better navigation within and to the plastisphere, allowing these genera to spread (colonize) and survive with greater speed. Additionally, an electronegatively charged anions as capsular colanic acid or LPS, also known as lipopolysaccharide is produced by bacterium that are gram-negative. These compounds improve the bacterial electrostatic, molecular or covalent bond and the gas hydrogen's relationships to the polymer material, thereby improving the bacterium's adherence.

In this regard, glycosphingolipids, an increasingly anionic charge substitute for LPS (Wedeking and van Echten-Deckert 2007), are expressed by the AlphaproteobacteriaSphingomonas (Zettler et al. 2013; Dussud et al. 2018). This may account for the bacteria's abundance in the plastisphere. While the effect of the structure of cytoplasmic membrane on attachment to plastispheres has not been explicitly studied, conclusions about the most successful adhesion tactics employed by the plastisphere can be drawn from the strategies employed by its most prevalent bacteria.

Gammaproteobacteria is the predominant bacterial class found in the plastisphere, and the species belonging to Vibrionales, Alteromonadales, Oceanospirillales, and Pseudomonadales, are characteristic of this class (Dussud et al. 2018; Delacuvellerie et al. 2019). Vibrionales and Pseudomonadales additionally produces type IV pili (Tfp), which facilitates their adherence to the plastic; conclusively, gamma proteobacterium are diderm (gram-negative) aerobic bacterium that secrete LPS, which enables them stick to plastic (Chao and Zhang 2011). The creation of expanding capilliform compositions by Actinobacteria, developed specifically for adhesion to substrates, could result in the recurrent coupling of Gram-positive Actinobacteria, Gram-negative Planctomycetes and Cyanobacteria, and mixed-Gram Firmicutes (Pelicic 2019). Though less frequent than Bacteroidetes and Proteobacteria, they constitute several of the platisphere's most prolific microbes (Oberbeckmann et al. 2021; Delacuvellerie et al. 2022a, b). Extracellular polymeric substances, sometimes known as EPS is specifically secreted by cyanobacteria to promote microbial aggregation and protection (Schlundt et al. 2020a, b; Lagarde et al. 2016).

The diversity of the plastisphere is also greatly influenced by a variety of protists, cnidaria, sponges, nematodes, and arthropods (Zettler et al. 2013; Kirstein et al. 2018; Bryant et al. 2016). It is unknown how many eukaryotes still exist that adhere to plastic, such as the protozoan Radiolaria (Zettler et al. 2013). Given their capacity to attach to substrates via hyphae and archaella (Pelicic 2019; Kirstein et al. 2018; Bryant et al. 2016), fungi and archaea are also sporadically detected (Woodall et al. 2018; Latva et al. 2022). For eukaryotes and archaea, cell shape thus appears to be less important in plastisphere adhesion than it is in bacteria. Notably, bacteria are usually the initial colonials, whereas archaea, and eukaryotes are auxillary (secondary) colonials (Pollet et al. 2018; Bryant et al. 2016). Therefore, the chemical, physical, and molecular structure of the plastisphere that constitutes the initial prokaryotes is prone to be influenced by position and arrangement of cells than the plastisphere's maturity.

1.2.4 Impact of Plastisphere on Marine Organisms and Ecosystems

The region of the Mediterranean Sea has the highest average density of plastics in the world. (Munari et al. 2021; Pedrotti et al. 2022; van der Hal et al. 2017). Numerous studies have indicated that misplaced or unreported polymers (plastic), organic plastic non-biodegradable detritus may be engulfed by coastlines such as seagrass beds, coral reefs, and mangrove forests (Lebreton and Andrady 2019; Isobe and Iwasaki 2022; Bond et al. 2018). This could result in physical harm to the habitats (such as abrasion and shearing) and obstruct sediment oxygenation, which could lead to mangrove suffocation and possible leaf loss (van Bijsterveldt et al. 2021); sedimentation and decrease of seagrass cover (Menicagli et al. 2021). Coral disorders such as black band disease, white syndrome, and skeletal eroding band disease have become more common when there is a correlation between the debris and the corals (Lamb et al. 2018).

Plastic marine waste poses a severe hazard to the marine ecosystem worldwide, with dire effects for human health, wildlife, and the ocean. There is a great diversity of microorganisms adhering to plastic, and the main microbial assemblages found on plastic often consist of prokaryotic and eukaryotic phototrophs as well as bacteria that break down hydrocarbons. The addition of endocrine disruptor (Godswill and Godspel 2019; Lyche 2017; Ugoeze et al. 2021; Paluselli et al. 2018; Aransiola et al. 2023a) results in plastic wastes that cause serious injuries, deformities, or intoxication (Kühn et al. 2015; Alomar et al. 2020; Güven et al. 2017). Plastics can lower an organism's capacity for reproduction as well as its rates of growth and survival once ingested (Foley et al. 2018).

The waste from plastic are hydrophobic in nature, it can assimilate or take up dangerous alloys (metals) and persisting biotic or natural contaminants in the maritime environment (da Costa et al. 2016). A number of rafting species can also cling to the plastic waste and use it as an aquatic conveyance to reach new, alien ecosystems. These species include macrobenthos, microalgae, bryozoa, and insects (Keswani et al. 2016).

The buildup of marine trash, particularly plastic litter, can have detrimental impacts on the ecosystem, including the formation of anoxic conditions in the sediment, a reduction in organic matter and primary production, and an impact on biodiversity (Green et al. 2015; Rios-Fuster et al. 2023). These plastic wastes include toxic elements that saturate and poison the ecosystem (Ugoeze et al. 2021).

Plastic has the potential to act as a carrier for dangerous microbes. Plastic can travel great distances in the marine environment because of its extended residence duration, which helps colonists survive. Vibrio, a genus that contains dangerous species on plastic fragments in the Atlantic, and a possibly pathogenic Vibrio parahaemolyticus on many microplastic particles from the North Sea and Baltic Sea were found to be present, according to Zettler et al. (2013) and Kirstein et al. Couet et al. (2018) discovered transient cysts of the dinoflagellate Alexandriumtaylori, a species known to cause harmful algae blooms (HABs), adhered to plastic waste discovered in Mediterranean Sea coastal and open ocean waters, even in areas of the sea that have never seen HAB episodes. The Mediterranean Sea has potentially dangerous poisonous species, including Ostreopsis and Coolia spp.

Plastic surfaces are abundant in potentially hazardous cyanobacteria. Marine species, especially shellfish, may become poisoned in these environments because to the biological accumulation of cyanotoxins produced by these cyanobacteria, such as Lyngbya (Mazard et al. 2016). The ecotoxicological implications of the plastisphere cannot be disregarded, even if it is unknown if marine creatures in these settings eat colonized plastic trash.

An increasingly serious environmental concern and a growing menace to both humans and ecosystems, plastic has been found recently in a variety of habitats, including human tissues. Eventually, the disturbance of physical and chemical processes in the water causes the plastic waste to break into microplastics (MPs) (Morfin 2021). Animals can also consume (micro) plastics, which can cause digestive system lesions that can result in false satiation, malnutrition, and frequently the death of the organisms (Gesamp 2016). Microplastics can enter the food chain and eventually find their way into the human digestive system through the ingestion of microplastics by zooplankton and other small marine creatures (Keswani et al. 2016).

Pathogenic organisms and toxics can enter the food chain through the consumption of (micro)plastics because plastic can serve as a vector for both chemical and biological compounds (Keswani et al. 2016). Extensive studies have revealed the fact that tiny polymeric materials, known as nanoplastics, can penetrate through the protective membranes of cells. Nanoplastics were discovered in Oryziaslatipes's blood, liver, and even brains following exposure to the plastics, suggesting that the polymers could pass through the blood-brain barrier (GESAMP 2016; Almutairi et al. 2016).

1.3 Methods for Studying Plastispheres

1.3.1 Sampling Techniques for Collecting Plastisphere Samples

Sampling strategies that are both efficient and capture the diversity and distribution of plastisphere in marine environments are essential for comprehending its composition and dynamics. The sampling method employed, primarily depends on the environment being sampled as well as the size limits of the microplastic to be collected (Duis and Coors 2016). Nevertheless, the sampling techniques have been said to be subject to evolution and adaptation as no standard practice has been established for the procedure of sampling micro plastics (Adomat and Grischek 2021).

The majority of water bodies contains microplastics and the method employed to gather water samples for microplastic analysis depends on the kind of aquatic environment being examined. The physiochemical characteristics of microplastics, such as their shape, density, and chemical composition variations as well as the degree of biofouling, and their distribution in aqueous environments allow for their discovery in surface waters, deep within the ocean, or suspended in the water column. The location of the microplastics in the water dictates the sampling technique to be used, which can be either vertical sampling through the water column or horizontal sampling along the water's surface. The weather and the availability of boats and equipment are examples of factors that should be considered early in the planning stages of sample collection because the sampling technique selected will have a big impact on the outcome.

1.3.1.1 Sampling Along the Water's Surface

Neutson nets which have been used in 28 out of 33 studies of microplastic sampling are the most widely used type of nets in surface water sampling. The horizontal sampling of both hyponeuston—organisms which exist below the water's surface and epineuston—organisms that exist on the water's surface film are usually performed using neuston nets. The net aperture of rectangular neuston nets which were reported were found to range between 3 and 200 cm (Hidalgo-Ruz et al. 2012).

Undoubtedly, surface sampling is most effective in calm, level waters. However, in more wide areas, where the net's depth can change significantly, it is advisable to employ a more stable manta trawl or catamaran (Fossi et al. 2016). The microplastic samples obtained from this method depends largely on the net's design, aperture, mesh size and length.

1.3.1.2 Sampling through the Water Column

The sampling of water column can be both vertically and horizontally. In the case of vertical microplastic sampling, the net is either drawn up from a predetermined depth to provide a top-down view of the entirewater column. A bongo net is pulled upwards to achieve this. Another sampling methodfor water column vertically is to open and close at specific depth levels (Kukulka et al. 2012). In a similar vein, microplastics are sampled horizontally using a bongo net. The net is lowered to a predetermined depth, over the bottom, and hauled there for a predetermined period of time at a predetermined speed before being recovered. Although, the net aperture is sometimes not disclosed, but when it is, circular bongo nets were reported to be between 79 and 158 cm (Hidalgo-Ruz et al. 2012).

1.3.2 Analytical Methods for Studying Plastisphere Composition

Researchers utilize diverse analytical techniques to examine microbial communities and their relationships with plastic surfaces after obtaining plastisphere samples. The first step usually involves the physical characterization of the microplastics predicated on a number of microscopies, including scanning electron, fluorescence, and optical microscopy (Gniadek and Dąbrowska 2019). The plastisphere have been shown to be a crowded, surface-based microecosystem that contains a wide variety of microorganisms, as shown by imaging technologies likescanning electron microscopy (SEM), spectral imaging fluorescence in situ hybridization (CLASI-FISH) and combinatorial labeling (Zhao et al. 2021; Schlundt et al. 2020a, b; Wang et al. 2022).

Scanning electron microscopy (SEM) is an effective technology for identifying minute structural details, such as defects, and living things that have colonized surfaces of plastic wastes (Zettler et al. 2013; Gniadek and Dąbrowska 2019).

The chemical composition of identified substances can be confirmed by subjecting it to spectroscopic analyses (Wang et al. 2017). The FTIR and Raman spectroscopic techniques have been found to be the most effective and widely applied methods for the detection and characterization of microplastics (Käppler et al.

Table 1.1 Analytical methods employed in the identification and characterization of microplastics

Method	Advantages
Optical microscopy	Fast, easy and inexpensive
FTIR	Fast data acquisition and analysis; non-destructive; single point measurement with good specific resolution; surface map of a large sample
Raman spectroscopy	Dependable; non-destructive; single point measurement with good specific resolution; spectra as a source of both quantitative and qualitative data; surface map of a large sample
GC-MS	Dependable and sensitive; no solvent required; automatic registration and analysis

Gniadek and Dąbrowska (2019)

2016). These two methods are said to be complementary since the bands which appear strong in one are absent or weak in the other (Gniadek and Dąbrowska 2019). Zettler et al. 2013 employed the Raman spectroscopy in the identification of the resin composition of plastic pieces by comparing with reference scans of known composition. Using mass spectrometry (MS) with specialized interfaces, polymers, degradation products, and additives can be identified. These interfaces include thermal (pyrolysis, py) and chromatographic (GC and LC) techniques for more accurate characterization of microplastics (Hale et al. 2022). Table 1.1 provides a comparison of the analytical methods employed in the identification and characterization of microplastics.

1.3.3 Advances in Technology and Methodologies

The study of plastispheres has evolved through recent technological developments, which have made it possible to study this complex microbial environment using new tools and techniques. Some of these advancements in technology include the SEM-EDS spectroscopy, amplicon and metagenomics sequencing. Detailed information on the elemental composition of microplastics and additional information about the inorganic additives they contain can be obtained by combining scanning electron microscopy and energy-dispersive X-ray spectroscopy (SEM-EDS). Due to the simultaneous collection of images and elemental mapping, the use of SEM-EDS facilitates the further differentiation of natural components that produce microplastics (Gniadek and Dąbrowska 2019). Pan et al. 2019, discovered debris made of polypropylene, polyethylene, and polystyrene may show strong nitrogen peaks on their surface, which may indicate the presence of biomass. The high concentration of Nitrogen in the microplastic samples indicates bioaccumulation as it shows interactions between biota and microplastics.

Next-generation sequencing (NGS) methods have become increasingly popular in recent years for the investigation of microbial populations (Amaral-Zettler et al. 2020; De Tender et al. 2017a, b). Amplicon and metagenomic sequencing have been utilized in the majority of plastisphere microbiome surveys. In microbiome research, polymerase chain reaction (PCR) combined with taxonomic marker gene sequencing is widely used to determine the taxonomic profiles of microbial communities at a low cost. For the purpose of identifying bacterial communities within the plastisphere, the most often utilized marker is the 16S rRNA gene. Sequencing variable sections from the conserved 16S ribosomal RNA (16S rRNA) is the primary method employed to explore changes in species diversity (Knapp et al. 2017; Chen et al. 2019; Learman et al. 2019). The bacterial communities on microplastics taken from two Chinese urban waterways were examined by Wang et al. 2017 using 16S rRNA gene sequencing. Likewise, by using Illumina sequencing of 16S rRNA gene libraries, Ogonowski et al. 2018 investigated the Baltic bacterioplankton and plastisphere and revealed substrate-driven selection.

In recent investigations, multiomics, including, metabolomic, metatranscriptomics and proteogenomic approaches—have been applied to enable a full understanding of the active members and acting functions of the plastisphere, following significant advances in technology and analysis. These techniques hold great promise for examining the functional microorganisms within the plastispheres, including possible bacteria or enzymes involved in the breakdown of plastics.

1.3.4 Challenges and Limitations in Plastisphere Research

The field of plastisphere research is limited and confronted with various obstacles, even with notable advancements. One significant barrier is the heterogeneity in plastisphere composition and function under varying environmental conditions, which makes it difficult to draw broad conclusions and develop universal principles for sampling and analysis.

Research on the fate, distribution and effects of plastic waste in the environment has been hampered thus far by the absence of sampling and analytical techniques that are able to fully characterize its diversity. There exists a range of sizes of plastic debris in the environment, and plastics break down over time in the environment, at different rates based on the content of the polymers and the surrounding circumstances. Size distributions therefore change throughout time. A subset of microplastics, usually those larger than 300 μm, are the focus of the majority of reported approaches (Hale et al. 2022). Smaller particles might not be reported as the majority of methods use sampling equipment (such as plankton nets) with apertures larger than 300 μm. Although harder to find, smaller microplastics may, paradoxically, be more prevalent in environmental samples (Enders et al. 2015).

Effective sample purification has been discovered to have a major impact on the quality of spectroscopic results. The subject of microplastic sample contamination across the whole assessment process, from sampling to analysis, is critical and requires the application of standardized contamination prevention methods (Löder and Gerdts 2015).

1.4 Influence of Plastisphere on Marine Organisms

The influence of the plastisphere on marine organisms is complex and multifaceted, with both direct and indirect effects on individual organisms, populations, and ecosystems. Understanding these interactions is crucial for assessing the ecological impacts of plastic pollution and developing strategies to mitigate its effects on marine biodiversity and ecosystem functioning. In the last few decades, the accumulation of plastics in the sea has increased drastically. A larger portion of these plastics submerged in water because they are denser than water (such as Polyvinyl Chloride, PVC), after some period of time, as a result of fouling by microorganism, most positively buoyant plastic (floating plastic) tends to adsorb or absorb water and sink. However, plastic items on the surface of waterways usually photodegrade by oxidation when exposed to UV radiation from sunlight (Gewert et al. 2015). Physical forces like wind, waves, and sand abrasion cause the plastic debris to break down and erode over time. Biological processes and chemical weathering also play a role. The toxicity of plastic polymers and their additives—which are frequently purposefully added to products during production or processing to enhance performance; for marine life (through ingestion and other mechanisms) has become a significant worry (Gewert et al. 2015).

Micro plastics as well as damaged plastic pieces are similar to phytoplankton which are consumed by some marine organism such as fish, crustaceans, shear waters, albatrosses, or petrels etc. Ingested plastic debris has been shown to impair digestion, hinder growth, injure internal organs, and result in intestinal blockages (Wilcox et al. 2016). In certain cases, plastic clogging of fishing nets or many other ring-shaped objects can lead to drowning, reduced feeding capacity, and strangulation. According to estimates, plastic waste poses a threat to at least 25% of aquatic mammal species, 30% of seabird species, and 74% of sea turtle species worldwide (Das and Dash 2020). Figure 1.1 shows the influence of plastisphere formation and function.

1.4.1 Impact on Microbial Communities

Plastisphere exerts significant influence on microbial communities in aquatic environments, with implications for ecosystem functioning, biogeochemical cycling, and ecological interactions. Plastic debris provides a substrate for microbial attachment and growth, altering the composition and dynamics of microbial communities in aquatic ecosystems (Aransiola et al. 2023b). This can lead to shifts in the balance

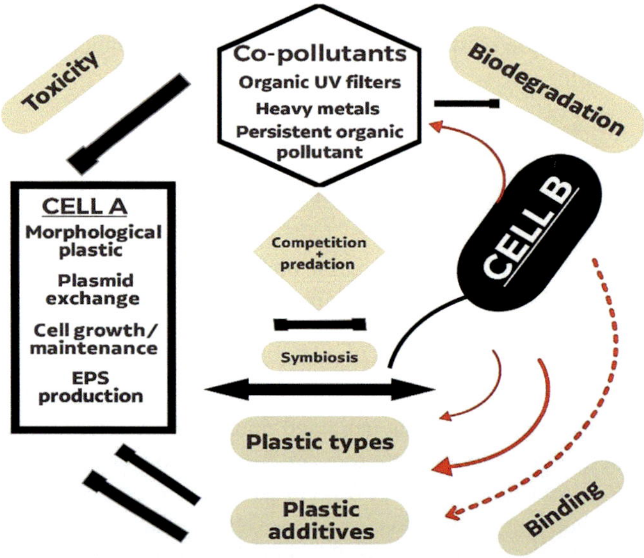

Fig. 1.1 Influence of plastisphere formation and function

of microbial populations, potentially favoring certain species over others. Microbial communities in the plastisphere release metabolites that impact the nearby ecosystem and are more resilient to human intervention than bacteria in the surrounding water (Bryant et al. 2016). Moreover, plastic waste encourages pathogen infiltration. Food security and human health may be at risk from compounds produced by microorganisms that colonize plastic waste. These microorganisms may include known pathogens (Amaral-Zettler et al. 2020). Some microorganisms within the plastisphere possess the ability to degrade certain types of plastics.

These plastic-degrading microorganisms produce enzymes capable of breaking down the polymers in plastics (Fig. 1.2), leading to physical and chemical changes in the plastic material. This process can influence the composition and metabolic activities of microbial communities within the plastisphere. Plastics can be broken down by microbes like fungi and bacteria, and numerous reviews of the literature have updated the list of these microbes (Jadaun et al. 2022; Kale et al. 2015; Pathak 2017). In addition, plastic has the potential to harbor antibiotic resistance genes (ARGs), serving as a source of ARGs in aquatic environments. ARG transfer rate may be accelerated and a substrate for gene exchange could be provided by microbial biofilm formation in the plastisphere of aquatic ecosystems (Li et al. 2020). Microorganisms in the plastisphere often form biofilms, complex communities of microorganisms encased in a matrix of extracellular polymeric substances. These biofilms facilitate microbial attachment to plastic surfaces and can modify local nutrient cycling, microbial interactions, and ecosystem dynamics. The biosynthesis of extracellular polysaccharides (EPS) on the surface of microplastic particles can facilitate the formation of biofilms by a number of bacterial taxa. Microplastics

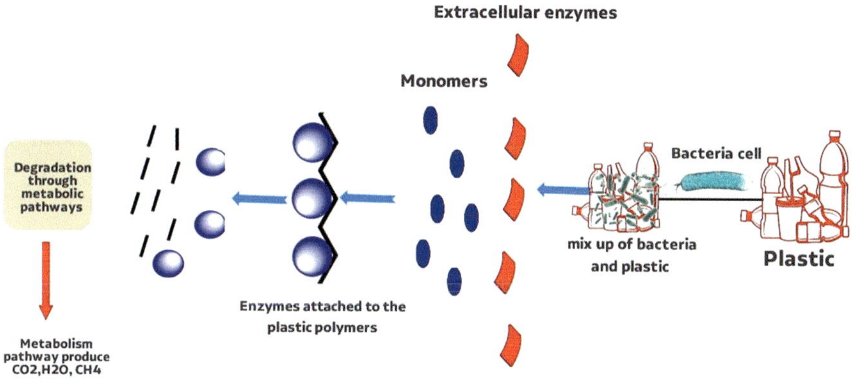

Fig. 1.2 Mechanism of plastic degradation through microbial enzymes

fragment and biodegrade as a result of biofilm formation on their surface (Zhou et al. 2022). The nutrients present in the surrounding seawater can have a significant impact on the composition of biofilm communities. It was discovered that many bacterial species attach to the substrate surface when nutrient levels are low, but many microorganisms' ability to attach is diminished when nutrient levels are high, which is detrimental to the formation of specialized biofilms (Stanley and Lazazzera 2004).

Deep water's low temperature and lack of sunlight change the microbial communities on their surface, and bacteria and other microorganisms primarily break down microplastics through biodegradation (Rummel et al. 2017). Plastics can absorb and accumulate organic pollutants and nutrients from the surrounding environment. Microorganisms within the plastisphere may interact with these chemicals, either metabolizing them as energy sources or being negatively impacted by their presence. These chemical interactions can shape the composition and function of microbial communities within the plastisphere.

1.4.2 Behavioral and Physiological Responses to Plastisphere Presence

The presence of the plastisphere can elicit various behavioral and physiological responses in organisms within the ecosystems. Organisms may exhibit altered feeding behaviors in the presence of plastic debris colonized by the plastisphere. Some species may mistake plastic particles for food, leading to ingestion and potential digestive issues or blockages. Others may avoid feeding near plastic debris altogether, altering their foraging patterns and potentially impacting their nutritional intake (Naidoo et al. 2020). Organisms may exhibit altered behaviors in response to physical interactions with plastic debris. For example, marine animals may become entangled in discarded fishing nets or other plastic materials, leading to injuries, reduced mobility, or even death. Additionally, the accumulation of plastic debris on the seafloor can modify habitat structure, affecting species that rely on benthic environments for shelter and substrate (Fang et al. 2018).

Plastics can create oxygen-depleted microenvironments on the surface of marine debris, which may affect the respiratory physiology of organisms living in close proximity to these surfaces Some chemicals released from plastic debris have been found to mimic hormones in organisms, disrupting endocrine function. This can lead to developmental abnormalities, reproductive dysfunction, and altered behavior. Organisms that come into contact with plastic debris may experience inflammatory responses as a result of the interaction with microorganisms in the plastisphere. This can lead to immune system activation and potentially chronic inflammatory conditions (Li et al. 2020).

1.5 Ecological Consequences of Plastisphere on Marine Organisms

Ecological consequences of the plastisphere on marine organisms can be significant and wide-ranging. The ocean's ecology and food chain are impacted by the significant amount of microplastics in the water. Microplastics in the ocean have the potential to impact the ocean's carbon stock, slow down zooplankton swimming and hence decrease their reproductive efficiency, and affect the photosynthesis and growth of phytoplankton (Sjollema et al. 2016; Lim 2021; Wu et al. 2020a, b). Furthermore, some pathogenic microorganisms may use floating microplastics as a means of transportation, which would hasten the spread of infectious diseases (Keswani et al. 2016). The microbiota that drifting microplastics carry can be transferred to new habitats and even across oceans when they are dispersed by waves and currents (Keswani et al. 2016). Researchers have also discovered significant concentrations of microplastics in the sea surface and sediments of the Arctic Ocean, which has been less impacted by human activity (Ramasamy et al. 2021; Collard et al. 2021). It's likely that the food chain and food web have enriched microplastics. They are contaminating the food web to a far greater extent than we currently realize because they have been discovered in fish, shellfish, Antarctic krill, and oceanic coral (Hipfner et al. 2018; Savinelli et al. 2020; Gonçalves et al. 2019; Dawson et al. 2018). Most common micro-plastics type is mainly blue-colored (Mattan-Moorgawa et al. 2021). Because marine animals may mistake these colorful microplastics for food and consume them directly, they are especially dangerous in aquatic environments (Mattan-Moorgawa et al. 2021). Microplastics that penetrate biological cells may affect living things more severely because they can disrupt energy redistribution, cause localized inflammation, and cause weight loss (Li et al. 2020; Sussarellu et al. 2016). Apart from the potential risks posed by the micro-plastics themselves, the organic pollutants that are adsorbed on their surface, like diazinon, nonylphenol, and phenanthrene, can also seriously harm the environment (Seidensticker et al. 2018). Microplastics can adsorb and accumulate various toxic pollutants from seawater, including heavy metals, pesticides, and persistent organic pollutants (POPs). When marine organisms ingest microplastics, they may also be exposed to these harmful chemicals, leading to toxicity, bioaccumulation, and biomagnification through the food chain. Exposure to microplastics and associated pollutants can induce genetic and physiological stress in marine organisms, affecting their reproductive success, growth rates, immune responses, and overall fitness. In addition, the accumulation of microplastics and their associated effects can lead to ecological imbalances in marine ecosystems, with potential cascading effects on ecosystem structure and function.

1.6 Conclusion, Challenges and Future Prospect

The plastisphere is a unique microbial ecosystem that lives on and within plastic debris in the marine environment. This study has shed important light on the prevalence and distribution of this ecosystem. It has made clear the complex interactions between microbiological ecology and plastic pollution. It has been possible to obtain important insights into the genesis, dynamics, and ecological significance of the plastisphere by examining the literature and empirical data. The present review highlights the pressing necessity of implementing coordinated measures to tackle the widespread hazard of plastic pollution and its correlated effects on marine environments.

To improve our knowledge and lessen its ecological effects, it is important to address critical issues that has to do with the plastisphere. Understanding the intricate relationships that exist within the plastisphere and how they affect marine biodiversity and ecosystem functioning is one of the main challenges. Furthermore, it is extremely difficult to determine the prevalence and distribution of the plastisphere in various habitats and geographical areas due to the heterogeneous nature of plastic debris and the dynamic nature of marine environments. Also. For reliable data collection and comparison, standardized techniques for researching the plastisphere in various marine environments must be developed. Moreover, little is known about the possible effects of the plastisphere on human health, underscoring the necessity of more study in this important field.

In light of the aforementioned, efforts in research should concentrate on resolving these issues and deepening our understanding of the ecological dynamics and implications of the plastisphere. Deciphering the intricacies of the plastisphere and its interactions with marine ecosystems will require integrating interdisciplinary approaches, such as those from microbiology, ecology, oceanography, and environmental science. Furthermore, utilizing cutting-edge technologies will improve our capacity to investigate the plastisphere at different temporal and spatial scales. Examples of these technologies include remote sensing, imaging methods, and genomic sequencing. Thorough grasp of the plastisphere's function in the marine environment can be obtained by tackling these issues and following these future research avenues. This information can help develop policies that reduce plastic pollution, encourage biodegradation, and safeguard the wellbeing of marine environments.

References

Adomat Y, Grischek T (2021) Sampling and processing methods of microplastics in river sediments – a review. Sci Total Environ 758:143691. https://doi.org/10.1016/j.scitotenv.2020.143691

Almutairi MM, Gong C, Xu YG, Chang Y, Shi H (2016) Factors controlling permeability of the blood–brain barrier. Cell Mol Life Sci 73:57–77

Alomar C, Estarellas F, Deudero S (2016) Microplastics in the Mediterranean Sea: deposition in coastal shallow sediments, spatial variation and preferential grain size. Mar Environ Res 115:1–10

Alomar C, Deudero S, Compa M, Guijarro B (2020) Exploring the relation between plastic ingestion in species and its presence in seafloor bottoms. Mar Pollut Bull 160:111641

Amaral-Zettler LA (2022) Colonization of plastic marine debris: the known, the unknown, and the unknowable. In: Andrady AL (ed) Plastics and the ocean: origin, characterization, fate, and impacts. Wiley, Hoboken, pp 301–316

Amaral-Zettler LA, Zettler ER, Slikas B, Boyd GD, Melvin DW, Morrall CE et al (2015) The biogeography of the plastisphere: implications for policy. Front Ecol Environ 13(10):541–546

Amaral-Zettler LA, Erik RZ, Tracy JM (2020) Ecology of the plastisphere. Nat Rev Microbiol 18:139–151. https://doi.org/10.1038/s41579-019-0308-0

Anand TP, Chellaram C, Shanthini F (2013) Competitive dominance of potential bacteria from marine organisms. J Pharm Sci Res 5(6):137

Anjana K, Hinduja M, Sujitha K, Dharani G (2020) Review on plastic wastes in marine environment – biodegradation and biotechnological solutions. Mar Pollut Bull 150:110733

Aransiola SA, Ikhumetse AA, Babaniyi BR, Abioye OP, Oyedele OJ, Falade NO (2023a) Phytoaccumulation of micro and nanoplastics: root uptake. In: Maddela NR, Reddy KV, Ranjit P (eds) Micro and nanoplastics in soil. Springer, Cham. https://doi.org/10.10007/978-3-031-21195-9_8

Aransiola SA, Victor-Ekwebelem MO, Ajiboye AE, Leh-Togi Zobeashia SS, Ijah UJJ, Oyedele OJ (2023b) Ecological impacts and toxicity of micro and nanoplastics in agroecosystem. In: Maddela NR, Reddy KV, Ranjit P (eds) Micro and nanoplastics in soil. Springer, Cham. https://doi.org/10.10007/978-3-031-21195-9_10

Auta SH, Abioye OP, Aransiola SA, Bala JD, Chukwuemeka VI, Hassan A, Aziz A, Fauziah SH (2022) Enhanced microbial degradation of PET and PS microplastics under natural conditions in mangrove environment. J Environ Manag 304(2022):114273. https://doi.org/10.1016/j.jenvman.2021.114273

Bakir A, Rowland SJ, Thompson RC (2014) Enhanced desorption of persistent organic pollutants frommicroplastics under simulated physiological conditions. Environ Pollut (Barking, Essex: 1987) 185:16–23

Bolaños LM, Choi CJ, Worden AZ, Baetge N, Carlson CA, Giovannoni S (2021) Seasonality of themicrobial community composition in the North Atlantic. Front Mar Sci 8:624164

Bond T, Ferrandiz-Mas V, Felipe-Sotelo M, Van Sebille E (2018) The occurrence and degradation of aquatic plastic litter based on polymer physicochemical properties: a review. Crit Rev Environ Sci Technol 48(7–9):685–722

Boucher J, Friot D (2017) Primary microplastics in the oceans: a global evaluation of sources. International Union for Conservation of Nature (IUCN), Gland

Brandon J, Goldstein M, Ohman MD (2016) Long-term aging and degradation of microplastic particles: comparing in situ oceanic and experimental weathering patterns. Mar Pollut Bull 110(1):299–308

Bryant JA, Clemente TM, Viviani DA, Fong AA, Thomas KA, Kemp P, Karl DM, White AE, DeLong EF (2016) Diversity and activity of communities inhabiting plastic debris in the North Pacific gyre. mSystems 1(3):e00024-00016

Buchan A, LeCleir GR, Gulvik CA, González JM (2014) Master recyclers: features and functions of bacteria associated with phytoplankton blooms. Nat Rev Microbiol 12(10):686–698

Cai L, Wu D, Xia J, Shi H, Kim H (2019) Influence of physicochemical surface properties on the adhesion of bacteria onto four types of plastics. Sci Total Environ 671:1101–1107. https://doi.org/10.1016/j.scitotenv.2019.03.434

Chao Y, Zhang T (2011) Probing roles of lipopolysaccharide, type 1 fimbria, and colanic acid in the attachment of Escherichia coli strains on inert surfaces. Langmuir 27(18):11545–11553

Chen J, Li J, Zhang H, Shi W, Liu Y (2019) Bacterial heavy-metal and antibiotic resistance genes in a copper tailing dam area in northern China. Front Microbiol 10:1916. https://doi.org/10.3389/fmicb.2019.01916

Cheng J, Jacquin J, Conan P, Pujo-Pay M, Barbe V, George M, Fabre P, Bruzaud S, Ter Halle A, Meistertzheim A-L, Ghiglione J-F (2021) Relative influence of plastic debris size and shape, chemicalcomposition and phytoplankton-bacteria interactions in driving seawater plastisphere abundance, diversity andactivity. Front Microbiol 11:610231

Collard F, Husum K, Eppe G, Malherbe C, Hallanger IG, Divine DV, Gabrielsen GW (2021) Anthropogenic particles in sediment from an Arctic fjord. Sci Total Environ 772:145575

Coons AK, Busch K, Lenz M, Hentschel U, Borchert E (2021) Biogeography rather than substrate type determines bacterial colonization dynamics of marine plastics. PeerJ 9:e12135

Corcoran PL, Norris T, Ceccanese T, Walzak MJ, Helm PA, Marvin CH (2015) Hidden plastics of Lake Ontario, Canada and their potential preservation in the sediment record. Environ Pollut 204:17–25

Corradini F, Meza P, Eguiluz R, Casado F, Huerta-Lwanga E, Geissen V (2019) Evidence of microplastic accumulation in agricultural soils from sewage sludge disposal. Sci Total Environ 671:411–420

Couet D, Pringault O, Bancon-Montigny C, Briant N, Poulichet FE, Delpoux S et al (2018) Effects of copper and butyltin compounds on the growth, photosynthetic activity and toxin production of two HAB dinoflagellates: the planktonic Alexandriumcatenella and the benthic Ostreopsis cf. ovata. Aquat Toxicol 196:154–167

Crawford CB, Quinn B (2017a) 8 – Microplastic collection techniques. In: Microplastic pollutants. Elsevier, pp 179–202. https://doi.org/10.1016/B978-0-12-809406-8.00008-6

Crawford CB, Quinn B (2017b) 4 – Physiochemical properties and degradation. In: Crawford CB, Quinn B (eds) Microplastic pollutants. Elsevier, Amsterdam, pp 57–100

da Costa JP, Santos PS, Duarte AC, Rocha-Santos T (2016) (Nano)plastics in the environment – sources, fates and effects. Sci Total Environ 566:15–26

Dang H, Lovell CR (2016) Microbial surface colonization and biofilm development in marine environments. Microbiol Mol Biol Rev 80(1):91–138

Das MR, Dash A (2020) Impact of plastic pollution on aquatic ecosystem. PalArch's J Archaeol Egypt/Egyptol 1s7(6):5201–5209

Datta MS, Sliwerska E, Gore J, Polz MF, Cordero OX (2016) Microbial interactions lead to rapidmicro-scale successions on model marine particles. Nat Commun 7(1):11965

Dawson A, Huston W, Kawaguchi S, King C, Cropp R, Wild S et al (2018) Uptake and depuration kinetics influence microplastic bioaccumulation and toxicity in Antarctic krill (Euphausia superba). Environ Sci Technol 52(5):3195–3201

De Tender CA, Devriese LI, Haegeman A, Maes S, Ruttink T, Dawyndt P (2015) Bacterial communityprofiling of plastic litter in the Belgian part of the North Sea. Environ Sci Technol 49(16):9629–9638

De Tender C, Schlundt C, Devriese LI, Mincer TJ, Zettler ER, Amaral-Zettler LA (2017a) A review of microscopy and comparative molecular-based methods to characterize "plastisphere" communities. Anal Methods 9:2132–2143. https://doi.org/10.1039/C7AY00260B

De Tender C, Devriese LI, Haegeman A, Maes S, Vangeyte J, Cattrijsse A et al (2017b) Temporal dynamics of bacterial and fungal colonization on plastic debris in the North Sea. Environ Sci Technol 51(13):7350–7360

Delacuvellerie A, Cyriaque V, Gobert S, Benali S, Wattiez R (2019) The plastisphere in marine ecosystemhosts potential specific microbial degraders including Alcanivoraxborkumensis as a key player for the low-densitypolyethylene degradation. J Hazard Mater 380:120899

Delacuvellerie A, Ballerini T, Frère L, Matallana-Surget S, Dumontet B, Wattiez R (2022a) From rivers tomarine environments: a constantly evolving microbial community within the plastisphere. Mar Pollut Bull 179:113660

Delacuvellerie A, Géron A, Gobert S, Wattiez R (2022b) New insights into the functioning and structure ofthe PE and PP plastispheres from the Mediterranean Sea. Environ Pollut (Barking, Essex: 1987) 295:118678

Du Y, Liu X, Dong X, Yin Z (2022) A review on marine plastisphere: biodiversity, formation, and role in degradation. Comput Struct Biotechnol J 20:975–988

Duis K, Coors A (2016) Microplastics in the aquatic and terrestrial environment: sources (with a specific focus on personal care products), fate and effects. Environ Sci Eur 28:1–25

Dussud C, Hudec C, George M, Fabre P, Higgs P, Bruzaud S, Delort A-M, Eyheraguibel B, Meistertzheim A-L, Jacquin J, Cheng J, Callac N, Odobel C, Rabouille S, Ghiglione J-F (2018) Colonization of non-biodegradable and biodegradable plastics by marine microorganisms. Front Microbiol 9:1571

Eerkes-Medrano D, Thompson RC, Aldridge DC (2015) Microplastics in freshwater systems: a review of the emerging threats, identification of knowledge gaps and prioritisation of research needs. Water Res 75:63–82

Enders K, Lenz R, Stedmon CA, Nielsen TG (2015) Abundance, size and polymer composition of marine microplastics ≥10 μm in the Atlantic Ocean and their modelled vertical distribution. Mar Pollut Bull 100(1):70–81. https://doi.org/10.1016/j.marpolbul.2015.09.027

Eriksson C, Burton H, Fitch S, Schulz M, van den Hoff J (2013) Daily accumulation rates of marine debris on sub-Antarctic Island beaches. Mar Pollut Bull 66:199–208

Fang C, Zheng R, Zhang Y, Hong F, Mu J, Chen M, Song P, Lin L, Lin H, Le F (2018) Microplastic contamination in benthic organisms from the Arctic and sub-Arctic regions. Chemosphere 209:298–306

Fazey FM, Ryan PG (2016) Biofouling on buoyant marine plastics: an experimental study into the effect of size on surface longevity. Environ Pollut 210:354–360

Fish KE, Osborn AM, Boxall J (2016) Characterising and understanding the impact of microbial biofilms and the extracellular polymeric substance (EPS) matrix in drinking water distribution systems. Environ Sci Water Res Technol 2(4):614–630

Foley CJ, Feiner ZS, Malinich TD, Höök TO (2018) A meta-analysis of the effects of exposure to microplastics on fish and aquatic invertebrates. Sci Total Environ 631:550–559

Fossi MC, Marsili L, Baini M, Giannetti M, Coppola D, Guerranti C, Caliani I, Minutoli R, Lauriano G, Finoia MG, Rubegni F, Panigada S, Bérubé M, Urbán Ramírez J, Panti C (2016) Fin whales and microplastics: the Mediterranean Sea and the Sea of Cortez scenarios. Environ Pollut 209:68–78

Free CM, Jensen OP, Mason SA, Eriksen M, Williamson NJ, Boldgiv B (2014) High-levels of microplastic pollution in a large, remote, mountain lake. Mar Pollut Bull 85(1):156–163

Frère L, Maignien L, Chalopin M, Huvet A, Rinnert E, Morrison H, Kerninon S, Cassone A-L, Lambert C, Reveillaud J, Paul-Pont I (2018) Microplastic bacterial communities in the bay of brest: influence of polymer type and size. Environ Pollut (Barking, Essex: 1987) 242(Pt A):614–625

Gesamp G (2016) Sources, fate and effects of microplastics in the marine environment: part two of a global assessment. IMO, London, p 220

Gewert B, Plassmann MM, MacLeod M (2015) Pathways for degradation of plastic polymers floating in the marine environment. Environ Sci: Processes Impacts 17(9):1513–1521
Gniadek M, Dąbrowska A (2019) The marine nano- and microplastics characterisationby SEM-EDX: the potential of the method in comparison with various physical and chemical approaches. Mar Pollut Bull 148:210–216. https://doi.org/10.1016/j.marpolbul.2019.07.067
Godswill AC, Godspel AC (2019) Physiological effects of plastic wastes on the endocrine system (Bisphenol A, Phthalates, Bisphenol S, PBDEs, TBBPA). Int J Bioinf Comput Biol 4:11–29
Gonçalves C, Martins M, Sobral P, Costa PM, Costa MH (2019) An assessment of the ability to ingest and excrete microplastics by filter-feeders: a case study with the Mediterranean mussel. Environ Pollut 245:600–606
Gong M, Yang G, Zhuang L, Zeng EY (2019) Microbial biofilm formation and community structure on low-density polyethylene microparticles in lake water microcosms. Environ Pollut 252:94–102
Green DS, Boots B, Blockley DJ, Rocha C, Thompson R (2015) Impacts of discarded plastic bags on marine assemblages and ecosystem functioning. Environ Sci Technol 49(9):5380–5389
Gregory SV, Swanson FJ, McKee WA et al (1991) An ecosystem perspective of riparian zones. Bioscience 41(8):540–551. https://doi.org/10.2307/1311607
Güven O, Gökdağ K, Jovanović B, Kıdeyş AE (2017) Microplastic litter composition of the Turkish territorial waters of the Mediterranean Sea, and its occurrence in the gastrointestinal tract of fish. Environ Pollut 223:286–294
Hale RC, Seeley ME, King AE, Yu LH (2022) Microplastic in the environment: pattern and process, Environmental contamination remediation and management. Springer, Cham
Harrison JP, Schratzberger M, Sapp M, Osborn AM (2014) Rapid bacterial colonization of low-density polyethylene microplastics in coastal sediment microcosms. BMC Microbiol 14(1):1–15
Harrison JP, Hoellein TJ, Sapp M, Tagg AS, Ju-Nam Y, Ojeda JJ (2018) Microplastic-associated biofilms: a comparison of freshwater and marine environments. In: Freshwater microplastics: emerging environmental contaminants? Springer, Cham, pp 181–201
Hidalgo-Ruz V, Gutow L, Thompson RC, Thiel M (2012) Microplastics in the marine environment: a review of the methods used for identification and quantification. Environ Sci Technol 46:3060–3075
Hipfner JM, Galbraith M, Tucker S, Studholme KR, Domalik AD, Pearson SF et al (2018) Two forage fishes as potential conduits for the vertical transfer of microfibres in northeastern Pacific Ocean food webs. Environ Pollut 239:215–222
Horton AA, Walton A, Spurgeon DJ, Lahive E, Svendsen C (2017) Microplastics in freshwater and terrestrial environments: evaluating the current understanding to identify the knowledge gaps and future research priorities. Sci Total Environ 586:127–141
Hossain MR, Jiang M, Wei Q, Leff LG (2019) Microplastic surface properties affect bacterial colonizationin freshwater. J Basic Microbiol 59(1):54–61
Huang D, Xu Y, Lei F, Yu X, Ouyang Z, Chen Y et al (2021) Degradation of polyethylene plastic in soil and effects on microbial community composition. J Hazard Mater 416:126173
Isobe A, Iwasaki S (2022) The fate of missing ocean plastics: are they just a marine environmental problem? Sci Total Environ 825:153935
Jackson SA, Crossman L, Almeida EL, Margassery LM, Kennedy J, Dobson AD (2018) Diverse andabundant secondary metabolism biosynthetic gene clusters in the genomes of marine sponge derived Streptomycesspp. isolates. Mar Drugs 16(2):67
Jadaun JS, Bansal S, Sonthalia A, Rai AK, Singh SP (2022) Biodegradation of plastics for sustainable environment. Bioresour Technol 347:126697
Jambek JR, Geyer R, Wilcox C et al (2015) Plastic waste inputs from land into the ocean. Science 347(6223):768–771
Jones JB Jr, Smock LA (1991) Transport and retention of particulate organic matter in two low-gradient headwater streams. J N Am Benthol Soc 10(2):115–126

Kale SK, Deshmukh AG, Dudhare MS, Patil VB (2015) Microbial degradation of plastic: a review. J Biochem Technol 6(2):952–961

Käppler A et al (2016) Analysis of environmental microplastics by vibrational micro-spectroscopy: FTIR, Raman or both? Anal Bioanal Chem 408:8377–8391

Keswani A, Oliver DM, Gutierrez T, Quilliam RS (2016) Microbial hitchhikers on marine plastic debris: human exposure risks at bathing waters and beach environments. Mar Environ Res 118:10–19

Kettner MT, Oberbeckmann S, Labrenz M, Grossart HP (2019) The eukaryotic life on microplastics in brackish ecosystems. Front Microbiol 10:538

Kirstein IV, Wichels A, Krohne G, Gerdts G (2018) Mature biofilm communities on synthetic polymers in seawater – specific or general? Mar Environ Res 142:147–154

Knapp CW, Callan AC, Aitken B, Shearn R, Koenders A, Hinwood A (2017) Relationship between antibiotic resistance genes and metals in residential soil samples from Western Australia. Environ Sci Pollut Res 24:2484–2494. https://doi.org/10.1007/s11356-016-7997-y

Kooi M, Nes EHV, Scheffer M, Koelmans AA (2017) Ups and downs in the ocean: effects of biofouling on vertical transport of microplastics. Environ Sci Technol 51(14):7963–7971

Kowalski N, Reichardt AM, Waniek JJ (2016) Sinking rates of microplastics and potential implications of their alteration by physical, biological, and chemical factors. Mar Pollut Bull 109(1):310–319

Krueger MC, Harms H, Schlosser D (2015) Prospects for microbiological solutions to environmental pollution with plastics. Appl Microbiol Biotechnol 99:8857–8874

Kühn S, Bravo Rebolledo EL, Van Franeker JA (2015) Deleterious effects of litter on marine life. In: Marine anthropogenic litter. Springer, Cham, pp 75–116

Kukulka T, Proskurowski G, Morét-Ferguson S, Meyer DW, Law KL (2012) The effect of wind mixing on the vertical distribution of buoyant plastic debris. Geophys Res Lett 39:7601

Lagarde F, Olivier O, Zanella M, Daniel P, Hiard S, Caruso A (2016) Microplastic interactions with freshwater microalgae: hetero-aggregation and changes in plastic density appear strongly dependent on polymer type. Environ Pollut (Barking, Essex: 1987) 215:331–339

Lamb JB, Willis BL, Fiorenza EA, Couch CS, Howard R, Rader DN et al (2018) Plastic waste associated with disease on coral reefs. Science 359(6374):460–462

Latva M, Dedman CJ, Wright RJ, Polin M, Christie-Oleza JA (2022) Microbial pioneers of plastic-colonisation in coastal seawaters. Mar Pollut Bull 179:113701

Learman DR, Ahmad Z, Brookshier A, Henson MW, Hewitt V, Lis A et al (2019) Comparative genomics of 16 microbacterium spp. that tolerate multiple heavy metals and antibiotics. PeerJ 6:e6258. https://doi.org/10.7717/peerj.6258

Lebreton L, Andrady A (2019) Future scenarios of global plastic waste generation and disposal. Palgrave Commun 5(1):1–11

Li S, Wang P, Zhang C, Zhou X, Yin Z, Hu T et al (2020) Influence of polystyrene microplastics on the growth, photosynthetic efficiency and aggregation of freshwater microalgae Chlamydomonas reinhardtii. Sci Total Environ 714:136767

Lim X (2021) Microplastics are everywhere—but are they harmful. Nature 593(7857):22–25

Löder MGJ, Gerdts G (2015) Methodology used for the detection and identifcation of microplastics—a critical appraisal. In: Bergmann M, Gutow L, Klages M (eds) Marine anthropogenic litter. Springer, Cham, pp 201–227

Lorite GS, Rodrigues CM, de Souza AA, Kranz C, Mizaikoff B, Cotta MA (2011) The role of conditioning film formation and surface chemical changes on Xylella fastidiosa adhesion and biofilm evolution. J Colloid Interface Sci 359(1):289–295

Lyche JL (2017) Phthalates. In: Reproductive and developmental toxicology. Academic, London, pp 829–856

Ma L-S, Hachani A, Lin J-S, Filloux A, Lai E-M (2014) Agrobacterium tumefaciens deploys a superfamily of type VI secretion dnase effectors as weapons for interbacterial competition in planta. Cell Host Microbe 16(1):94–104

Mammo FK, Amoah ID, Gani KM, Pillay L, Ratha SK, Bux F, Kumari S (2020) Microplastics in the environment: interactions with microbes and chemical contaminants. Sci Total Environ 743:140518
Mattan-Moorgawa S, Chockalingum J, Appadoo C (2021) A first assessment of marine meso-litter and microplastics on beaches: where does Mauritius stand? Mar Pollut Bull 173:112941
Mayali X (2018) Metabolic interactions between bacteria and phytoplankton. Front Microbiol 9:727
Mazard S, Penesyan A, Ostrowski M, Paulsen IT, Egan S (2016) Tiny microbes with a big impact: the role of cyanobacteria and their metabolites in shaping our future. Mar Drugs 14(5):97
McCormick AR, Hoellein TJ (2016) Anthropogenic litter is abundant, diverse, and mobile in urban rivers: insights from cross-ecosystem analyses using ecosystem and community ecology tools. Limnol Oceanogr 61(5):1718–1734
Menicagli V, Balestri E, Vallerini F, De Battisti D, Lardicci C (2021) Plastics and sedimentation foster the spread of a non-native macroalga in seagrass meadows. Sci Total Environ 757:143812
Miao L, Wang C, Adyel TM, Wu J, Liu Z, You G, Meng M, Qu H, Huang L, Yu Y, Hou J (2020) Microbial carbon metabolic functions of biofilms on plastic debris influenced by the substrate types and environmental factors. Environ Int 143:106007
Michels J, Stippkugel A, Lenz M, Wirtz K, Engel A (2018) Rapid aggregation of biofilm-covered microplastics with marine biogenic particles. Proc R Soc B 285(1885):20181203
Mincer TJ, Zettler ER, Amaral-Zettler LA (2019) Biofilms on plastic debris and their influence on marine nutrient cycling, productivity, and hazardous chemical mobility. In: Hazardous chemicals associated with plastics in the marine environment. Springer, Cham, pp 221–233
Moresco V, Oliver DM, Weidmann M, Matallana-Surget S, Quilliam RS (2021) Survival of human enteric and respiratory viruses on plastics in soil, freshwater, and marine environments. Environ Res 199:111367
Morfin APS (2021) Microbial biofilms on microplastics: a look into the estuarine plastisphere of the Chesapeake Bay. Doctoral dissertation, University of Maryland, College Park
Munari C, Scoponi M, Sfriso AA, Sfriso A, Aiello J, Casoni E, Mistri M (2021) Temporal variation of floatable plastic particles in the largest Italian river, the Po. Mar Pollut Bull 171:112805
Musa IO, Auta HS, Ilyasu US, Aransiola SA, Makun HA, Adabara NU, Abioye OP, Aziz A, Jayanthi B, Maddela NR, Prasad R (2024) Micro and nanoplastics in environment—degradation, detection, ecological impact. Int J Environ Res 18:1. https://doi.org/10.1007/s41742-023-00551-9
Naidoo T, Thompson RC, Rajkaran A (2020) Quantification and characterisation of microplastics ingested by selected juvenile fish species associated with mangroves in KwaZulu-Natal, South Africa. Environ Pollut 257:113635
Newbold JD, Mulholland PJ, Elwood JW et al (1982) Organic carbon spiralling in stream ecosystems. Oikos 38(3):266–272. https://doi.org/10.2307/3544663
Nguyen J, Lara-Gutiérrez J, Stocker R (2021) Environmental fluctuations and their effects on microbial communities, populations and individuals. FEMS Microbiol Rev 45(4):fuaa068
Oberbeckmann S, Loeder MGJ, Gerdts G, Osborn AM (2014) Spatial and seasonal variation in diversity and structure of microbial biofilms on marine plastics in northern European waters. FEMS Microbiol Ecol 90(2):478–492
Oberbeckmann S, Löder MG, Labrenz M (2015) Marine microplastic-associated biofilms – a review. Environ Chem 12(5):551–562
Oberbeckmann S, Osborn AM, Duhaime MB (2016) Microbes on a bottle: substrate, season and geography influence community composition of microbes colonizing marine plastic debris. PLoS One 11(8):e0159289
Oberbeckmann S, Bartosik D, Huang S, Werner J, Hirschfeld C, Wibberg D, Heiden SE, Bunk B, Overmann J, Becher D, Kalinowski J, Schweder T, Labrenz M, Markert S (2021) Genomic and proteomic profiles of biofilms on microplastics are decoupled from artificial surface properties. Environ Microbiol 23(6):3099–3115
Ogonowski M, Asa M, Karolina I, Eva H, Zandra G, Klas IU, Zoltan B, Elena G (2018) Evidence for selective bacterial community structuring on microplastics. Environ Microbiol 20:2796–2808. https://doi.org/10.1111/1462-2920.14120

Osborn AM, Stojkovic S (2014) Marine microbes in the plastic age. Microbiol Aust 35(4):207–210
Paluselli A, Aminot Y, Galgani F, Net S, Sempere R (2018) Occurrence of phthalate acid esters (PAEs) in the northwestern Mediterranean Sea and the Rhone River. Prog Oceanogr 163:221–231
Pan Z et al (2019) Microplastics in the northwestern Pacific: abundance, distribution and characteristics. Sci Total Environ 650:1913–1922
Park SY, Kim CG (2019) Biodegradation of micro-polyethylene particles by bacterial colonization of a mixed microbial consortium isolated from a landfill site. Chemosphere 222:527–533
Pathak VM (2017) Review on the current status of polymer degradation: a microbial approach. Bioresour Bioprocess 4(1):1–31
Pedrotti ML, Lombard F, Baudena A, Galgani F, Elineau A, Petit S et al (2022) An integrative assessment of the plastic debris load in the Mediterranean Sea. Sci Total Environ 838:155958
Pelicic V (2019) Monoderm bacteria: the new frontier for type IV pilus biology. Mol Microbiol 112(6):1674–1683
Pollet T, Berdjeb L, Garnier C, Durrieu G, Le Poupon C, Misson B, Briand J-F (2018) Prokaryotic community successions and interactions in marine biofilms: the key role of Flavobacteriia. FEMS Microbiol Ecol 94(6):fiy083
Quero GM, Luna GM (2017) Surfing and dining on the “plastisphere”: microbial life on plastic marine debris. Adv Oceanogr Limnol 8(2). https://doi.org/10.4081/aiol.2017.7211
Ramasamy EV, Sruthy S, Harit AK, Mohan M, Binish MB (2021) Microplastic pollution in the surface sediment of Kongsfjorden, Svalbard, Arctic. Mar Pollut Bull 173:112986
Rios-Fuster B, Compa M, Alomar C, Deudero S (2023) Stranded and floating marine debris detected along the coastline of Cabrera National Park (Balearic Islands). Mar Pollut Bull 194:115288
Rummel CD, Jahnke A, Gorokhova E, Kühnel D, Schmitt-Jansen M (2017) Impacts of biofilm formation on the fate and potential effects of microplastic in the aquatic environment. Environ Sci Technol Lett 4(7):258–267
Sadri SS, Thompson RC (2014) On the quantity and composition of floating plastic debris entering and leaving the Tamar Estuary, Southwest England. Mar Pollut Bull 81(1):55–60
Savinelli B, Fernández TV, Galasso NM, D'Anna G, Pipitone C, Prada F et al (2020) Microplastics impair the feeding performance of a Mediterranean habitat-forming coral. Mar Environ Res 155:104887
Schlundt C, Jessica LM, Anna MK, Erik RZ, Linda AA (2020a) Spatial structure in the “plastisphere”: molecular resources for imaging microscopic communities on plastic marine debris. Mol Ecol Resour 20:620–634. https://doi.org/10.1111/1755-0998.13119
Schlundt C, Mark Welch JL, Knochel AM, Zettler ER, Amaral-Zettler LA (2020b) Spatial structure in the “plastisphere”: molecular resources for imaging microscopic communities on plastic marine debris. Mol Ecol Resour 20(3):620–634
Schwabl P, Köppel S, Königshofer P, Bucsics T, Trauner M, Reiberger T, Liebmann B (2019) Detection of various microplastics in human stool: a prospective case series. Ann Intern Med 171(7):453–457
Seidensticker S, Grathwohl P, Lamprecht J, Zarfl C (2018) A combined experimental and modeling study to evaluate pH-dependent sorption of polar and non-polar compounds to polyethylene and polystyrene microplastics. Environ Sci Eur 30:1–12
Shim WJ, Thompson RC (2015) Microplastics in the ocean. Arch Environ Contam Toxicol 69:265–268
Sjollema SB, Redondo-Hasselerharm P, Leslie HA, Kraak MH, Vethaak AD (2016) Do plastic particles affect microalgal photosynthesis and growth? Aquat Toxicol 170:259–261
Solano C, Echeverz M, Lasa I (2014) Biofilm dispersion and quorum sensing. Curr Opin Microbiol 18:96–104
Stanley NR, Lazazzera BA (2004) Environmental signals and regulatory pathways that influence biofilm formation. Mol Microbiol 52:917–924. https://doi.org/10.1111/j.1365-2958.2004.04036.x

Stubbins A, Law KL, Munoz SE, Bianchi TS, Zhu L (2021) Plastics in the earth system. Science 373:51–55. https://doi.org/10.1126/science.abb0354

Sussarellu R, Suquet M, Thomas Y, Lambert C, Fabioux C, Pernet MEJ et al (2016) Oyster reproduction is affected by exposure to polystyrene microplastics. Proc Natl Acad Sci 113(9):2430–2435

Tang B-L, Yang J, Chen X-L, Wang P, Zhao H-L, Su H-N, Li C-Y, Yu Y, Zhong S, Wang L, Lidbury I, Ding H, Wang M, McMinn A, Zhang X-Y, Chen Y, Zhang Y-Z (2020) A predator-prey interaction between a marine Pseudoalteromonas sp. and gram-positive bacteria. Nat Commun 11(1):285

Tank JL, Rosi-Marshall EJ, Griffiths NA, Entrekin SA, Stephen ML (2010) A review of allochthonous organic matter dynamics and metabolism in streams. J N Am Benthol Soc 29(1):118–146

Tobias-Hünefeldt SP, Wenley J, Baltar F, Morales SE (2021) Ecological drivers switch from bottom–up to top–down during model microbial community successions. ISME J 15(4):1085–1097

Tu C, Chen T, Zhou Q, Liu Y, Wei J, Waniek JJ, Luo Y (2020) Biofilm formation and its influences on the properties of microplastics as affected by exposure time and depth in the seawater. Sci Total Environ 734:139237

Ugoeze KC, Amogu EO, Oluigbo KE, Nwachukwu N (2021) Environmental and public health impacts of plastic wastes due to healthcare and food products packages: a review. J Environ Sci Public Health 5(1):1–31

van Bijsterveldt CE, van Wesenbeeck BK, Ramadhani S, Raven OV, van Gool FE, Pribadi R, Bouma TJ (2021) Does plastic waste kill mangroves? A field experiment to assess the impact of macro plastics on mangrove growth, stress response and survival. Sci Total Environ 756:143826

van der Hal N, Ariel A, Angel DL (2017) Exceptionally high abundances of microplastics in the oligotrophic Israeli Mediterranean coastal waters. Mar Pollut Bull 116(1–2):151–155

Wang Z-M et al (2017) SEM/EDS and optical microscopy analyses of microplastics in ocean trawl and fish guts. Sci Total Environ 603–604:616–626

Wang J, Xiao Q, Jiabao G, Weiqian J, Qian W, Mengjun Z, Yi H (2020) Evidence of selective enrichment of bacterial assemblages and antibiotic resistant genes by microplastics in urban Rivers. Water Res 183:116113. https://doi.org/10.1016/j.watres.2020.116113

Wang J, Peng C, Li H, Zhang P, Liu X (2021) The impact of microplastic-microbe interactions on animal health and biogeochemical cycles: a mini-review. Sci Total Environ 773:145697

Wang C, Liuwei W, Yong SO, Daniel CW, Deyi H (2022) Soil plastisphere: exploration methods, influencing factors, and ecological insights. J Hazard Mater 430:128503. https://doi.org/10.1016/j.jhazmat.2022.128503

Webster JR, Benfield EF, Ehrman TP, Schaeffer MA, Tank JL, Hutchens JJ, D'angelo DJ (1999) What happens to allochthonous material that falls into streams? A synthesis of new and published information from Coweeta. Freshw Biol 41(4):687–705

Wedeking A, van Echten-Deckert G (2007) Glycosphingolipid structure and function in membranes. Curr Org Chem 11(7):579–589

Wilcox C, Mallos NJ, Leonard GH, Rodriguez A, Hardesty BD (2016) Using expert elicitation to estimate the impacts of plastic pollution on marine wildlife. Mar Policy 65:107–114

Woodall LC, Jungblut AD, Hopkins K, Hall A, Robinson LF, Gwinnett C, Paterson GLJ (2018) Deep-sea anthropogenic macrodebrisharbours rich and diverse communities of bacteria and archaea. PLoS One 13(11):e0206220

Wright RJ, Erni-Cassola G, Zadjelovic V, Latva M, Christie-Oleza JA (2020) Marine plastic debris: anew surface for microbial colonization. Environ Sci Technol 54(19):11657–11672

Wu N, Zhang Y, Zhao Z, He J, Li W, Li J et al (2020a) Colonization characteristics of bacterial communities on microplastics compared with ambient environments (water and sediment) in Haihe Estuary. Sci Total Environ 708:134876

Wu Y, Tam WS, Chau HF, Kaur S, Thor W, Aik WS et al (2020b) Solid-phase fluorescent BODIPY–peptide synthesis via in situ dipyrrin construction. Chem Sci 11(41):11266–11273

You X, Xu N, Yang X, Sun W (2021) Pollutants affect algae-bacteria interactions: a critical review. Environ Pollut (Barking, Essex: 1987) 276:116723

Zettler ER, Mincer TJ, Amaral-Zettler LA (2013) Life in the "plastisphere": microbial communities on plastic marine debris. Environ Sci Technol 47(13):7137–7146. https://doi.org/10.1021/es401288x

Zhang S-J, Zeng Y-H, Zhu J-M, Cai Z-H, Zhou J (2022) The structure and assembly mechanisms ofplastisphere microbial community in natural marine environment. J Hazard Mater 421:126780

Zhao S, Erik RZ, Linda AA, Tracy JM (2021) Microbial carrying capacity and carbon biomass of plastic marine debris. ISME J 15:67–77. https://doi.org/10.1038/s41396-020-00756-2

Zhou Y, Kumar M, Sarsaiya S, Sirohi R, Awasthi SK, Sindhu R et al (2022) Challenges and opportunities in bioremediation of micro-nano plastics: a review. Sci Total Environ 802:149823

Zhu J, Wang C (2020) Biodegradable plastics: Green hope or greenwashing? Mar Pollut Bull 161:111774

Chapter 2
The Effects of Marine White Pollution on Higher Animals, Plants, and the Food Chain

Innocent Ojeba Musa, Abdulhakeem Idris Abdulhakeem, Mustapha Abdulsalam, Miracle Uwa Livinus, Olaitan Lateefat Salam, Shehu-Alimi Elelu, Auwal Sagir Muhammad, Job Oloruntoba Samuel, and Adamu Mustapha

Abstract Marine white pollution, also known as marine plastic pollution, presents a substantial danger to ecosystems on a global scale, affecting higher organisms, plants, and the entire food web. Marine pollution refers to the buildup of plastic waste in aquatic habitats, such as oceans, seas, and coastal regions. This can lead to obstructions in the gastrointestinal tract, malnourishment, and finally fatality. In addition, animals may become ensnared in plastic waste, which can hinder their mobility, feeding, and reproductive capabilities, so further jeopardising their survival. Furthermore, the presence of plastic trash in marine habitats can affect the behavior and distribution of species, leading to alterations in predator-prey dynamics and competition for resources. This can disturb the functioning of marine ecosystems and undermine their ability to offer essential ecosystem services, such as

---

I. O. Musa (✉) · M. Abdulsalam
Department of Microbiology, Skyline University Nigeria, Kano, Kano State, Nigeria

A. I. Abdulhakeem
Department of Fisheries and Aquaculture, Federal University Dustin-Ma, Dustin-Ma, Kastina State, Nigeria

M. U. Livinus
Department of Biochemistry, Skyline University Nigeria, Kano, Kano State, Nigeria

O. L. Salam
Department of Chemistry and Molecular Biology, University of Gothenburg, Gothenburg, Sweden

S.-A. Elelu
Department of Chemistry, Howard University, Washington, DC, USA

A. S. Muhammad
Department of Computer Science, Xiamen University, Xiamen, China

J. O. Samuel · A. Mustapha
Department of Microbiology, Federal University of Technology Minna, Minna, Nigeria

S. A. Aransiola et al. (eds.), *White Pollution: Biodiversity and hazards in marine plastisphere*, Environmental Contamination Remediation and Management,
https://doi.org/10.1007/978-3-031-95547-1_2

fisheries production and carbon sequestration. Marine white pollution has large and extensive repercussions on higher animals, vegetation, and the food chain in marine environments. Urgent and concentrated efforts are needed to ameliorate this rising environmental concern by actions such as lowering plastic production and use, improving waste management procedures, and promoting the use of alternative materials. By tackling marine white pollution, we can maintain the health and integrity of marine ecosystems for present and future generations.

Keywords Marine · Pollution · Ecosystem · Food chain

2.1 Introduction

Marine environment White pollutants have a number of diverse consequences on the marine environment. They alter the non-living environment, creating physical and chemical changes. White contaminants can create changes in water clarity and color and can interfere with many physical qualities of the water. They can get into the sediments and alter the benthic ecosystem. White contaminants are delivered to many isolated places of the ocean by air transport. Most issues produced by white pollution are indirect; it is sometimes impossible to pinpoint particular cause and effect correlations. This is because white pollution generally impacts the ecological level. An example of a direct consequence would be if a fish were to ingest a deadly pellet mistaking it for food. The pellet may kill the fish. An example of an indirect impact would be if an otter were to eat a crab that had absorbed harmful chemicals. The chemicals may subsequently be passed to the otter with lethal results. Consequences on humans also have adverse consequences on the marine ecosystem. Humans are responsible for white pollution and so it has direct consequences on humans. It also impacts the fish sector, which in many nations is economically crucial.

White contaminants have become an integral component of the ecology and have produced many difficulties in the marine environment. These pollutants are so-called because of their hue, although they are not genuinely white. White pollutants are complex combinations of synthetic organic compounds, metals, and detergents. Certain cosmetics and medications are also covered. They normally do not dissolve in water and are not commonly detected in the air or water. Instead, they collect on the soil where they create greatest damage. They were important contaminants from sewage before the introduction and widespread usage of sewage treatment systems. White contaminants are now acknowledged to be one of the greatest risks to the health of the earth's marine ecosystem. Close to 200,000 distinct synthetic compounds are created in extremely large amounts, and 1000 new chemicals are released each year. High hazardous compounds like DDT are no longer utilised, but many chemicals like PCBs and heavy metals are persistent and all are potential contaminants. White contaminants are present in all marine ecosystems from the poles to the equator. This form of pollution is not simply local; it impacts the entire oceans.

A considerable number of white contaminants is observed near coastal and estuary locations. White pollutants have varied consequences on the environment. These include toxic effects on higher animals, long-term consequences through accumulation, changes in species diversity, habitat alterations, effects on the ecosystem level, and effects on people. The most important concern produced by white pollution is the harmful impact on higher animals.

Styrofoam has one of the greatest environmental costs amongst polymers. As Styrofoam and plastic are both relatively low specific gravity materials, their movement within the marine environment is not always predictable. It is fairly typical for white pollution to travel throughout the world and might be washed up into coasts many years after it was gone. Often the first sign of trash is the unattractive appearance of a local place. Beaches and coral reefs are of particular importance as a biodiversity hotspot, and white pollution may operate as a nuisance for local tourism income. Littering of the seafloor tries to cause confusion and harm or disturbance to marine ecosystems, mainly coral reefs.

Plastic is believed to be a 'wonder material'—affordable and long-lasting. These two features are what make it so destructive to the environment. It may be argued to be a similar circumstance to the use of CFCs or pesticides, however unintended owing to a lack of awareness or disrespect for the repercussions. A little-known truth is that at both extremities of the temperature range, plastic becomes exceedingly brittle and will shatter and shard quite quickly (Auta et al. 2022). This implies that even within a plastic bag, exposure to light and heat in the environment from the sun will cause the plastic to break up into small fragments. Plastic does not just 'decompose' like other materials. It photodegrades, breaking up smaller and smaller without breaking into simpler compounds. This results in an exceptionally high concentration of minute plastic particles, which regard the top few hundred millimeters of the water column to be the last resting place. This section of the maritime ecosystem is most biologically active, and the fragmentation of plastic will make its way into marine food chains at a vast scale. High quantities of plastic pieces will find their way onto beaches, often dangerous to humans and sometimes hard to clear using trash picking machines owing to the size of the waste (Aransiola et al. 2023a; Musa et al. 2024).

Maritime white pollution encompasses maritime debris composed of lightweight materials, including plastics and Styrofoam. There is no distinct demarcation between the act of littering and the problem of maritime pollution. 'Litter' refers to any discarded item, such as general waste, cans, packaging, bottles, and other such materials. Garbage of this nature may be found virtually anywhere on the planet. While littering is usually a deliberate and personal choice, marine pollution, also known as white pollution, is frequently seen as unintentional or indirect. White pollution, also known as plastic pollution, has a specific definition and a substantial impact on the marine environment. It causes extensive ecological damage due to the presence of common, everyday objects made of plastic.

2.1.1 Ecological Consequences

Marine white pollution, which refers to the widespread occurrence of plastic waste in marine environments, presents significant obstacles to maintaining ecological stability (Jambeck et al. 2015). The detrimental effects of this type of pollution on biodiversity, ecological services, and human well-being have received greater attention in recent years (Wilcox et al. 2016).

The ingestion and entanglement of plastic pose significant threats to marine creatures, such as seabirds, turtles, and marine mammals (Gall and Thompson 2015). Plastic consumption can lead to internal injury, hunger, and mortality among marine species (Schuyler et al. 2016), while entanglement can impede mobility and reproductive success (Laist 1997).

Marine plants, particularly those inhabiting coastal environments, are vulnerable to the physical and chemical effects of plastic pollution (Kooi et al. 2016). Plastic trash may choke and degrade critical ecosystems such as coral reefs and seagrass beds, reducing ecological stability and biodiversity (Lamb et al. 2018).

Plastic pollution disrupts marine food webs through mechanisms such as toxin transmission and changed predator-prey dynamics (Rochman et al. 2013). Plastic waste can release toxic substances and serve as vectors for pollution accumulation, offering risks to animals across trophic levels (Wright et al. 2013). This can eventually impair human health through contaminated seafood ingestion (Li et al. 2016).

2.2 Effects on Higher Animals

Marine white pollution, mostly consisting of plastic trash, exerts major impacts for higher organisms in marine ecosystems. The ingestion and entanglement of plastic waste pose major threats to the health and survival of numerous marine species.

Plastic ingestion is a widespread hazard for higher species such as seabirds, turtles, seals, and marine mammals (Gall and Thompson 2015). These animals frequently misinterpret plastic things for prey, resulting to ingestion. Once consumed, plastic debris can produce internal blockages, gastrointestinal injury, and malnutrition, ultimately leading to illness and mortality (Schuyler et al. 2016).

Entanglement with plastic trash offers another major hazard for higher animals. Animals may become stuck in abandoned fishing gear, plastic bags, or other debris, which can restrict their mobility, feeding capacity, and reproductive efficacy (Laist 1997). Entangled animals frequently suffer from lacerations, infections, and reduced movement, making them more prone to predation and other risks.

In addition, plastic ingestion and entanglement can have cascading repercussions on entire populations and ecosystems. For example, the reduction in numbers of apex predators, such as sharks and marine mammals, owing to plastic pollution might disrupt trophic cascades and impact ecosystem dynamics (Wilcox et al.

2015). Additionally, the loss of essential species can lead to imbalances in food webs, altering the amount and distribution of other organisms within the ecosystem.

The consequences of plastic pollution on higher animals extend beyond individual organisms to destroy entire groups and societies. For example, seabirds who dine on seafood contaminated with plastic particles may have decreased reproductive success and population reductions (Schuyler et al. 2016). Similarly, marine species that consume plastic rubbish may suffer from decreased reproductive rates and higher susceptibility to infections.

2.2.1 *Impact on Marine Mammals*

Marine creatures, notably whales, dolphins, seals, and sea lions, are substantially injured by marine white pollution, mostly via ingestion, entanglement, and habitat disruption.

Plastic ingestion is a common hazard among marine creatures, with many species accidently eating plastic waste when scavenging for food (Schuyler et al. 2016). These organisms may ingest plastic bags, bottles, shards, and other debris, mistaking them for prey items like as fish or squid. Once eaten, plastic debris can create gastrointestinal blockages, perforations, and internal injuries, leading to starvation, malnutrition, and death (Laist 1997).

Entanglement in plastic debris provides another major peril to aquatic species. Abandoned fishing gear, such as nets and lines, and other sorts of plastic rubbish can entangle marine creatures, compromising their ability to swim, feed, and breathe (Gall and Thompson 2015). Entangled animals may suffer from serious wounds, infections, and restricted mobility, ultimately leading to fatigue, predation, or drowning.

Moreover, marine white pollution can impact the habitats and feeding grounds of marine creatures. Plastic garbage can amass in coastal regions, estuaries, and offshore ecosystems, impacting marine animal movement patterns, breeding grounds, and feeding environments (Lamb et al. 2018). For example, plastic pollution in maritime environments could interfere with echolocation and communication signals used by cetaceans, decreasing their capacity to find and locate prey.

The repercussions of marine white pollution on marine animals extend beyond individual individuals to affect whole populations and ecosystems. Plastic ingestion and entanglement may result in poor reproductive success while also lowering the survival of offspring and poet decreases. In addition to the above dangers, the presence of plastic waste in marine ecosystems exposes dangerous substances and toxins that damage the food chain, posing additional risks to marine animal health and wellness.

2.2.2 *Impact on Sea Birds*

Marine white pollution, characterized by the development of plastic rubbish in marine ecosystems, presents major threats to sea birds, changing their health, behavior, and population dynamics.

The most visible impact of marine white pollution on sea birds is plastic ingestion. Sea birds often take the floating plastic garbage for food and consume objects like plastic parts, bottle caps, and angling lines. Contingent on their size and structure, these articles can make inside damage, blockages, and radiant destitution and, as such, will starve to death.

Entanglement in plastic waste is another critical danger to marine birds. Abandoned fishing tackle like nets and connect lines and other types of plastic garbage may likewise wrap marine birds and limit their ability to move, taking off, and searching for food. Entangled birds are at a greater exclusive danger of having wounds, sickness, and trouble; which additionally makes them more vulnerable to predation and different hazards.

Moreover, plastic pollution in maritime ecosystems could influence the foraging behaviours and feeding patterns of sea birds. Floating plastic waste can simulate natural food items, compelling marine birds to spend valuable foraging time attempting to ingest non-nutritive materials (Ryan 2015). This can result in poorer feeding efficiency, lowered energy reserves, and poor reproductive success among afflicted bird populations.

The repercussions of marine white pollution on sea birds extend beyond individual individuals to affect whole populations and ecosystems. Plastic ingestion and entanglement can lead to lower reproductive success, greater mortality rates, and population declines (Schuyler et al. 2016). Furthermore, the absorption of plastic waste can introduce harmful chemicals and toxins into the food chain, posing significant concerns to marine bird health and ecological integrity.

2.2.3 *Disruption of Fish and Reptile Habitats*

Marine white pollution, mostly consisting of plastic rubbish, damages the habitats of fish and reptiles in marine settings, contributing to ecological imbalances and harmful effects on these species.

Plastic waste collection impacts the physical structure and composition of ecosystems utilised by fish and reptiles. Coastal habitats, estuaries, and coral reefs are particularly prone to plastic pollution, where rubbish may suffocate and destroy important organisms (Lamb et al. 2018). Such habitat degradation affects the availability of adequate shelter, nesting areas, and food supplies for fish and reptiles, leading to declines in population abundance and variation.

Furthermore, plastic contamination in maritime environments could interfere with the behavior and natural migrations of fish and reptiles. Floating plastic garbage can impair swimming and foraging behaviours, decreasing the capacity of these animals to gather food, escape predators, and participate in reproductive practices (Gall and Thompson 2015). This perturbation of typical behaviors might have cascading implications on population dynamics and ecological functioning.

Plastic ingestion offers a direct threat to fish and reptiles in marine ecosystems. Many animals may misinterpret plastic items for prey or swallow them inadvertently while feeding, resulting to internal damage, digestive blockages, and hunger (Schuyler et al. 2016). Chronic eating of plastic garbage can reduce development, reproductive performance, and overall health in fish and reptiles, ultimately affecting population survival.

Additionally, entanglement in plastic debris presents a major hazard to fish and reptiles. Abandoned fishing gear, such as nets and lines, and other kinds of plastic debris can entangle these species, resulting in damage, reduced movement, and hypoxia (Laist 1997). Entangled individuals may experience restricted mobility and more vulnerability to predation, exacerbating population declines.

The disruption of fish and reptile habitats due to marine white pollution might have greater effects for marine ecosystems. Fish and reptiles serve crucial roles in food webs, nutrient cycling, and ecological stability. Population declines or local extinctions of these species can lead to imbalances in predator-prey interactions, disturbances in community structure, and decreased resilience to environmental changes.

2.3 Effects on Plants

Marine white pollution, typified by the buildup of plastic rubbish in marine environments, also impacts plants in numerous ways, although these repercussions are typically indirect.

One key effect on marine plants is habitat alteration. Plastic waste may smother and cover the bottom, seagrass meadows, and coral reefs, decreasing the development and survival of marine plants (Lamb et al. 2018). Seagrasses, for example, which constitute crucial habitat and nursery grounds for many marine species, could be particularly vulnerable to plastic pollution. The presence of plastic garbage might limit light penetration and hamper nutrient exchange, resulting to decreased photosynthetic activity and growth of marine plants.

Moreso, plastic pollution impairs nutrient cycling and ecosystem functioning in marine environments, which can indirectly impact marine plants. Plastic waste can function as substrates for the growth of algae and other microorganisms, influencing nutrient dynamics and microbial communities in marine environments (Dussud et al. 2018). Changes in food supply and microbial activity might effect the competitive interactions between marine plants and other species, possibly leading to modifications in community composition and structure.

Plastic pollution also creates hazards to marine plants by the introduction of toxic chemicals and toxins into the marine environment. Plastic trash can absorb and concentrate contaminants such as heavy metals, herbicides, and persistent organic pollutants from seawater (Teuten et al. 2009). These compounds can leach from plastic garbage into the surrounding water, possibly harming the health and physiology of marine plants.

Additionally, plastic trash can transfer invasive species and diseases, which can have harmful impacts on native marine plant communities. Floating plastic garbage can operate as a vector for the introduction of non-native species to new settings, where they may outcompete native plants and alter ecosystem dynamics (Barnes et al. 2009; Aransiola et al. 2023b). This can lead to the loss of biodiversity and ecological resilience in marine plant communities.

2.4 Impact on the Food Chain

One of the primary impacts of marine white pollution on the food chain is through the ingestion of plastic garbage by marine organisms. Plastics can be mistaken for food by a broad range of marine creatures, including fish, seabirds, turtles, and marine mammals (Rochman et al. 2013). Once eaten, plastic trash can develop in the digestive systems of these species, resulting to physical damage, reduced feeding efficiency, and interference with nutritional absorption.

As plastic garbage lingers in the marine environment, it can undergo physical and chemical degradation, releasing dangerous chemicals and toxins into the surrounding water (Wright et al. 2013). These contaminants can bioaccumulate in the tissues of marine creatures, becoming increasingly concentrated as they advance up the food chain. Consequently, predators at higher trophic levels, including humans, may be exposed to heightened concentrations of toxins and pollutants through the eating of contaminated seafood.

Moreover, plastic pollution can influence the behavior and distribution of marine species, resulting to variations in predator-prey dynamics and competition for resources. For example, the presence of plastic litter could modify the foraging behavior of marine predators, such as fish and seabirds, by influencing food availability and accessibility (Lavers et al. 2019). This can disturb the balance of the food chain and alter population dynamics across the ecosystem.

Additionally, plastic pollution can act as a conduit for the transfer of exotic species and viruses, which can further disrupt food webs and ecosystem functioning (Barnes et al. 2009). Invasive species introduced via plastic garbage may outcompete native species for resources, disturb habitat structure, and contribute to the reduction of vulnerable populations within the food chain.

2.5 Sources of Marine White Pollution

Marine white pollution, usually referred to as marine plastic pollution, arises from numerous sources, including both land-based and ocean-based activities (Jambeck et al. 2015). These sources contribute to the building of plastic waste in marine environments, leading to serious environmental impacts.

1. Land-based sources

   Improper waste management: Inadequate garbage disposal and unsuitable waste management procedures contribute considerably to marine white pollution (Geyer et al. 2017). Plastic waste from land-based sources, such as metropolitan centres, industrial sites, and recreational spots, can be transported by wind, rivers, and stormwater runoff into waterways and eventually into the ocean.

   Littering: Direct littering by humans on beaches, along coasts, and in rivers can transport plastic rubbish into marine environments (Derraik 2002). Items like plastic bottles, food packaging, and single-use plastics are often dumped carelessly, adding to marine white pollution.

   Industrial activities: Industrial operations and manufacturing facilities may dump plastic pellets (nurdles), microplastics, and other plastic garbage straight into rivers, affecting marine environments (Horton et al. 2017).
2. Ocean-based sources:

   Fishing and aquaculture: Lost or discarded fishing gear, consisting nets, lines, and traps (known as ghost gear), adds considerably to marine white pollution (Wilcox et al. 2018). Abandoned or abandoned fishing gear continues to imprison and kill marine species, and eventually breaks down into microplastics.

   Shipping and maritime activities: Shipping vessels, cargo ships, and recreational boats may unintentionally or actively dump plastic garbage into marine habitats (Lippiatt et al. 2013). Discharge of plastic trash, including packaging materials, fishing gear, and cargo spills, can occur during transportation and loading/unloading processes.

   Tourism and leisure activities: Tourism and recreational activities in coastal locations produce plastic garbage, including food and beverage containers, packaging materials, and recreational equipment (Lebreton et al. 2017). Items left behind or inadequately disposed of contribute to marine white pollution.
3. Wastewater and sewage runoff: Urban runoff: Urban areas produce substantial amounts of plastic rubbish, which can be carried into rivers via stormwater runoff (Muthukumar et al. 2017). Plastic garbage may emerge from numerous urban sources, including residential areas, business districts, and municipal landfills.

   Sewage systems: Plastic microfibers from synthetic textiles, personal care items containing microbeads, and other plastic-based elements can penetrate marine environments through sewage systems (Browne et al. 2011). Wastewater treatment plants may not adequately filter out all microplastics, resulting to their release into aquatic habitats.

4. At-sea activities: Maritime operations: Oil and gas exploration and extraction activities, offshore drilling platforms, and marine research operations can contribute to marine white pollution through unintentional spills, discharge of plastic debris, and equipment failure (Thompson et al. 2004).

   Military activities: Military drills, naval operations, and the disposal of military equipment at sea can seep plastic waste into marine habitats, affecting ocean ecosystems (Lavers et al. 2019).

2.5.1 *Plastic Waste and Microplastics*

Plastic waste and microplastics are important contributors to marine white pollution, coming from diverse sources and posing pervasive risks to marine ecosystems and biodiversity.

1. Plastic Waste: Plastic waste originates from land-based sources such as metropolitan regions, industrial facilities, and agricultural activities (Geyer et al. 2017). Improper waste management practices, including littering and inadequate recycling infrastructure, result in the accumulation of plastic debris in terrestrial and aquatic environments.

   Single-use plastics, including bottles, bags, and packaging materials, contribute greatly to plastic waste pollution due to their pervasive consumption and disposal (Jambeck et al. 2015). These creatures frequently find up in streams via stormwater runoff and can linger in the ecosystem for hundreds of years.
2. Microplastics: Microplastics are small plastic particles less than 5 mm in size, resulting from the fragmentation of larger plastic items or as primary microplastics from personal care products and industrial abrasives (Cole et al. 2011). Microplastics can also emerge from the degradation of synthetic textiles and fishing gear.

   These small plastic particles are abundant in marine environments, damaging coastal sediments, surface waters, and even isolated oceanic regions (Van Cauwenberghe and Janssen 2014). Microplastics are ingested by a wide range of marine creatures, from zooplankton to apex predators, posing dangers to marine life and human health.

Plastic waste and microplastics have detrimental repercussions on marine ecosystems:

Physical harm: Marine creatures may become entangled in larger plastic debris, resulting to damage, suffocation, and death (Laist 1997). Ingestion of plastic garbage, especially microplastics, can cause internal harm, digestive blockages, and malnutrition.

Chemical contamination: Plastic garbage may absorb and concentrate dangerous substances such as heavy metals and persistent organic pollutants from seawater (Teuten et al. 2009). These contaminants can seep from plastic particles, leading to chemical exposure and bioaccumulation in marine organisms.

Ecological disruption: Plastic garbage and microplastics can affect marine habitats, nitrogen cycling, and ecosystem functioning (Wright et al. 2013). Microplastics can be mistaken for food by marine organisms, leading to reduced eating efficiency and reproductive success. Furthermore, plastic pollution might facilitate the transmission of alien species and diseases, damaging natural ecosystems and biodiversity.

2.5.2 Chemical Pollutants

Chemical pollutants in marine environments, including heavy metals, pesticides, and industrial chemicals, pose substantial hazards to marine ecosystems and human health. These toxins occur from different sources and can accumulate in marine animals, leading to ecological disruptions and negative health repercussions.

1. Heavy Metals: Heavy metals such as mercury, lead, cadmium, and arsenic are dumped into marine environments by industrial operations, mining activities, and atmospheric deposition (Mason et al. 2018). These metals may persist in the environment for lengthy durations and accumulate in sediments, water, and marine organisms.

   Marine species, including fish, shellfish, and marine mammals, can bioaccumulate heavy metals through the food chain (Storelli and Marcotrigiano 2018). Chronic exposure to significant levels of heavy metals can disrupt physiological activity, change reproductive systems, and produce developmental problems in marine organisms.
2. Pesticides: Pesticides, including insecticides, herbicides, and fungicides, are applied in agriculture, aquaculture, and urban pest management (Crain et al. 2020). Runoff from agricultural fields and urban areas can transport pesticides into streams and coastal zones, harming marine ecosystems.

   Pesticides can bioaccumulate in marine organisms and biomagnify in the food chain, resulting to greater quantities in apex predators (Guzzetti et al. 2021). Exposure to pesticides can trigger brain damage, reproductive problems, and immune system suppression in marine creatures.
3. Industrial Chemicals: Industrial chemicals such as polychlorinated biphenyls (PCBs), polycyclic aromatic hydrocarbons (PAHs), and per- and polyfluoroalkyl substances (PFAS) are introduced into marine environments through industrial activities, waste disposal, and urban runoff (Sarkar et al. 2020).

   These chemicals are persistent organic pollutants (POPs) that can bioaccumulate in marine animals and biomagnify in the food chain (Costa and Giordano 2007). Chronic exposure to industrial pollution can lead to reproductive issues, immune system dysfunction, and cancer in marine creatures.
4. Plastic Additives: Plastic additives, including plasticizers, flame retardants, and UV stabilizers, can seep from plastic waste into marine environments (Rochman

2018). Plastic additives may include toxic compounds that can accumulate in marine organisms and affect ecosystem health.

Exposure to plastic additives has been connected to reproductive disorders, endocrine disruption, and developmental problems in marine animals (Bergmann et al. 2020).

2.5.3 Oil Spills and Petroleum Products

Oil spills and petroleum products constitute considerable environmental threats to maritime habitats, creating immediate and long-term ecological devastation, as well as economic and societal ramifications. These spills can occur owing to mishaps involving oil tankers, offshore drilling rigs, pipeline breaks, or unauthorised discharge of petroleum products.

1. Immediate Environmental Impacts:

   Physical Coating: When oil is spilled into marine ecosystems, it generates a slick on the sea surface, which may spread fast owing to wind and currents. This oil slick can cover coastal ecosystems such as beaches, mangroves, and coral reefs, suffocating and killing marine species (Graham et al. 2020).

   Toxicity: Petroleum products include a complex mixture of hydrocarbons and hazardous chemicals, including polycyclic aromatic hydrocarbons (PAHs) and heavy metals. These toxins can harm marine organisms, change physiological functions, and hinder reproductive and developmental processes (Incardona et al. 2019).
2. Long-term Environmental Consequences:

   Bioaccumulation: Oil spills can lead to the accumulation of petroleum-derived compounds in marine animals through ingestion, absorption, and inhalation. These compounds can bioaccumulate and biomagnify in the food chain, resulting in chronic exposure and long-term repercussions for ecosystem health (Socolofsky et al. 2019).

   Habitat Degradation: Oil spills can cause long-term devastation to coastal habitats, including seagrass meadows, salt marshes, and estuaries. Oil may sink into sediments and persist for years, hindering plant growth, disrupting nutrient cycling, and modifying ecosystem dynamics (Michel et al. 2020).
3. Economic and Social Impacts:

   Loss of Livelihoods: Oil spills may harm coastal communities that rely on marine resources for their livelihoods, including fishing, tourism, and recreation. Contamination of fisheries and coastal regions can lead to loss of income, employment, and cultural identity for local inhabitants (Bastiaens et al. 2019).

   Cleaning Costs: The cleanup and rehabilitation of oil spills are expensive and might take large efforts and resources. Costs associated with containment, recovery, and restoration of afflicted regions can run into billions of dollars, exerting a pressure on governments, enterprises, and taxpayers (NRC 2019).

4. Mitigation and Response:

   Prevention: Effective ways to prevent oil spills include tough regulation, safety procedures, and technology innovations in oil exploration, transportation, and storage (Hodges et al. 2020).

   Reaction: In the case of an oil spill, fast reaction and containment efforts are crucial to reduce environmental damage. Strategies similar as mechanical recovery, dispersion application, and shoreline protection can help lessen the impacts of oil spills on marine ecosystems (Lehr et al. 2019).

2.6 Measures to Reduce Marine White Pollution

Reducing marine white pollution, largely caused by plastic debris, demands a multi-faceted approach involving government regulations, industry actions, public awareness campaigns, and individual acts. Here are many solutions that can aid lessen this frequent environmental issue:

1. Regulatory Policies:

   Ban Single-Use Plastics: Implement bans or curbs on single-use plastics such as plastic bags, straws, utensils, and packaging materials to decrease their manufacture and usage (Geyer et al. 2017).

   Extended Producer Responsibility (EPR): Enforce EPR legislation that hold producers liable for the collection, recycling, and right disposal of plastic objects at the end of their lives (Jambeck et al. 2015).

   Polluter Pays idea: Apply the polluter pays concept to motivate enterprises to accept responsibility for the environmental effect of their plastic items, including the expenses of cleaning and restoration (G7 2019).
2. Waste Management Strategies:

   Improve Recycling Infrastructure: Invest in recycling facilities and technology to increase the collection, sorting, and processing of plastic garbage, thereby raising recycling rates and minimising landfilling and incineration (Geyer et al. 2017).

   Promote Circular Economy: Encourage the application of circular economy concepts, such as product redesign, reuse, and remanufacturing, to minimise the production of plastic waste and boost resource efficiency (Ellen MacArthur Foundation 2016).

   Minimise Plastic Packaging: Encourage merchants and manufacturers to adopt eco-friendly packaging alternatives, such as biodegradable materials, paper-based packaging, and reusable containers, to limit plastic packaging waste (PlasticsEurope 2020).
3. Education and Awareness: Environmental Education: Promote environmental education initiatives in schools, communities, and businesses to improve awareness about the implications of marine white pollution and drive behavior change

towards more sustainable consumption and waste management practices (UNEP 2020).

Citizen Science initiatives: Engage citizens in monitoring and cleaning efforts through citizen science programs and community-based projects to collect data on plastic pollution and empower local communities to take action (Jang et al. 2021).

Media Campaigns: Launch media campaigns, social media initiatives, and public outreach operations to enlighten and organise the public, corporations, and lawmakers on the necessity of minimising plastic waste and conserving coastal habitats (Surfrider Foundation 2021).

4. Innovation and Research:

Develop Alternative Materials: Invest in research and development of sustainable alternatives to standard plastics, such as biodegradable polymers, bio-based plastics, and innovative packaging materials created from renewable resources (European Commission 2020).

Innovative Cleanup Technologies: Support the research and application of innovative technologies for plastic pollution cleanup, including ocean cleaning systems, autonomous drones, and biodegradable cleanup materials (The Ocean Cleaning n.d.).

Plastic Tracing and Monitoring: Develop monitoring techniques and technology, such as satellite pictures, drones, and machine learning algorithms, to determine the sources, transit patterns, and accumulation hotspots of plastic pollution in marine environments (Maximenko et al. 2019).

By implementing these activities in a coordinated and collaborative fashion, stakeholders may assist towards minimising marine white pollution and sustaining the health and integrity of marine ecosystems for future generations.

2.7 Conclusion

In conclusion, eliminating marine white pollution, notably created by plastic debris, is a challenging yet critical issue demanding joint actions from governments, corporations, communities, and individuals globally. The remedies outlined above offer a complete method to fixing this major environmental issue and sustaining the health and integrity of marine ecosystems.

Regulatory actions, such as restricting single-use plastics and creating extended producer responsibility regimes, are crucial for limiting the manufacturing and use of plastic goods. Additionally, waste management measures, including enhancing recycling infrastructure and promoting circular economy principles, are crucial for reducing plastic pollution and encouraging sustainable resource use.

Education and awareness campaigns play a critical role in informing and mobilizing the public, corporations, and politicians about the importance of marine conservation and the need to adopt more sustainable consumption and waste management practices. Citizen science programmes and media campaigns can enable individuals and communities to actively participate in monitoring and cleanup activities.

Investments in innovation and research are vital to generate alternative materials to typical plastics, enhance cleanup technologies, and boost plastic tracking and monitoring capabilities. By fostering collaboration, innovation, and accountability, stakeholders may work together to mitigate the impacts of marine white pollution and protect the long-term health and resilience of marine ecosystems.

Eventually, fighting marine white pollution involves a collective commitment to sustainable development and appropriate care of our waters. By adopting the measures indicated above and taking a holistic approach to marine conservation, we can conserve marine species, aid coastal communities, and maintain the beauty and health of our oceans for future generations.

References

Aransiola SA, Ikhumetse AA, Babaniyi BR, Abioye OP, Oyedele OJ, Falade NO (2023a) Phytoaccumulation of micro and nanoplastics: root uptake. In: Maddela NR, Reddy KV, Ranjit P (eds) Micro and nanoplastics in soil. Springer, Cham. https://doi.org/10.10007/978-3-031-21195-9_8

Aransiola SA, Victor-Ekwebelem MO, Ajiboye AE, Leh-Togi Zobeashia SS, Ijah UJJ, Oyedele OJ (2023b) Ecological impacts and toxicity of micro and nanoplastics in agroecosystem. In: Maddela NR, Reddy KV, Ranjit P (eds) Micro and nanoplastics in soil. Springer, Cham. https://doi.org/10.10007/978-3-031-21195-9_10

Auta SH, Abioye OP, Aransiola SA, Bala JD, Chukwuemeka VI, Hassan A, Aziz A, Fauziah SH (2022) Enhanced microbial degradation of PET and PS microplastics under natural conditions in mangrove environment. J Environ Manag 304(2022):114273. https://doi.org/10.1016/j.jenvman.2021.114273

Bastiaens L, Hübner U, De Paepe J, Stojanov R, Meselhe EA (2019) Economic and social impacts of oil spill events and policies: international evidence. Ocean Coast Manag 170:1–10

Bergmann M, Gutow L, Klages M (2020) Marine anthropogenic litter. Springer, Cham

Barnes DK, Galgani F, Thompson RC, & Barlaz M (2009) Accumulation and fragmentation of plastic debris in global environments. Philos Trans R Soc B: B Biol Sci 364(1526):1985–1998

Browne MA, Crump P, Niven SJ, Teuten E, Tonkin A, Galloway T, Thompson R (2011) Accumulation of microplastic on shorelines woldwide: sources and sinks. Environ Sci Technol 45(21):9175–9179

Cole M, Lindeque P, Halsband C, Galloway TS (2011) Microplastics as contaminants in the marine environment: a review. Mar Pollut Bull 62(12):2588–2597

Costa LG, Giordano G (2007) Developmental neurotoxicity of polybrominated diphenyl ether (PBDE) flame retardants. Neurotoxicology 28(6):1047–1067

Crain CM, Kroeker K, Halpern BS (2020) Interactive and cumulative effects of multiple human stressors in marine systems. Ecol Lett 23(3):476–486

Dussud C, Meistertzheim AL, Conan P, Pujo-Pay M, George M, Fabre P, Ghiglione JF (2018) Evidence of niche partitioning among bacteria living on plastics, organic particles and surrounding seawaters. Environ Pollut 236:807–816

Derraik JG (2002) The pollution of the marine environment by plastic debris: a review. Mar Pollut Bull 44(9):842–852

Ellen MacArthur Foundation (2016) The new plastics economy: rethinking the future of plastics. Ellen MacArthur Foundation, Cowes

European Commission (2020) A European strategy for plastics in a circular economy. European Commission, Brussels

Geyer R, Jambeck JR, Law KL (2017) Production, use, and fate of all plastics ever made. Sci Adv 3(7):e1700782

Gall SC, & Thompson RC (2015) The impact of debris on marine life. Marine pollution bulletin, 92(1-2):170–179

Graham WM, Condon RH, Carmichael RH, D'Ambra I, Patterson HK, Linn LJ, Hernandez FJ (2020) Oil carbon entered the coastal planktonic food web during the Deepwater Horizon oil spill. PNAS 117(13):7031–7041

Guzzetti F (2021) Invited perspectives: Landslide populations–can they be predicted? Nat Hazards Earth Syst Sci 21(5):1467–1471

Hodges A, Lafferty KD, Moershel M, Stelzenmüller V, Merrie A (2020) A global review of oil spill governance: examining the policy landscape of oil spill response. Mar Policy 118:104018

Horton AA, Walton A, Spurgeon DJ, Lahive E, Svendsen C (2017) Microplastics in freshwater and terrestrial environments: Evaluating the current understanding to identify the knowledge gaps and future research priorities. Sci Total Environ 586:127–141

Incardona JP, Vines CA, Anulacion BF, Baldwin DH, Day HL, French BL et al (2019) Deepwater horizon crude oil toxicity to the developing hearts of large predatory pelagic fish. Proc Natl Acad Sci 116(5):1920–1925

Jambeck JR, Geyer R, Wilcox C, Siegler TR, Perryman M, Andrady A et al (2015) Plastic waste inputs from land into the ocean. Science 347(6223):768–771

Jang M, Shin S, Ha HK, Kim Y (2021) Citizen science for marine plastic litter management: opportunities and challenges. Sustainability 13(2):792

Kooi M, Reisser J, Slat B, Ferrari FF, Schmid MS, Cunsolo S, Koelmans AA (2016) The effect of particle properties on the depth profile of buoyant plastics in the ocean. Sci Rep 6(1):33882

Laist DW (1997) Impacts of marine debris: entanglement of marine life in marine debris including a comprehensive list of species with entanglement and ingestion records. In: Marine debris. Springer, New York, pp 99–139

Lamb JB, Willis BL, Fiorenza EA, Couch CS, Howard R, Rader DN, Harvell CD (2018) Plastic waste associated with disease on coral reefs. Science 359(6374):460–462

Lehr WJ, Simecek-Beatty D, Plourde JD (2019) Introduction to oil spill response: past, present, and future. In: Oil spill science and technology. Elsevier, Amsterdam, pp 1–23

Li J, Galley M, Brockett C, Spithourakis GP, Gao J, & Dolan B (2016) A persona-based neural conversation model. arXiv preprint arXiv:1603.06155

Lippiatt S, Opfer S, Arthur C (2013). Marine debris monitoring and assessment: recommendations for monitoring debris trends in the marine environment. NOAA technical memorandum NOS-OR&R-46

Lavers JL, Dicks L, Dicks MR, Finger A (2019) Significant plastic accumulation on the Cocos (Keeling) Islands, Australia. Sci Rep 9(1):7102

Lebreton LC, Van Der Zwet J, Damsteeg JW, Slat B, Andrady A, & Reisser J (2017) River plastic emissions to the world's oceans. Nat Commun 8(1):15611

Mason RP, Choi AL, Fitzgerald WF, Hammerschmidt CR, Lamborg CH, Soerensen AL, Sunderland EM (2018) Mercury biogeochemical cycling in the ocean and policy implications. Environ Res 163:116–133

Maximenko N, Corradi P, Law KL, Van Sebille E, Garaba SP, Lampitt RS et al (2019) Toward the integrated marine debris observing system. Front Mar Sci 6:447

Michel J, Owens EH, Zengel S, Graham A, Nixon Z, Allard T et al (2020) Extent and degree of shoreline oiling: Deepwater Horizon oil spill, Gulf of Mexico, USA. PloS One 15(2):e0226974

Musa IO, Auta HS, Ilyasu US, Aransiola SA, Makun HA, Adabara NU, Abioye OP, Aziz A, Jayanthi B, Maddela NR, Prasad R (2024) Micro and nanoplastics in environment—degradation, detection, ecological impact. Int J Environ Res 18:1. https://doi.org/10.1007/s41742-023-00551-9

Muthukumar M (2017) 50th anniversary perspective: a perspective on polyelectrolyte solutions. Macromolecules 50(24):9528–9560

National Research Council (NRC) (2019) The role of oil spill dispersants in the Deepwater Horizon oil spill. National Academies Press, Washington, DC

Plastics Europe (2020) Plastics: the facts 2020. An analysis of European plastics production, demand and waste data. Plastics Europe, Brussels
Rochman CM (2018) Microplastics research—from sink to source. Science 360(6384):28–29
Ryan PG (2015) Does size and buoyancy affect the long-distance transport of floating debris? Environ Res Lett 10(8):084019
Rochman CM, Hoh E, Kurobe T, & Teh SJ (2013) Ingested plastic transfers hazardous chemicals to fish and induces hepatic stress. Sci Rep 3(1):3263
Sarkar A, Ray BK, Soares AMVM, de Andrade VM (2020) Polycyclic aromatic hydrocarbons in marine environments: a review. Mar Pollut Bull 158:111396
Storelli MM, Marcotrigiano GO (2018) Heavy metal residues in fish tissues from the Mediterranean Sea (Italy). Bull Environ Contam Toxicol 100(4):541–546
Schuyler QA, Wilcox C, Townsend KA, Wedemeyer-Strombel KR, Balazs G, van Sebille E, Hardesty BD (2016) Risk analysis reveals global hotspots for marine debris ingestion by sea turtles. Glob Chang Biol 22(2):567–576
Socolofsky SA, Gros J, North E, Boufadel MC, Parkerton TF, & Adams EE (2019) The treatment of biodegradation in models of sub-surface oil spills: A review and sensitivity study. Mar Pollut Bull 143:204–219
Surfrider Foundation (2021) Rise above plastics campaign. Surfrider Foundation, San Clemente
Teuten EL, Saquing JM, Knappe DR, Barlaz MA, Jonsson S, Björn A et al (2009) Transport and release of chemicals from plastics to the environment and to wildlife. Philos Trans R Soc B Biol Sci 364(1526):2027–2045
The Ocean Cleanup (n.d.) Technology. Retrieved from https://theoceancleanup.com/technology/
Thompson RC, Olsen Y, Mitchell RP, Davis A, Rowland SJ, John AW, Russell AE (2004) Lost at sea: where is all the plastic?. Science 304(5672):838–838
UNEP (2020) Marine litter vital graphics. United Nations Environment Programme, Nairobi
Van Cauwenberghe L, Janssen CR (2014) Microplastics in bivalves cultured for human consumption. Environ Pollut 193:65–70
Wilcox C, Van Sebille E, Hardesty BD (2018) Threat of plastic pollution to seabirds is global, pervasive, and increasing. Proc Natl Acad Sci 115(25):201717594
Wilcox AC, Escauriaza C, Agredano R, Mignot E, Zuazo V, Otárola S, Mao L (2016) An integrated analysis of the March 2015 Atacama floods. Geophys Res Lett 43(15):8035–8043
Wright SL, Rowe D, Thompson RC, Galloway TS (2013) Microplastic ingestion decreases energy reserves in marine worms. Curr Biol 23(23):R1031–R1033

Chapter 3
Characterization of Biodegradable Plastic-Producing Bacteria Isolates from Atlantic Ocean

Scott O. Fayemi, Bunmi Kotun, O. A. Fatunbi, O. Oda, and J. Ayedun

Abstract The importance of biodegradable plastic producing bacteria is determined by the properties they conferred on the type of plastic(s) they degrade. Despite the known advantages of biocompatibility, similarity to the petrochemical plastics properties and complete biodegradability that bioplastics possesses, the characteristics of the producers of these bioplastics is of immense importance to bioplastic synthesis. This review explored the different types of bioplastics used in our daily activities, the conditions by which certain bacteria synthesized them and the conditions of their biodegradability. Revelations from this study shows that different bioplastics usefulness is determined by the inherent ability conferred by their producing microbes and in particularly bacteria. It also shows that most of the bioplastic synthesizing bacteria are indigent to the hypersaline environment, even though the possibility of survival for some outside this habitat exists. However, there are mechanisms and conditions of proliferation and synthesizing the bioplastic. It is therefore concluded that for any bacteria to exhibit synthesis of bioplastics, there are conditions for the synthesis and optimal production of same.

Keywords Biodegradable · Bioplastics · Synthesis · Hypersaline and inherent ability

S. O. Fayemi (✉) · B. Kotun · O. A. Fatunbi · O. Oda
Department of Microbiology and Industrial Biotechnology, College of Pure and Applied Sciences, Caleb University, Imota, Lagos, Nigeria
e-mail: scott.fayemi@calebuniversity.edu.ng

J. Ayedun
Department of Biological Sciences and Industrial Biotechnology, College of Pure and Applied Sciences, Imota, Ghana

S. A. Aransiola et al. (eds.), *White Pollution: Biodiversity and hazards in marine plastisphere*, Environmental Contamination Remediation and Management,
https://doi.org/10.1007/978-3-031-95547-1_3

3.1 Introduction

The characteristics of biodegradable plastic-producing bacteria has been proven to be pivotal to the strength, durability, buoyancy and widespread application in everyday life when compared to the synthetic plastics. The use of Plastics in food and other packaging materials as disposables (ending up in dumping sites, landfills and waterbodies including oceans) had led to its emergence everywhere thus, constituting nuisance and environmental hazards that if not controlled, will have immense adverse effects on humans' health. Howbeit, a sure solution to the imminent environmental hazards posed by plastics is its biodegradable characteristics when disposed into the environment. It is worthy to note that bioplastics are biodegradable while synthetic ones are not. The classification of biodegradable character is majorly based on how long it takes 90% of the plastic material (polymer) to degrade into its corresponding simpler compounds within six (6) months by the microbial metabolic activities microbes. Bioplastics are biopolymers compound of polyethylene, aliphatic polyesters, hybride plastics and polyhydroxyalkanoates (PHAs) which are relatively quickly degraded by bacteria (Thakur et al. 2018; Boey et al. 2021). They can be synthesized by fungi or bacteria under stress condition of nitrogenous non-availability, coupled with excess carbon in the environment. However, among stress conditions known as extremes are acidic ($pH < 5$), alkaline ($pH > 9$), pressurized (>0.1 MPa), hot (>40 °C), cold (<5 °C), dry ($a_w < 0.80$), high-radiation and hypersaline (salinity >35‰) environments.

PHAs are of commercial interest as they can be processed into bioplastics, and environmentally-friendly alternative to the conventional petrochemical-derived plastics. They have promising applications in medicine, material science and agriculture (Koller 2022). PHAs are used in food packaging, plastic films, surgical sutures, controlled drug delivery, bone plates, nails, screws, and in the treatment of Osteomyelitis (Pandey et al. 2022). They can be derived from renewable and inexpensive sources, and are completely biodegradable by many microorganisms. Due to aforementioned importance of PHA it is necessary to characterize PHA producing bacteria from hypersaline water bodies where they are found in abundance and diverse (Tan et al. 2017).

3.1.1 *Types of Biodegradable Plastics*

There are over twenty (20) groups of biodegradable plastics (Jia 2020) but, some of them are discussed viz.:.

3.1.1.1 Starch Blends

Starch are derived from cassava, potato and corn. Their application is usually in the packaging of food, production of cutlery bags and thin plastic film used for mulching. Starch is abundant in nature and relatively cheap (Jia 2020). While their characteristics exhibits low mechanical strength but have higher rate of biodegradability. They therefore have need of being combined with other plastic source(s) for effectiveness and hence the term adjective 'blend', signifying its need of having to be combined with other plastic source or sources for its functionality (Jia 2020). In an anaerobic composting condition of 58 °C, 85% of starch blend biodegraded in 90 days while in anaerobic composting condition of 22 °C and 55% moisture, 26.9% of starch blends decomposed in 27 hours. In a soil condition of 20 °C and 60% moisture, 14.2% of starch blend biodegraded in 110 days (Jia 2020).

3.1.1.2 Polybutylene-Based Polymers

They are fossil-based biodegradable plastics, which means their basal constituent is crude oil. This is therefore a group of abioplastic, consolidating the fact that biodegradable plastics are not necessarily bioplastics. There are some that are produced from bio based feedstock like sugarcane. There are two major representatives in this group namely polybutylene succinate (PBS) and Polybutylene adipate-co-teraphthalate (PBAT). They are used for producing disposable cutleries, mulching film and plastic bags (Jia 2020). Their Biodegradation of 90% occurred in anaerobic compost of 58–65 °C and 50–55% moisture for 160 days (Jia 2020).

3.1.1.3 Polylactic Acid (PLA)

They rely on plant feedstock such as corn, cassava, sugar cane, and potato like the starch blends does. Agricultural wastes seem to be potential feedstock for (PLA) production (Jia 2020). They are usually biodegraded by the industrial composting technique. For an environmental condition of 58 °C for 58 days, 84% biodegradation occurred by industrial composting. In 60% moisture at same temperature for 30 days, 60% biodegradation occurred (Jia 2020).

Other biodegradable plastic worthy of note includes:

3.1.1.4 Polyethylene

This type of plastic is commonly used in packaging materials, bottles, and containers. It's lightweight, flexible, and resistant to moisture (Tsironi et al. 2022).

3.1.1.5 Polypropylene

It's a versatile PHA are biodegradable polymers obtained from a wide range of gram-negative and gram-positive bacteria that accumulate them in the cytoplasm in order to act as storage materials such as carbon, energy, and reducing power; the manufacturing are mostly possible under by these organisms under unbalanced culture conditions (Urtuvia et al. 2014). An excess of carbon activates the PHA polymerase enzyme encoded by *PhaC*. If starvation occurs, PHA depolymerase encoded by *PhaZ* degrades PHA and R-hydroxyalkanoic acids, to facilitate their use as sources of carbon and energy (Tan et al. 2014).

The production of PHA by halophilic micro-organisms is viewed to be advantageous for industrial applications since the saline environment favours the growth of halophiles, but not other contaminating organisms (Quillaguamán et al. 2010). For example, *Halomonas* TD01 was reported to produce P(3HB) in a non-sterile condition (Choi and Lee 1999; Tan et al. 2014). Their plastic are used in various applications like food containers, automotive parts, and textiles. It's known for its high melting point and durability (Swetha et al. 2023).

3.2 Importance and Uses of Biodegradable Plastics

Most of the vast attention drawn towards PHAs is mainly ascribed to their properties that is similar to some petrochemical plastics, biocompatibility and complete biodegradability. Bacteria metabolic activity in the natural environment breaks down bioplastics to release carbon dioxide and water or methane under aerobic and anaerobic conditions, respectively. PHAs are biodegradable and of great importance in the applications of medicine, engineering, agriculture, telecommunications, transportation and household utilities (Aditi et al. 2015; Zhang et al. 2015; Manikandan and Senthilkumar 2017).

The present emergence of human daily lives and corresponding technological development has further championed the significance of plastics. It has also resulted into the continuous demand and production of the same throughout the world at large. Plastic materials, which have a long-lasting durability, buoyancy, widespread use, omnipresence, and accumulation in large proportions, represent a major source of waste since about 70% of the plastic waste generated is improperly managed, being deposited in landfills (Ncube et al. 2021) and aquatic ecosystems, including the open sea, coasts, rivers, beaches, and wetlands having severe impacts on the ecosystems and public health at a global level (Manisalidis et al. 2020).

3.2.1 Types of Biodegradable Plastics and Their Characteristics

Microbial degradation of plastics is a fascinating process where microorganisms, such as bacteria and fungi, break down plastic materials into simpler compounds. These microorganisms produce enzymes that can break the chemical bonds in plastics, allowing them to utilize the plastic as a source of carbon and energy.

The process of microbial degradation can be influenced by factors such as temperature, moisture, pH, and the presence of specific enzymes produced by the microorganisms. In some cases, researchers have even identified and isolated specific bacteria or fungi that are particularly effective at degrading certain types of plastics.

Microbial degradation of plastics has gained attention as a potential solution for plastic waste management and environmental sustainability. By harnessing the natural abilities of microorganisms, scientists are exploring ways to enhance the degradation process and develop more efficient and eco-friendly strategies for plastic disposal.

3.3 Macro and Micro Plastics

Regarding aquatic ecosystems, microplastics are extensively distributed in most types of marine and freshwater environments, and they have even been detected in preserved areas and polar regions in recent years. A number of studies have indicated that these microplastics exhibit toxicological effects on various aquatic organisms through uptake and accumulation; further, they pose a potential threat to human health by transmission along the food web (Sussarellu et al. 2016; Aransiola et al. 2023a). In addition, microplastics have been revealed to be indirect harmful factors, which may serve as vectors that selectively enrich aquatic pathogens to new environments compared with natural substrates (e.g., rock and leaf) and could provide footholds for pathogens to invade aquatic organisms e.g., coral; (Wu et al. 2022). Through in long-term exposure and migration by water flow in aquatic environments, microplastics can be potential carbon sources that gradually release organic compounds into the water phase, thus contributing to the prolonged survival and growth of surrounding bacterial communities and the selective colonization of microplastic-attached biofilms that are distinct from other natural substrates (Tomboulian et al. 2004; Rummel et al. 2017; Aransiola et al. 2023b).

Howbeit, the major problem is the presence of microplastic particles (MPs) (Koelmans 2022; Musa et al. 2024) with regular and irregular shapes, whose sizes range between 1 μm and 5 mm and which are insoluble in water and classified as primary and secondary microplastics. Primary microplastics are manufactured as micrometric-sized plastic particles, fibers, or powders, used in personal care products, synthetic textiles, detergents, and paints, while secondary microplastics are derived from the breaking of larger plastics, such as containers or bags, which are then released into the environment under physical, chemical, and biological activities (Lehtiniemi et al. 2018; Auta et al. 2022). MPs have toxic effects on aquatic organisms, due to their size, which is like that of their food source, being absorbed or accidentally eaten, generating representative changes in several biomarkers, at the cellular level; on the other hand, MPs tend to accumulate in sediments, affecting benthic species.

3.4 Microbial Synthesis of Plastics

PHAs are synthesized in nature through fermentation of sugar or lipids and may be stored as carbon or utilized by expended energy by the living cell. Also, they are bioplastics, thermoplastic or elastomeric materials of temperature range between 140 °C to 180 °C. More than 150 monomers can be combined within this family to give materials with extremely different properties (Valappil et al. 2007).

These characteristics make them superior to synthetic plastics because they can be naturally recycled and do not contribute to environmental pollution at the end of their use. Concerns about plastic pollution are greatly helped by biodegradable polymers, which are produced by microbial synthesis. These polymers are eco-friendly because microorganisms may decompose them into innocuous byproducts like water and carbon dioxide (Du et al. 2022).

According to Suzuki et al. (2020) and Sanchez-Hernandez et al. (2020), biodegradable polymers, such as polyhydroxyalkanoates (PHAs), produced by microbial synthesis, have demonstrated significant promise in mitigating the environmental risks linked to traditional plastics.

The significance of microbial procedures in plastic manufacture is highlighted by the critical role that microbial enzymes play in accelerating ester bond formation, an essential phase in the sustainable use of plastics (Lai et al. 2023).

Additionally, the environmentally friendly and sustainable characteristics of the microbial synthesis of succinic acid and 3-hydroxypropionic acid using renewable substrates have been acknowledged, underscoring the significance of microbial engineering in the creation of sustainable plastics (Wang et al. 2023; Wang 2021).

Microbial synthesis has the ability to produce plastics with desirable qualities, as demonstrated by the microbial production of polyesters such as polyhydroxyalkanoates (PHAs), which exhibit tensile and barrier properties similar to conventional plastics (Vizireanu et al. 2018).

Further demonstrating the potential of microbial processes to support the circular as well as sustainable economy of plastics is the utilization of microbial polyester hydrolases for the breakdown of polymers such as polyethylene terephthalate (PET) (Salvador et al. 2019).

One potential way to lessen the negative effects of plastic waste on the environment is to use microbiological processes to synthesize biodegradable plastic from renewable resources (Uday 2023).

Microbial synthesis of plastics is a feasible way to produce bio-based and biodegradable plastics. It provides a sustainable alternative to traditional plastic manufacturing processes. It is also a sustainable and environmentally friendly way to make polymers. By combining microorganisms with renewable raw materials, bio-based polymers could be manufactured with a lower environmental effect than those generated using conventional methods. By recovering waste resources for the production of plastic, using microbial techniques for plastic synthesis not just offers a more ecologically friendly alternative, but it also promotes the circular economy.

Recent research has made tremendous progress toward understanding the production of plastics. Numerous aspects of producing plastic with biological systems have been studied. Wang et al. (2023), for example, worked on building an *Escherichia coli* with redirected carboxylic acid reductase cascade for the production of the synthetic plastic monomer 1,6-diaminohexane. This reveals the possibility of producing plastic precursors using microbial systems.

Koller (2022) highlighted the industrial potential of these bioplastics by discussing the autotrophic production of microbial polyesters referred to as polyhydroxyalkanoates (PHAs), which are biodegradable, utilizing modified *Cupriavidus necator* strains. In order to increase production capabilities and develop downstream processing techniques for manufacturing that is more affordable, a study that evaluated PHAs and focused on the processes, metabolic pathways, as well as enzymes included in their biosynthesis was conducted (Choi et al. 2020).

The economic viability of biosynthesizing environmentally friendly substitutes for plastic straws was demonstrated by a study carried out with bacterial cellulose-based straws (Yang et al. 2021). The viability of creating sustainable substitutes for conventional plastics was shown by this study.

The manufacture of phenol, a crucial component of many plastics and polymers, in *Escherichia coli* was studied by Thompson et al. (2021), demonstrating the possibility enabling bio-based production of these vital plastic components. The study of plastic biosynthesis through biological systems shows promise for environmentally acceptable and sustainable manufacture of polymers and their precursors. These researches have been laying the groundwork for a more ecologically responsible method of producing plastic by utilizing microbial processes.

The manufacturing of polymers by microbial synthesis has attracted a lot of interest as a sustainable and eco-friendly method. Microbial techniques present a possible substitute for conventional petroleum-based plastic synthesis by employing renewable raw materials such glycerol, agri-food wastes, and acetate (Sheng et al. 2021). Compared to plastics made from fossil fuels, these bio-based polymers made by microbial synthesis have been not just scalable and affordable, but they also have less of an adverse influence on the environment (Visco et al. 2022).

3.5 Hypersaline Plastics Producing Microbes and Their Characteristics

Hypersaline water bodies are marine water body that are characterized by 3.5–30% (w/v) salt concentration (Elbaz and Pratt 2015), low oxygen (< 6.6 mg/L at 25 °C) concentration (Podell et al. 2014), High or low temperatures and most times alkaline (> pH 7.5). It is an extreme environment that integrates into some of the characters associated with the classified nine categories of extreme environments (Rosgen 1994) namely halophilic, psychrophilic, thermophilic and alkaliphilic, others are radioresistant, acidophilic, xerophilic, barophilic and endoliths.

Physicochemical of any medium really determines the types of life forms and microbes that can be found therein. They can be said to determine the ecological niche of any environment. For hypersaline aquatic environments, temperature is an indisputable element in the rate of 'salt holding' and dissolution capacity in the medium and these two factors determines the water activity of the aquatic environment. In a hypersaline aquatic environments literature reveals that there is always extreme cold or hot temperature at any given period of time.

Hypersaline water pH is majorly dependent on numerous substances dissolution and interaction that results from organismal metabolic activities. These include the rate of respiration, photosynthetic activity of saline aquatic plants, and decomposition of organic matter, dissolution of CO_2 and oxidation and reduction of saline water body (Mohanakrishna et al. 2018). Hypersaline water bodies are mostly alkaline and the importance of alkalinity are hinged on its role in CO_2 chemistry, trace metal and buffer capacity of water. Alkalinity is also a tracer for studying mixing processes between different water masses and useful additional variable to study hydrographical features. The processes that affects alkalinity are photosynthesis, respiration, nitrification, denitrification, sulfide oxidation, sulfide reduction, and $CaCO_3$ dissolution (Liu et al. 2021).

Halophilic Bacteria

Halophilic bacteria are one of the nine groups of extremophiles known as psychrophiles (extreme cold); thermophiles (extreme heat); radioresistant microbes (extreme radiation); alkaliphiles (extreme bases; high pH levels); acidophiles (extreme acids; low pH levels); xerophiles (extreme dryness; lack of water); barophiles (extreme pressure); and endoliths (extreme rockiness) (Yuhana and Hanselmann 2006; Wang 2015; Zhang et al. 2015). Halophiles are organisms that possess the ability to survive extremely salty environments (5–10 times saltier than ocean water). Examples of halophilic bacteria among others that lives in hypersaline environments are *Agrobacterium aurantiacum*, *Staphylococcus aureus*, *Chromobacterium violaceum*, *Serratia marcescens*, *Bacillus Spp*, *Flavobacterium sp*, (with corresponding colors of Pink–red, Golden Yellow, Purple, red, Creamy and yellow respectively). Halophiles have been found in Utah's Great Salt Lake, the Dead Sea between Israel and Jordan, Atlantic Ocean in SouthWestern Nigeria region (Fayemi et al. ??), even growing on saltine crackers (Vinothini et al. 2014).

3.5.1 Categories of Halophiles

In any environment of which determines the category of resident organisms, the acquisition of compensatory mechanisms or adaptability of cellular biochemistry must be exercised by the organism in question, for example the ability to cope with the low water availability of resident bacteria in an halophilic environment depicts the category of halophiles (Vera-Gargallo et al. 2023) therefore, the classified halophiles is based on their salt requirement and tolerance and they are: extreme halophiles (growing best at 2.5–5.2 M NaCl), borderline extreme halophiles (having their optimum at 1.5–4.0 M NaCl), moderate halophiles (preferring media with 0.5–2.5 M NaCl) and halotolerant microorganisms that do not require salt for growth but tolerate salt often in high concentrations. Terry and Oren (2012) also describes extreme and moderate halophiles as those that grow in environments with salt concentrations above 100 g L^{-1} (equivalent to 1.7 M NaCl) and salt saturation (around 5.2 M or 300 g L^{-1} NaCl respectively. He also identified that such halophiles can be found in each of the three domains of life of Archaea, Bacteria and Eukarya. He stated that the best known extreme halophiles are found within the domain Archaea (phylum Euryarchaeota) in the class Halobacteria, which contains the single order Halobacteriales, with a single family the Halobacteriaceae.

3.5.2 Mechanisms of Growth Displayed by Halophilic Bacteria in Hypersaline Environments

Hypersaline water bodies (less than 10% NaCl or dissolved salt contents) are characterized by low-water content or water activity (a_w) level due to the high-salt concentration. The implication is that as a_w decreases, there is less water for dissolving nutrients, general metabolic processes, and for the hydrating of proteins and nucleic acids by the bacteria found in this water bodies. However, halophilic bacteria are characterized for possessing mechanisms for the exhibition of physiological responses to salt (ref) and this includes osmo-regulatory mechanisms, increased intracellular ionic gradients, large vacuole, and also active-transport activity among others. The aforementioned characters enable them to survive, attain optimum growth and replication potentials in this harsh condition (DasSarma and DasSarma 2012).

Halophiles include prokaryotes and eukaryotes with the capacity to balance the osmotic pressure of the environment and resist the denaturing effects of salts. They are known to show considerable growth at salt concentrations typically within 0.2–5.1 mol L^{-1} and they grow optimally between 0.5 and 2.5 M salt concentration.

Therefore, in order for bacteria to survive hypersaline environment, any of the characteristic mechanisms of overcoming high salt effects needed are identified and discussed below:

3.5.2.1 Osmotic Pressure Balance

the permeability in water characteristic found in biological membranes is evident in a non-halophile that is in a high-salt environment, because water will exit the cell due to osmotic principle (plasmolysis), and the resultant process will leave behind a highly concentrated cytoplasm which will result to the loss of structure and functions in proteins, nucleic acids and other macromolecules. But, bacteria adapted to living at low water activity may possess a cytoplasm that is at the same osmotic pressure as the brines of its medium. The intracellular osmotic pressure should exceed that of the extracellular environment if the turgor pressure of the cell is to be maintained (as displayed by all cells but with the possible exception of some halophilic Halobacteriaceae family), McGenity and Oren 2012).

3.5.2.2 Accumulation of KCl to Molar Concentrations (Salt-In to Balance Salt-Out Strategy)

This type of strategy is also known as salt-in to balance salt-out strategy and found to be displayed by few groups of prokaryotes only. Also, the archaeal family Halobacteriaceae, bacteria aerobe Salinibacter and the anaerobic fermentative Halanaerobiales (Low-G + C branch of the Firmicutes) displays salt-in to balance salt-out strategy (Brown 1990; McGenity and Oren 2012). It is also observed that microbes that employ the use of salt-in to balance salt-out strategy are obligate extreme halophiles thus suggesting that they are unable to adapt to life where the salinity is low, because adequate molar concentration of KCl is necessary for the activity of their complete enzymatic machinery to function.

Proteins from obligate extreme halophiles are typically highly acidic, with a great excess of the negatively charged glutamate and aspartate over the positively charged lysine and arginine residues. Literature also reveals that the charge of the proteins may be responsible for creating negatively charged surface, which facilitates retention of a hydration shell around the protein. Researchers (Chen et al. 2023) on the other hand, argued that it may not be the negative charge that is important but the fact that negatively charged amino acids also have short side chain that will eventually decrease the surface area of the protein accessible to ions. But it must be noted that maintenance of cellular proteins in their native and active form requires necessary molar salt concentrations without which will denature the cellular proteins.

3.5.2.3 Accumulation of Organic Compatibles (Osmolyte Strategy)

Osmolyte strategy requires more energy than that of salt-in to balance salt-out, but this confers on the bacteria the action of flexibility and adaptations in case of changing concentrations of the salt contents of the environments because, there is no need for special protein structures. The mechanism makes use of the strategy of excluding salt from their cytoplasm as much as possible, and this implies that high concentrations of organic 'compatible' solutes is prevented from any great interference with the normal enzymatic activity of the cell thus, the intracellular saline concentrations can be regulated to meet up with the survivability of the bacteria since few adaptations of the cells' proteome are needed (Ventosa et al. 1998). Internal osmotic pressure is therefore balanced with that of the external osmotic pressure by making the two solutions isotonic, or close to it.

The mechanism of accumulation of compatible solutes in bacteria is the most commonly used method of salt resistance among halophiles. Inclusive in the solutes that are regulated in the cells' cytoplasm for the achievement of osmotic balance are glycerol, amino acid derivative such as glycine, betaine and ectoine (1,4,5,6-tetrahy dro-2-methyl-4-pyrimidine carboxylic acid); included sugars too are sucrose and trehalose and many other compounds. It is also noted that glucosylglycerol is almost exclusively found in moderately halophilic or halotolerant cynobacteria and that N∑-acethyl—(-lysine and N-acethylortine) have thus far been detected only in aerobic members of the Firmicutes (Terry and Oren 2012).

3.5.2.4 Multiple Extremes

The form of extreme condition in an hypersaline environment can be more than one, and so many bacteria found to survive in this environment are polyextremophiles therefore, additional forms of environmental extremes makes these organisms adapted not only to high salt but to one or more additional forms of environmental extremes. For example, reports of evaluating the growth in shallow hypersaline ponds that results in exposure to very high levels of ultraviolet light, which eventually leads to DNA damage and cell death in unadapted organisms (McGenity and Oren 2012). He also reported that *Halobacterium salinarum*, an archaeon that thrives in such environment which is among the most UV-resistant of any organism.

Hypersaline environments are found deep in the Earth's crust and at the bottom of the deep sea, high pressure is known to significantly influence metabolism (Zhang et al. 2015), Hypersaline acidic environments are rare, but Halobacteriaceae has been reportedly isolated from such habitat, the members are capable of growing between pH 4 and 6, the twin East Lake in Australia are both highly acidic (pH 3) and hypersaline (160 g L^{-1}), they harbours many bacteria, such as *Sphingomonads* that are not typically found in neutral lakes of equivalent salinity. The report of the detection of microbial activity in cold, hypersaline environments, and the culture of moderately halophilic psychrotolerant strains has been documented, also included is the report of psychrotolerant extreme halophile (*Halorubrum lacusprofundi*) from Deep Lake, Antarctica (Kyrpides 2016).

It is observed that the optimum temperature of this Deep Lake is 31–37 °C, yet bacteria growth is reported at temperatures as low as 4 °C, thus, it could grow during the short periods when the temperature of Deep Lake increases from subzero to a maximum of 11.5 °C (McGenity and Oren 2012). Many members of the Halobacteriaceae can grow up to 50 °C, the most thermophilic halophiles are bacteria. For example, *Halothermothrix orenii*, from a salt lake in Tunisia, grows anaerobically up to 68 °C (optimum 60 °C) and up to 200 g L^{-1} salt (Cayol et al. 1994). An isolate isolated from Wadi An Natrun, *Natranaerobius thermophilus*, of representative of a new order, the Natranaerobiales, also grows optimally at 54 °C (up to 56 °C) at pH 9.5, and at a total Na^+ concentration between 3.3 and 3.9 M (range: 3.1–4.9 M) (McGenity and Oren 2012). A better understanding of the compounding and in some cases off-setting effects of multiple extremes on organisms is needed if we are going to identify and even extend the frontiers of life.

3.5.2.5 Pigmentation

Microbes that inhabits and survive the process involved in resultant hypersaline water body environments do so, on the basis of their character to develop the mechanism of pigmentation (Edbeib et al. 2016). This is necessary because they found themselves in an abundant environment of sunlight which reaches them with doses of elevated temperatures and most especially ultraviolet radiations that can cause damage to their DNA if not protected. Therefore, rich array of carotenoid pigments are synthesized by some of these halophilic microbes that lives in hypersaline environments e.g. shallow evaporation pans).

Microbial pigmentation may be coupled with the character of good DNA repair ability of halophilic microbes that are frequently exposed to ultraviolet rays if they will have to survive the condition. It is also argued that these pigments acts as a protective barrier to the ultraviolet light rays from the sun because, conjugated double bonds that functions as free radical scavengers (that acts as protective shields) were contained in carotenoids (Valmalette et al. 2012; Vatansever et al. 2013; Udayan et al. 2017). Pigmented body surfaces are also known to absorb more heat from the sun than non-pigmented bodies (Alibardi 2015).

Halophilic archae contains among others two types of pigments namely bacteriordopsin and halorhodopsin. The former is a vehicle in the export of Hydrogen-hydrogen ions in the ATP generation process (this type of energy production does not involve chlorophyll as found in plant or photosynthetic microbes). Exposure to UV light is necessary for the activation of bacteriorhodopsin, a purple chromoprotein located within the cell membrane, which acts as a proton pump and drives ATP synthesis (Toth et al. 2020) but, the latter excretes hydrogen ions and import chlorine ions into the cells (Lange et al. 2013; Skulachev et al. 2013). For example bacteriorhodopsin that is reportedly found in the genus *Halobacterium*, allows for energy production under reduced oxygen (micro aerobic) or anaerobic conditions and high light intensities. Thus competitive advantage against total aerobic energy production mechanism, dependent organisms are seen where bacteriorhodopsin is present as found in *Halobacterium* spp. (Ventosa et al. 2014).

3.5.2.6 Halotaxism

This is a mechanism whereby sensitivity and response to salt concentration beyond halophilic bacteria adaptation is developed by migrating to preferred saline (McKew et al. 2011). For example filamentous cyanobacterium Coleofasciculus (*Microcoleus chthonoplastes*) is reported to migrate from a less salt concentration to a higher and required choice of salt concentration.

3.6 Characterization of Bioplastics Producing Bacteria

The process of microbial degradation can be influenced by factors such as temperature, moisture, pH, and the presence of specific enzymes produced by the microorganisms. In some cases, researchers have even identified and isolated specific bacteria or fungi that are particularly effective at degrading certain types of plastics.

In a recent unpublished study, the effects of temperature on the growth of bacterial isolates shows that 60 °C among other experimented temperatures is the general optimum growth index at 1 and 2 h of incubation. However, there was a complete decline at 3 h incubation but the optimal growth index of temperature obtained in this work for incubation periods from 4 to 36 h was 60 °C. In spite of the results obtained in this study, which is in total agreement with the findings of Oren (2013) which ascertain that hypersaline water bacteria survive or thrive in both high and low temperatures, the decline at 3 h incubation may be due to the initial release of CO_2. Hence, the release of CO_2 according to Steinweg et al. (2008) could be associated with the assimilation of the carbon source of plastic materials after degradation and fragmentation.

This same mentioned work above clearly shows growth resuscitation at temperature of 70 °C of incubation period for 1 h and 75 °C at 2 h. This may be attributed to the suggestion that there might be mechanisms that tend to repair damages incurred by the effects of temperature (Zion et al. 2006). Generally, the trends of these different hours of culture affects the bacteria growth in such a way that a repair of the effects of the damage that might have been caused by the intensity of the different temperatures are repaired/resuscitated on the pivot of time of exposure (incubation) depending on the genetic constitution of each bacterium in question (Yadav et al. 2018). It may also be deduced that the period of 'heat shock' and exposure can be recoverable at a maximum period of 24 h after which repair may be impossible. This is because at 36 h of incubation period, it was noticed that the general growth index tended towards bacterial inactivation. This finding was in agreement with the work that ascertained trends of recovery profile response of bacteria to temperature in spring environment even though the optimum growth temperature was 60 °C (Yadav et al. 2018).

The work on temperatures shows initiation of exponential period at 53 h from the effects of 3% salt concentration on the growth of bacterial isolates. This may be due to the contributions of temperature, osmotic pressure, atmospheric pressure, and moisture availability among other environmental factors. This may also be due to characteristic of enough salt concentration accumulation (known as lag period of which may also be termed period of growth adjustment) for proliferation initiation (cellular multiplication). Hence, genetic makeup may also be responsible for the bahaviour as ascertained in literature (Elabed et al. 2019). The delay in growth until this 53rd hour could also mean that, bacterial isolates had gone into dormancy (Aanderud et al. 2016) due to the demonstration of defence against unfavourable temperature and exposure time. This work shows increase in growth index at 12% salt concentration of bacterial incubation at 2–18 h. But, 'salt shock' was observed in *Lactobacillus delbrueckii* where growth index was low compared to the other bacterial isolates before 53rd hour of incubation where recovery and optimum growth recovery was obtained. This result is also suggesting that *Lactobacillus delbrueckii* has an acclimatization period of over 53 h for 3–30% concentration of salts (NaCl) medium. It also suggests that the severity of salt shock for *Lactobacillus delbrueckii* is dependent on the availability or deposition/accumulation of salt concentration required by the bacterial cell for proliferation as ascertained in literature (Jan et al. 2022). This result also suggests that pH may have more effects on the growth of bacteria once the required pH is attained; growth is only obtained at 30% salt concentration of the same incubation period of 53 h. Thus, the pH and not salt concentration over time may be responsible for preventing dormancy or its breakage in *Lactobacillus delbrueckii* (Bashar et al. 2018).

The significant difference displayed in the growth of the bacterial isolates at 6% and 9% (w/v) NaCl concentration mainly at 33 hours after incubation may be due to the different categories and adaptation abilities of halophiles that may be either halotolerant or halophilic, this is in agreement with the findings that time of exposure to NaCl is a function of the growth index displayed by different bacteria as documented in the literature (Ventosa et al. 1998).

Polyhydroxylalkanoates (PHAs) are biopolymers and may be produced by bacteria in order to overcome environmental stresses that are characteristic of hypersalinity. The vast attention towards PHAs is mainly attributed to their properties that resemble some petrochemical plastics, as well as its biocompatibility and complete biodegradability. Microbial activity in the natural environment breaks down such bioplastics to release carbon dioxide and water or methane under aerobic and anaerobic conditions, respectively. PHAs are biodegradable and of immense importance in the applications of medicine, engineering, agriculture, telecommunications, transportation and household utilities (Aditi et al. 2015; Zhang et al. 2015; Manikandan and Senthilkumar 2017).

3.7 Importance of Halophilic Bacteria

Apart from the global warming and its effects of burning petroleum-based plastics, it also emits toxic fumes that are inimical to human and animal's health of which may eventually disrupt the ecosystem. Therefore, the aforementioned fact necessitates the search for environmentally-compatible alternatives to petrochemical-based plastics.

However, PHAs are a class of linear polyesters with properties comparable to petrochemical-based plastics, they possess the additional advantages of biocompatibility and biodegradability. Production and degradation of PHAs can be carried out by microorganisms of which bacteria is not left out.

Bacteria found in hypersaline environments are known as halophiles which are of great industrial potentials. Halophilic bacteria have been categorized into four, based on their optimal growth in different salt concentration. This include: non-halophilic bacteria, slight halophilic bacteria, moderate halophilic bacteria, and extreme halophilic bacteria (Kamekura et al. 1988). Non-halophilic bacteria require less than 1% sodium chloride (NaCl). But despite the fact that this group of bacteria do not need high salt concentration for their optimal growth, they may tolerate it and so, they are considered as halo-tolerant bacteria. Slight halophiles (for example, marine bacteria) grow best in media with 1–6% NaCl, whereas the optimal salt concentrations for moderate halophiles range between 7% and 15% NaCl; on the other hand, extreme halophiles show optimal growth in media containing 15% to 30% NaCl (Kushner 2020). The ability of bacteria display of tolerability or requirement of different levels of salinity in growth is controlled by specific genes, and these genes can be switched on or off by different environmental factors that the bacteria may find itself (Thattai and Van Oudenaarden 2004; West-Eberhard 2005) and thus, man having being armed with this knowledge is able to exploit the conversion of such beneficial technology for the advancement of agriculture, engineering, food Industry, medicine, waste management (Uzogara 2000; Tilman et al. 2002) of which optimum production of and degradation of bioplastics are not left out. The gene manipulations also extend to planet conservation and alternate power generation (Zhang et al. 2009).

The different group characterized by these halophiles is the factors responsible for the diversity and characteristics of their products or metabolites. The survival and or optimal growth of halophilic bacteria necessitates the production of one or more of the followings; PHAs, enzymes, metabolites, exopolysaccharides, endoploysaccharides, pigments and compatible solutes (Rothschild and Mancinelli 2001) of which they have high commercial value. The biomolecules produced by halophilic bacteria are stable due to their ability to live under extreme conditions, and thus this is shown in their exhibition of significance and potential needs, or in actual use in different industries such as agriculture, cosmetic, chemical, environmental, pharmaceutical, etc. Therefore the isolation, identification and characterization of bacteria that produces PHAs in the hypersaline environments are of great interest.

Some microorganisms which are moderately or extremely halotolerant do have the ability to produce PHAs. These PHA-producing microorganisms can be isolated and put to use in the treatment of effluents which have a similar salt concentration to the environment in which they flourish. Also, high salt concentrations prevent growth of other mesophilic organisms, so there is no need for strict sterile conditions and repeated sterilization of the equipment or media used. With such organisms, a dual purpose of effluent treatment as well as cost-effective production processes due to reduction in the use of energy can be achieved (Bashar et al. 2018). The reduction in cost will be greatly beneficial to mankind and will encourage the manufacture and sale of biocompatible plastics.

In a bid to adapt or tolerate extreme environmental conditions, halophiles produces compatible solutes such as glycine, betaines, and ectoines which are used as a stress protectant against high salinity, thermal denaturation, desiccation and freezing as well as stabilizers of enzymes, nucleic acids, membranes and entire cells. Also, trehalose, glycerol, proline, sugars, and hydroxyectoine are produced (Manikandan and Senthilkumar 2017).

The significance of halophilic bacteria to biotechnology is hinged on their synthesis of enzymes and eventual production of useful metabolites for mankind. It employs the application of cellular and bio molecular harness processes for the development of technologies and products with the intention to improve our lives and the health of our planet. Enzymes such as amylase and protease that are produced by halophilic bacteria (Manikandan and Senthilkumar 2017) are vital in biotechnological advancement. Amylase is involved in the biotechnological production of pharmaceuticals; textile, detergent, food technology as well as in paper and chemical industries (Manikandan and Senthilkumar 2017). Harvests of *Bacillus sp* have been reported to produce the largest amylase activity despite protease enzymatic applications are used in medicine, detergent and food processing industries (Zhang and Kim 2012).

The report of halophilic enzymes' diverse potential and usage in biomedical science and chemical industries has been credited to the Gram-negative genera known as, while Gram-positive, representatives of the genera identified as *Bacillus* has been documented to be among the bacteria that produce wide diversity of enzymes namely; llipolytic, pproteases, amylases, nucleases (DNA), cellulase and xylanases (Manikandan and Senthilkumar 2017).

Halophilic bacteria produces cellulases which are important for cotton processing, paper recycling, detergent formulation, juice extraction, and animal feed additives; agricultural biotechnology and production of bio-energy and bio based products (Venkatachalam et al. 2014; Manikandan and Senthilkumar 2017). Macrolactins, polypeptides, polyketides, lipopeptides, and isocoumarins are among the bioactive produced by halophilic bacteria as secondary metabolite (Surovy et al. 2022). These metabolites and compatible solutes are bimolecular or stress protective agents during metabolic processeses (Waditee-Sirisattha et al. 2016).

Halophilic microbes are producers of surfactants (Fariq and Yasmin 2020) of which some have been documented to have broad-spectrum antibiotic characteristic. It is employed in the laundry industries as it also reduces the surface tension in order to form stable emulsions.

The capability of halophilic bacterial growth under harsh environmental conditions makes it an excellent microbe for the bioremediation of salty environment (Ghafoori et al. 2016). Oil spillages have been reportedly cleaned up by the action of halotolerant and halophilic bacteria; they are also known to generally tolerate conspicuous amounts of toxic metals in their environment (Margesin and Schinner

2001). Halophiles may be demonstrated to be a good industrial manufacturer's tool, producing desired metabolites because of the fact that some halophiles are able to grow in high pH and high NaCl containing medium under higher temperature, allowing fermentation processes to run continuously and unhindered under such harsh conditions (Manikandan and Senthilkumar 2017).

Halophilic bacteria have been known to be candidates of potentiality for genes of desirable characters. Hence, desirable genes may be transferred and manipulated in other microbes in order to produce a 'super producer' machine for the industrial application.

Medical applications of PHA materials are visibly found in surgical pins, sutures, staples, swabs, wound dressings, bone replacement and plate and blood vessel replacement by piezoelectric properties (Zamierowski 2008).

PHA are been reported to be a useful source of stereo regular compounds which can serve as a chiral precursors for the chemical synthesis of optically active compounds, particularly in the synthesis of some drugs or insect pheromones. These substances are biologically active in the correct stereo chemical configuration (Anjum et al. 2016).

PHA is used for implants like heart valves, stents, sutures and bone scaffolding (Williams et al. 2005).

3.8 Factors Responsible for the Biodegradability of Bioplastic Synthesized by Bacteria

The biodegradation of bioplastics occurs in three separate phases. The first stage is called bio-deterioration, during which microorganisms infiltrate the material's surface and alter the polymers' chemical, mechanical, and physical properties. The porosity has a big impact on this step. In the second stage, referred to as bio-fragmentation, microbial activity breaks down polymers into oligomers and monomers. The third stage explains the use of bioplastics, in which biomass, CO_2, and H_2O are produced as a result of microorganisms absorbing the bio-fragmented molecules (Folino et al. 2020). In order to gain energy for their growth, microorganisms bind to the surface of polymers and secrete enzymes that break down the polymers (Danso et al. 2019). Bioplastics are biodegradable due to the actions of microorganisms such as bacteria and fungi that can absorb and metabolize these bioplastics as a source of energy. Hence, there are certain conditions required to proceed with the biodegradation process. These conditions would be reviewed in this section.

3.8.1 Factors Responsible for Biodegradation Includes

Bioplastic-degrading enzymes: Plastics can be broken down by microorganisms and their enzymes to depolymerize waste petro-plastics into monomers for recycling or to mineralize them into carbon dioxide, water, and new biomass while simultaneously producing higher-value bioproducts. The process of breaking down plastics involves the microorganism excreting extracellular enzymes, attaching the enzyme to the plastic's surface, hydrolyzing it to short polymer intermediates, and finally assimilated by microbial cells as a source of carbon dioxide for the release of CO_2. These are derived from organisms from the esterase, cutinase, or proteinase families, these enzymes degrade bioplastics that are not their naturally occurring substrates. They break large polymers down into smaller monomers and these are then taken up by the microbial cells. The breakdown of bioplastics gives microorganisms a source of carbon. These bioplastic-degrading enzymes are secreted (Jyoti 2021). Numerous enzymes that are known to degrade various types of plastic materials have been isolated, including lignolytic enzymes, exopolysaccharides from algae, laccases, cutinases, carboxylesterases from actinomycetes, PETase, Alkane hydroxylase, cutinase from bacteria, depolymerases, peroxidases, laccases, and cutinases from fungi (Urbanek et al. 2020).

Recent research has demonstrated that the esterase enzyme from the yeast Pseudozyma antarctica speeds up the breakdown of plastic mulch films stored in lab settings. Biodegradable plastic used to make mulch films demonstrated a rapid rate of breakdown when the strain was cultivated with xylose present. The plastic films were incubated with the enzyme for 12 weeks, and the SEM analysis showed that the smooth surface of the films had holes pierced through it, resulting in an uneven surface. Additionally, Adetunji and Olaniran (2021) reported that the amorphous portion of the films deteriorated more quickly than the crystalline portion. The yeast Cryptococcus sp. MTCC 5455, which was cultivated on leftover agricultural waste, produced lipase enzyme that showed a tremendous amount of hydrolyzing polybutylene succinate (PBS) and polybutylene succinate-co-adipate (PBSA). For PBS, the full deterioration of the plastic materials was seen after 72 h, whereas for PBSA, it took 16 hours (Thirunavukarasu et al. 2016).

3.8.1.1 Environmental Conditions

The rate of biodegradation is also significantly influenced by environmental factors. While plastics tend to be retained longer in colder, drier climates, deterioration is aided by both increased temperatures and wetness. As a result of higher microbial proliferation in warmer, more humid environments. Biodegradation of bioplastics is dependent on their surroundings. In uncontrolled environments, biodegradation may be less efficient; in natural ecosystems, like soil or water, the process may occur more slowly. Conversely, non-biodegradable bioplastics, despite coming from plants, do not break down easily in the environment. One example of a non-biodegradable bioplastic is polylactic acid (PLA) plastic, which has gained popularity because it can be produced from renewable resources like corn flour. Although PLA is compostable in industrial composting facilities, it does not readily biodegrade in natural environments. Tanadchangsaeng and Pattanasupong (2022) tested polyhydroxybutyrate (PHB) in saltwater and found that there was a significant 61.20% weight reduction. In contrast, polyhydroxyalkanoate (PHA) showed a high level of biodegradability in terrestrial settings. In a 30-day period, Patil et al. (2023) observed an astounding 98.62% and 89.75% weight loss for PHA in sewage and soil, respectively. Temperature, PH, oxygen availability, polymer size and structure, and other environmental conditions all have an impact on the biodegradation of bioplastic by bacteria. In a different study, the biodegradability of three fungal strains—*Fusarium* sp. B30 M, *Geomyces* sp. B10I, and *Sclerodinia* sp. B11IV—was evaluated at 4 °C, 20 °C, and 28 °C. The strains of *Sclerotinia* sp. B11IV and *Fusarium* sp. B30 M biodegraded most effectively at 20 °C after 4 weeks of incubation, with biodegradability rates of 45.99% and 49.65%, respectively, and 49.68% (PBSA) and 33.7% (PCL). *Geomyces* sp. B10I biodegraded PCL and PBSA at 11.34% and 4.46%, respectively, at 20 °C, whereas at 14 °C, the rates were 25.67% and 5.71%. The least efficient biodegradation occurred at 28 °C (Jia et al. 2021).

3.8.1.2 Structure/Properties of Bioplastic

The physical and chemical properties of plastics influence their biodegradation process. Significant roles are played by the

(a) surface conditions that is the surface area, hydrophilic and hydrophobic properties.
(b) high order structures such as the glass transition temperature, melting temperature, modulus of elasticity, crystallinity, and crystal structure.
(c) and first order structures which includes the chemical structure, molecular weight, and molecular weight distribution of polymers in the biodegradation processes.

Polyesters having side chains are often less digested than polyesters without side chains. Since the molecular weight affects the polymer's numerous physical characteristics, it is also significant for biodegradability. Tokiwa et al. (2009) reported that the degradability of the polymer was reduced when its molecular weight increased. Rhizopus delemar lipase (endo-cleavage type) broke down PCL with a greater molecular weight (Mn > 4000) more slowly than PCL with a lower Mn. Furthermore, the rates of biodegradation of polymers are significantly influenced by their appearance. Since enzymes primarily target a polymer's amorphous regions, the degree of crystallinity has a significant impact on biodegradability. Because the molecules in the amorphous region are not tightly packed, it is more prone to breaking down.

3.8.1.3 Kind of Bioplastic

Studies like those by Wicaksono et al. (2022) and Abe et al. (2021) on starch-based bioplastics showed that they were highly degradable in standard soil conditions and compost soil, losing 74% and 69% of their weight over 120 days, respectively. Additionally, Abe et al.'s (2021) investigation of a starch–polyhydroxyurethane hybrid revealed an 88% weight loss in soil during the same time period. Polylactic acid (PLA), which is made from tapioca, corn starch, or sugarcane roots and is frequently used in pharmaceuticals, is another biodegradable bio-plastic. PLA in soil can be broken down by *B. licheniformis* and *Amycolatopsis* sp. (Khan and Majeed 2019). The degradability of polylactic acid (PLA) has been studied in a variety of settings. In PLA, Brdlík et al. (2021) found that after 60 days of composting, there was a 16% weight reduction. This is consistent with the results of Karamanlioglu et al. (2017), who found that PLA lost 20% of its weight in soil over a 20-month period. Notably, PLA completely degraded in 60 days with controlled composting and bacterial treatment. PLA and PHAs both maintain their weight for a considerable amount of time about 12 weeks, whereas starch-based polymers exhibit a reduction in weight and a faster rate of breakdown in field. After 100 days, cellulose-based bioplastics showed the best biodegradability (80–100%) (Bilo et al. 2018). Drawing from the kinetic constants of the three blends under investigation mixes with PHAs had the highest kinetic constant in soil, followed by mixes with PLAs (Kalita et al. 2021). Different biodegradation paths can be followed by biodegradable plastics; some, like PHAs, can be broken down by microbes directly, while the breakdown of other types of plastics is aided and enhanced by environmental elements like heat, oxygen, and UV light. Oxidative-degradable polymers break down more quickly when exposed to heat or ultraviolet radiation that causes oxidation. Because oxygen-containing components can absorb UV radiation and use it to start a primary breakdown, UV radiation can break polymer chains. These polymers are referred to as photodegradable polymers. Since oxygen-containing components can absorb UV radiation and start a primary breakdown, UV radiation can break polymer chains. These polymers are referred to as photodegradable polymers. Both molar mass and crystal structure are impacted by photodegradation. The term "hydro-biodegradable" refers to plastics that can break down via hydrolytic processes, such as biopolymers consisting of cellulose, starch, and polyesters like PHA.

3.8.1.4 Microbial Diversity

Widespread distribution of bioplastics can have an impact on the ecosystems that degrade them. The kinetics of biodegradation can be influenced by the interactions between various microbial species that contribute to distinct stages of the degradation process. These microbial populations could adapt to new conditions, including the introduction of bioplastics. Therefore, comprehending the diversity and dynamics of microbial communities is essential to developing effective techniques for biodegrading bioplastics in these environments (Ruggero et al. 2023). Among the several microbial species that are able to break down bioplastics are firmicutes, proteobacteria, ascomycetes, and basidiomycetes. The fact that these microbes are dispersed throughout a variety of habitats demonstrates how diverse microbial metabolisms can be from one another. When cultivated in a lab, many of the biodegradable species that have been reported thus far are frequently found in growth conditions similar to the stationary phase. Cells produce proteins at this growth stage to help them scavenge and overcome nutritional stress (such as proteinases, which provide amino acids to cells so they can continue to grow) (Aguilar-paredes et al. 2023). Bioplastics can be broken down by a wide variety of bacteria and fungus. Strains, such as *Bacillus pumilus* A-1, *Paenibacillus amyloyticus* TB-13, *Actinomadura* sp. TF1, *Streptomyces* sp. APL3, *Laceyella* sp. TP4, and Aspergillus fumigatus NKCM1706 (I), have been reported to biodegrade various kinds of bioplastics. Varied types of bioplastics undergo varied breakdown processes due to varying microorganisms and environmental conditions. Research shows that petroleum-based bioplastics, such as PLA, break down similarly to regular polymers. Studies show that the breakdown of PLA is facilitated by microbes such Pseudomonas and Bacillus MKY1 and MKY2 (Ali et al. 2023). When exposed to Bacillus sp. JY14, polyhydroxybutyrate (PHB) exhibits remarkable biodegradability, losing a significant 98% of its weight in just 14 days (Urbanek et al. 2021). This demonstrates how well microbial contact works to speed up the breakdown of PHB. In contrast, other plastics like Polybutylene Succinate (PBS), Poly(butylene succinate-co-butylene adipate) (PBSA), Poly(ε-caprolactone) (PCL), and Poly(butylene adipate-co-terephthalate) (PBAT) showed different levels of biodegradation. The strain *Sclerotinia* sp. B11IV, B11IV, and *Fusarium* sp. B3′M, respectively, showed roughly 50%, 33.70%, and 50.0% of weight loss for PBSA, PCL, and PBAT. Interestingly, when subjected to *Pseudomonas mendocina* and *Arthrobacter elegans* over a five-day period, the blend of Polylactic Acid/Polybutylene Adipate-co-Terephthalate (PLA/PBAT) showed lesser biodegradability, with only 12.94% and 9.27% of weight loss, respectively (Ali 2023).

3.9 Conclusion

It is imperative to state that this study opens the importance of bacteria biodiversity in bioplastic synthesis. It also beams lights on the characteristic exhibition of bioplastic producers search to start from the hypersaline environments. The search for the right purposeful bioplastic producers are encouraged to properly study the producing organism(s) most especially of bacteria origin.

References

Aanderud ZT, Vert JC, Lennon JT, Magnusson TW, Breakwell DP, Harker AR (2016) Bacterial dormancy is more prevalent in freshwater than hypersaline lakes. Front Microbiol 7:853

Abe MM, Branciforti MC, Brienzo M (2021) Biodegradation of hemicellulose-cellulose-starch-based bioplastics and microbial polyesters. Recycling 6:22

Adetunji AI, Olaniran AO (2021) Production strategies and biotechnological relevance of microbial lipases: a review. Braz J Microbiol 52(3):1257–1269. https://doi.org/10.1007/s42770-021-00503-5

Aditi T, Suman S, Sundaresha S, Nidhi S, Shukla PK, Debasis P et al (2015) Artificial microRNA mediated gene silencing of phytophthora infestans single effector Avr3a gene imparts moderate type of late blight resistance in potato. Plant Pathol J (Faisalabad) 14(1):1–12

Ali S, Isha, Chang Y-C (2023) Ecotoxicological impact of bioplastics biodegradation: a comprehensive review. Processes 11(12):3445. https://doi.org/10.3390/pr11123445

Anjum S, Arora A, Alam MS, Gupta B (2016) Development of antimicrobial and scar preventive chitosan hydrogel wound dressings. Int J Pharm 508(1–2):92–101

Aransiola SA, Ikhumetse AA, Babaniyi BR, Abioye OP, Oyedele OJ, Falade NO (2023a) Phytoaccumulation of micro and nanoplastics: root uptake. In: Maddela NR, Reddy KV, Ranjit P (eds) Micro and nanoplastics in soil. Springer, Cham. https://doi.org/10.10007/978-3-031-21195-9_8

Aransiola SA, Victor-Ekwebelem MO, Ajiboye AE, Leh-Togi Zobeashia SS, Ijah UJJ, Oyedele OJ (2023b) Ecological impacts and toxicity of micro and nanoplastics in agroecosystem. In: Maddela NR, Reddy KV, Ranjit P (eds) Micro and Nanoplastics in soil. Springer, Cham. https://doi.org/10.10007/978-3-031-21195-9_10

Auta SH, Abioye OP, Aransiola SA, Bala JD, Chukwuemeka VI, Hassan A, Aziz A, Fauziah SH (2022) Enhanced microbial degradation of PET and PS microplastics under natural conditions in mangrove environment. J Environ Manag 304(2022):114273. https://doi.org/10.1016/j.jenvman.2021.114273

Bashar R, Gungor K, Karthikeyan KG, Barak P (2018) Cost effectiveness of phosphorus removal processes in municipal wastewater treatment. Chemosphere 197:280–290

Bilo F, Pandini S, Sartore L, Depero LE, Gargiulo G, Bonassi A, Federici S, Bontempi EA (2018) Sustainable bioplastic obtained from rice straw. J Clean Prod 200:357–368

Boey JY, Mohamad L, Khok YS, Tay GS, Baidurah S (2021) A review of the applications and biodegradation of polyhydroxyalkanoates and poly (lactic acid) and its composites. Polymers 13(10):1544

Brdlík P, Borůvka M, Běhálek L, Lenfeld P (2021) Biodegradation of poly (lactic acid) biocomposites under controlled composting conditions and freshwater biotope. Polymers 13:594

Chen W, He H, Wang J, Wang J, Chia-en AC (2023) Uncovering water effects in protein–ligand recognition: importance in the second hydration shell and binding kinetics. Phys Chem Chem Phys 25(3):2098–2109

Choi S, Cho I, Lee Y, Kim Y, Kim K, Lee S (2020) Microbial polyhydroxyalkanoates and nonnatural polyesters. Adv Mater 32(35):e1907138. https://doi.org/10.1002/adma.201907138

Danso D, Chow J, Streit WR (2019) Plastics: environmental and biotechnological perspectives on microbial degradation. Appl Environ Microbiol 85:e01095-19

Du Y, Liu X, Dong X, Yin Z (2022) A review on marine plastisphere: biodiversity, formation, and role in degradation. Comput Struct Biotechnol J 20:975–988

Edbeib MF, Wahab RA, Huyop F (2016) Halophiles: biology, adaptation, and their role in decontamination of hypersaline environments. World J Microbiol Biotechnol 32:1–23

Elabed H, González-Tortuero E, Ibacache-Quiroga C, Bakhrouf A, Johnston P, Gaddour K et al (2019) Seawater salt-trapped Pseudomonas aeruginosa survives for years and gets primed for salinity tolerance. BMC Microbiol 19:1–13

Folino A, Karageorgiou A, Calabrò PS, Komilis D (2020) Biodegradation of wasted bioplastics in natural and industrial environments: a review. Sustainability 12(15):6030. https://doi.org/10.3390/su12156030
Ghafoori H, Askari M, Sarikhan S (2016) Purification and characterization of an extracellular haloalkaline serine protease from the moderately halophilic bacterium, Bacillus iranensis (X5B). Extremophiles 20:115–123
Jan G, Tarnaud F, do Carmo FLR, Illikoud N, Canon F, Jardin J et al (2022) The stressing life of Lactobacillus delbrueckii subsp. bulgaricus in soy milk. Food Microbiol 106:104042
Jia MZ (2020) Biodegradable plastics breaking down facts: production, composition and environmental impacts. Greenpeace East Asia, Hong Kong, pp 1–54
Jia H, Zhang M, Weng Y, Li C (2021) Degradation of polylactic acid/polybutylene adipate-co-terephthalate by coculture of *Pseudomonas mendocina* and *Actinomucor elegans*. J Hazard Mater 403:123679
Kalita NK, Hazarika D, Kalamdhad A, Katiyar V (2021) Biodegradation of biopolymeric composites and blends under different environmental conditions: approach towards end-of-life panacea for crop sustainability. Bioresour Technol Rep 15:100705
Karamanlioglu M, Preziosi R, Robson GD (2017) Abiotic and biotic environmental degradation of the bioplastic polymer poly (lactic acid): a review. Polym Degrad Stab 137:122–130
Khan A, Majeed T (2019) Biodegradation of synthetic and natural plastics by microorganisms: a mini review. J Nat Appl Sci Pak 1:180–184
Koller M (2022) Advances in polyhydroxyalkanoate (pha) production. Bioengineering 9(7):328
Kushner DJ (2020) Growth and nutrition of halophilic bacteria. In: The biology of halophilic bacteria. CRC Press, Boca Raton, pp 87–103
Lai J, Huang H, Lin M, Xu Y, Li X, Sun B (2023) Enzyme catalyzes ester bond synthesis and hydrolysis: the key step for sustainable usage of plastics. Front Microbiol 13. https://doi.org/10.3389/fmicb.2022.1113705
Lehtiniemi M, Hartikainen S, Näkki P, Engström-Öst J, Koistinen A, Setälä O (2018) Size matters more than shape: ingestion of primary and secondary microplastics by small predators. Food Webs 17:e00097
Liu Y, Jiao JJ, Liang W, Santos IR, Kuang X, Robinson CE (2021) Inorganic carbon and alkalinity biogeochemistry and fluxes in an intertidal beach aquifer: implications for ocean acidification. J Hydrol 595:126036
Manisalidis I, Stavropoulou E, Stavropoulos A, Bezirtzoglou E (2020) Environmental and health impacts of air pollution: a review. Front Public Health 8:505570
McGenity TJ, Oren A (2012) Hypersaline environments. CABI, pp 402–437. doi:https://doi.org/10.1079/9781845938147.0402
Musa IO, Auta HS, Ilyasu US, Aransiola SA, Makun HA, Adabara NU, Abioye OP, Aziz A, Jayanthi B, Maddela NR, Prasad R (2024) Micro and nanoplastics in environment—degradation, detection, ecological impact. Int J Environ Rese 18:1. https://doi.org/10.1007/s41742-023-00551-9
Ncube LK, Ude AU, Ogunmuyiwa EN, Zulkifli R, Beas IN (2021) An overview of plastic waste generation and management in food packaging industries. Recycling 6(1):12
Pandey A, Adama N, Adjallé K, Blais JF (2022) Sustainable applications of polyhydroxyalkanoates in various fields: a critical review. Int J Biol Macromol 221:1184–1201
Patil PB, Sarkar D, Poddar K, Gu JD, Sarkar A (2023) Degradation profiling of in-vitro-produced polyhydroxyalkanoate synthesized by the soil bacterium Bacillus sp. PhNs9 under different microenvironments. Int Biodeterior Biodegrad 181:105615
Ruggero F, Roosa S, Onderwater R, Delacuvellerie A, Lotti T, Gori R, Lubello C, Wattiez R (2023) Characterization of bacterial communities responsible for bioplastics degradation during the thermophilic and the maturation phases of composting. J Mater Cycles Waste Manag 25:3270–3285
Salvador M, Abdulmutalib U, González J, Kim J, Smith A, Faulon J, Jiménez J (2019) Microbial genes for a circular and sustainable bio-pet economy. Genes 10(5):373

Sanchez-Hernandez J, Capowiez Y, Ro K (2020) Potential use of earthworms to enhance decaying of biodegradable plastics. ACS Sustain Chem Eng 8(11):4292–4316
Sheng H, Jing Y, An N, Shen X, Sun X, Yan Y, Yuan Q (2021) Extending the shikimate pathway for microbial production of maleate from glycerol in engineered escherichia coli. Biotechnol Bioeng 118(5):1840–1850
Surovy MZ, Rahman S, Dame ZT, Islam T (2022) Discovery of bioactive natural products from bacillus species: chemistry, biosynthesis and biological activities. In: Bacilli in agrobiotechnology: plant stress tolerance, bioremediation, and bioprospecting. Springer, Cham, pp 47–87
Suzuki M, Tachibana Y, Kasuya K (2020) Biodegradability of poly(3-hydroxyalkanoate) and poly(ε-caprolactone) via biological carbon cycles in marine environments. Polym J 53(1):47–66
Swetha TA, Ananthi V, Bora A, Sengottuvelan N, Ponnuchamy K, Muthusamy G, Arun A (2023) A review on biodegradable polylactic acid (PLA) production from fermentative food waste-its applications and degradation. Int J Biol Macromol 234:123703
Tanadchangsaeng N, Pattanasupong A (2022) Evaluation of biodegradabilities of biosynthetic polyhydroxyalkanoates in Thailand seawater and toxicity assessment of environmental safety levels. Polymers 14:428
Thakur S, Chaudhary J, Sharma B, Verma A, Tamulevicius S, Thakur VK (2018) Sustainability of bioplastics: opportunities and challenges. Curr Opin Green Sustain Chem 13:68–75
Thompson B, Machas M, Abed O, Nielsen D (2021) Applying a 'metabolic funnel' for phenol production in *Escherichia coli*. Fermentation 7(4):216
Toth A, Meyrat A, Stoldt S, Santiago R, Wenzel D, Jakobs S et al (2020) Kinetic coupling of the respiratory chain with ATP synthase, but not proton gradients, drives ATP production in cristae membranes. Proc Natl Acad Sci 117(5):2412–2421
Tsironi TN, Chatzidakis SM, Stoforos NG (2022) The future of polyethylene terephthalate bottles: challenges and sustainability. Packag Technol Sci 35(4):317–325
Uday J (2023) Molecular docking analysis of PET with MHET. Bioinformation 19(3):255–259
Urbanek AK, Mirończuk AM, García-Martín A, Saborido A, Mata I, Arroyo M (2020) Biochemical properties and biotechnological applications of microbial enzymes involved in the degradation of polyester-type plastics. BBA Proteins Proteom 1868:140315
Vera-Gargallo B, Hernández M, Dumont MG, Ventosa A (2023) Thrive or survive: prokaryotic life in hypersaline soils. Environ Microbiome 18(1):17
Visco A, Scolaro C, Facchin M, Brahimi S, Belhamdi H, Gatto V, Beghetto V (2022) Agri-food wastes for bioplastics: European prospective on possible applications in their second life for a circular economy. Polymers 14(13):2752
Vizireanu S, Panaitescu D, Nicolae C, Frone A, Chiulan I, Ionita M, Dinescu G (2018) Cellulose defibrillation and functionalization by plasma in liquid treatment. Sci Rep 8(1):15473. https://doi.org/10.1038/s41598-018-33687-2
Wang H (2021) Directional bio-synthesis and bio-transformation technology using mixed microbial culture. Microb Biotechnol 15(1):26–28
Wicaksono JA, Purwadaria T, Yulandi A, Tan WA (2022) Bacterial dynamics during the burial of starch-based bioplastic and oxo-low-density-polyethylene in compost soil. BMC Microbiol 22:309
Wu X, Hou H, Liu Y, Yin S, Bian S, Liang S, Wan C, Yuan S, Xiao K, Liu B et al (2022) Microplastics affect rice (Oryza sativa L.) quality by interfering metabolite accumulation and energy expenditure pathways: a field study. J Hazard Mater 422:126834. [CrossRef] [PubMed]
Yang H, Liu Z, Yin C, Han Z, Guan Q, Zhao Y, Yu S (2021) Edible, ultrastrong, and microplastic-free bacterial cellulose-based straws by biosynthesis. Adv Funct Mater 32(15). https://doi.org/10.1002/adfm.202111713

Chapter 4
Interconnectedness of Plastisphere, Marine Snow and Blue Carbon: Ecological Implications and Policy Perspectives

Ojo Emmanuel Ige, Adesuyi Emmanuel Omole, and Babafemi Raphael Babaniyi

Abstract Plastisphere, marine snow, and blue carbon represent integral components of marine ecosystems, each playing a crucial role in shaping environmental dynamics and influencing ecosystem health. This study provides a comprehensive overview of the dynamics, ecological implications, and policy perspectives surrounding these interconnected phenomena. The plastisphere, a microbial community associated with plastic debris, alters microbial community composition and function, influencing nutrient cycling and species interactions. Meanwhile, marine snow facilitates carbon sequestration and nutrient cycling, serving as a conduit for organic carbon transport from surface waters to the deep ocean. Blue carbon habitats, including mangroves, seagrasses, and salt marshes, sequester atmospheric carbon dioxide and provide essential ecosystem services such as coastal protection and biodiversity conservation. Despite their ecological significance, plastisphere, marine snow, and blue carbon ecosystems face threats from plastic pollution, climate change, and habitat degradation. Plastic pollution disrupts microbial processes, alters carbon dynamics, and compromises ecosystem integrity, highlighting the urgent need for effective mitigation strategies. Policy and management implications for addressing plastic pollution, blue carbon conservation, and marine snow dynamics are discussed. Recommendations include strengthening regulations, promoting circular economy approaches, enhancing monitoring and enforcement, and fostering international cooperation to address these interconnected environmental challenges comprehensively. By integrating these recommendations into policy and management frameworks, governments, stakeholders, and communities can work

O. E. Ige (✉)
Department of Chemistry, Federal University of Technology, Akure, Nigeria

A. E. Omole
Department of Marine Biology, University of Padova, Padova, Italy

B. R. Babaniyi
Bioresources Development Center, National Biotechnology Research and Development Agency, Abuja, Nigeria

S. A. Aransiola et al. (eds.), *White Pollution: Biodiversity and hazards in marine plastisphere*, Environmental Contamination Remediation and Management,
https://doi.org/10.1007/978-3-031-95547-1_4

together to tackle plastic pollution while promoting the conservation of blue carbon and marine snow habitats, safeguarding marine ecosystem health and resilience for generations to come.

Keywords Carbon sequestration · Mitigation strategies · Plastisphere · Marine snow · Blue carbon

4.1 Introduction

Aquatic ecosystems, encompassing both freshwater and marine environments, host a multitude of intricate processes and ecological phenomena, each playing a pivotal role in shaping the health and functioning of our planet's waters. Among these phenomena are the plastisphere, marine snow, and blue carbon ecosystems, each offering unique insights into the interconnectedness of aquatic environments and their global significance.

The plastisphere, initially conceptualized within marine ecology, delineates the microbial ecosystems thriving on plastic surfaces in aquatic settings (Eriksen et al. 2014). While extensive research has elucidated marine plastisphere dynamics, attention to freshwater ecosystems remains disproportionately limited, despite escalating plastic pollution levels (Zettler et al. 2013). Plastic debris infiltrating freshwater bodies, including rivers, lakes, and streams, undergoes degradation, providing a platform for microbial colonization and the formation of distinctive plastisphere communities (Oberbeckmann et al. 2016a, b).

Parallel to the complexities of the plastisphere, marine snow represents a fundamental process in oceanic systems, involving the continuous descent of organic and inorganic particles through the water column (Macreadie et al. 2019). This phenomenon facilitates nutrient cycling, carbon sequestration, and energy transfer within marine ecosystems, shaping their structure and function on local and global scales (Mazarrasa et al. 2018). The exploration of marine snow spans numerous decades, marked by pivotal discoveries and milestones that have enriched our comprehension of this phenomenon. Originating from studies in the late nineteenth and early twentieth centuries, driven by a fascination with the vertical transportation of organic material within the ocean, research into marine snow intensified during the mid-twentieth century. Seminal contributions from researchers like D.B. Cushing, who observed the vertical migration of organic particles in the water column and coined the term "marine snow," laid the foundation for subsequent investigations. The 1980s and 1990s witnessed significant advancements in imaging and sampling technologies, facilitating more precise and detailed studies of marine snow.

In concert with these phenomena, blue carbon ecosystems, including mangroves, seagrasses, and salt marshes, emerge as crucial players in carbon sequestration and storage (Sasmito et al. 2020). These ecosystems serve as substantial reservoirs of organic carbon, contributing significantly to global climate change mitigation efforts (McLeod et al. 2011). The carbon stored within blue carbon habitats is primarily derived from the accumulation of organic matter produced by photosynthesis. Mangroves, with their extensive root systems, sequester carbon in coastal sediments, while seagrasses and salt marshes trap carbon in their biomass and sediment layers. The preservation and restoration of blue carbon ecosystems are essential not only for mitigating climate change but also for maintaining coastal resilience and biodiversity. These habitats provide critical ecosystem services, such as coastal protection from storms and erosion, nursery grounds for fish and other marine species, and support for livelihoods and cultural practices of coastal communities.

Understanding the interconnectedness of plastisphere, marine snow, and blue carbon ecosystems is crucial for effective environmental management and sustainability. Plastic pollution not only threatens marine life directly but also indirectly affects carbon cycling and nutrient dynamics through its interactions with the plastisphere. Similarly, alterations in marine snow dynamics, such as changes in particle sinking rates or composition, can have cascading effects on carbon sequestration and nutrient cycling in marine ecosystems. By examining these interconnected phenomena holistically, this review aims to provide insights into the complex interactions shaping aquatic ecosystems and their responses to environmental changes. Through an integrated approach to research and management, we can develop strategies to mitigate plastic pollution, enhance carbon sequestration in blue carbon habitats, and preserve the integrity of marine ecosystems for the benefit of current and future generations.

4.1.1 Plastisphere, Marine Snow, and Blue Carbon

In recent years, the significance of marine ecosystems in mitigating climate change and maintaining global carbon balance has gained considerable attention. Three key concepts central to this discussion are plastisphere, marine snow, and blue carbon. This literature review provides an overview of these concepts, their ecological importance, and their implications for carbon sequestration in marine environments.

4.1.1.1 Plastisphere

The plastisphere refers to the ecosystem that develops on the surface of plastic debris in marine environments. This phenomenon has garnered increasing attention due to the widespread presence of plastic pollution in oceans worldwide. The plastisphere is composed of a diverse array of microorganisms, including bacteria, archaea, algae, and other microorganisms, which colonize the surface of plastic debris (Zettler et al. 2013).

Studies have highlighted the significance of the plastisphere in marine ecosystems, demonstrating its role in facilitating the transport of microorganisms across vast distances and serving as a vector for the spread of potentially harmful species

(Zettler et al. 2013). Moreover, the plastisphere can alter the biogeochemical properties of marine environments by influencing nutrient cycling and the microbial degradation of plastic debris.

Environmental factors, such as plastic composition, surface texture, and exposure to sunlight and seawater, can influence the composition and dynamics of the plastisphere. Additionally, the presence of pollutants and leachates associated with plastic debris can impact the microbial communities within the plastisphere and have broader ecological implications for marine ecosystems (Oberbeckmann et al. 2016a, b).

The plastisphere represents a novel ecological niche in marine environments, with implications for ecosystem dynamics, biogeochemical cycling, and the fate of plastic pollution. Further research into the composition, function, and ecological consequences of the plastisphere is essential for developing effective strategies to mitigate the impacts of plastic pollution on marine ecosystem. Microbial communities within the plastisphere may interact with surrounding marine organisms, potentially affecting the structure and function of marine food webs. Furthermore, the presence of plastic debris and the plastisphere may influence the distribution and abundance of marine species, including those that interact directly with plastic debris or rely on related organisms for food or habitat. Understanding the composition, function, and ecological consequences of the plastisphere is essential for developing effective strategies to mitigate the impacts of plastic pollution on marine ecosystems. Future research should focus on elucidating the interactions between microorganisms within the plastisphere, as well as the broader ecological implications of plastic debris and associated microbial communities on marine biodiversity, ecosystem functioning, and human health. The emergence of the plastisphere, a term denoting the complex ecosystem thriving on ocean plastics, has garnered significant attention due to its potential ecological and economic implications. Lacerda et al. (2024), in their research, delve into this phenomenon by investigating the microbial communities inhabiting floating plastics in the Northwestern Mediterranean Sea, with a particular emphasis on viruses. Using shotgun DNA sequencing, the team uncovered a diverse range of microbial life associated with the plastics, encompassing bacteria, eukaryotes, viruses, and archaea. Notably, they observed a prevalence of Proteobacteria, alongside other groups such as Ascomycota fungi and Bacteroidetes. Interestingly, the researchers found that neither the composition of the plastic nor its shape significantly influenced the community structure, indicating the resilience of plastisphere organisms to varying environmental conditions. Of notable concern were the abundant pathogenic Vibrio species detected, suggesting potential health risks associated with plastic pollution. Moreover, the significance of comprehending the dynamics of the plastisphere is highlighted by the existence of viruses such as the White Spot Syndrome virus (WSSV), notorious for instigating detrimental outbreaks in aquaculture. The research conducted by Lacerda et al. (2024) underscores the pressing necessity for expanded inquiries into the ecological and economic ramifications of plastisphere organisms within marine ecosystems. These investigations are pivotal for formulating efficient strategies to alleviate the repercussions of plastic pollution on environmental health and human welfare. The notion of the plastisphere, delineating the multifaceted ecosystem burgeoning on ocean plastics, has garnered escalating interest due to its potential ecological and economic implications.

Delving into this phenomenon, Lacerda et al. (2024) meticulously explore the microbial communities thriving on floating plastics in the Northwestern Mediterranean Sea, with a special emphasis on viruses. Employing shotgun DNA sequencing, the researchers unveil a diverse spectrum of microbial life associated with the plastics, encompassing bacteria, eukaryotes, viruses, and archaea. Noteworthy is the prevalence of Proteobacteria, alongside other groups like Ascomycota fungi and Bacteroidetes. Interestingly, the composition and shape of the plastic appeared to have little effect on the community structure, suggesting that plastisphere organisms demonstrate resilience to environmental variables. Abundant pathogenic Vibrio species were detected, raising potential health risks associated with plastic pollution. Furthermore, the presence of viruses such as the White Spot Syndrome virus (WSSV), renowned for triggering devastating outbreaks in aquaculture, underscores the criticality of comprehending plastisphere dynamics.

The findings of Lacerda et al. (2024) underscore the pressing imperative for further exploration into the ecological and economic repercussions of plastisphere organisms in marine habitats. Such investigations are pivotal for formulating efficacious strategies to mitigate the impact of plastic pollution on both ecosystems and human well-being.

4.1.1.2 Marine Snow

Marine snow refers to the continuous shower of organic particles and aggregates sinking from the surface waters to the deep sea. These particles consist of various organic and inorganic materials, including phytoplankton, zooplankton fecal pellets, detritus, and mineral particles (Alldredge and Silver 1988). Marine snow plays a fundamental role in transporting carbon and nutrients from the surface ocean to the deep sea, thus influencing carbon sequestration and nutrient cycling in marine ecosystems.

Studies have highlighted the importance of marine snow in the biological pump, a process by which carbon is transported from the surface ocean to the deep sea through the aggregation and sinking of organic particles (Simon et al. 2002). As marine snow settles through the water column, it provides a source of organic carbon for deep-sea organisms and forms sedimentary deposits, contributing to long-term carbon storage in marine sediments (Lutz et al. 2007). Moreover, changes in oceanic conditions, such as temperature, pH, and nutrient availability, can affect the production, composition, and flux of marine snow, with potential implications for carbon cycling and ecosystem functioning (Buesseler et al. 2007). Marine snow comprises a complex matrix of organic aggregates and particles that sink through the water column, serving as a conduit for the vertical transport of carbon and nutrients in the ocean. Recent research has emphasized the role of marine snow in shaping microbial communities, nutrient cycling, and the biological pump, a process by which carbon is transferred from surface waters to the deep sea (Passow 2002). The composition and flux of marine snow vary spatially and temporally, influenced by factors such as phytoplankton blooms, zooplankton grazing, and physical mixing processes (Alldredge and Gotschalk 1989).

Studies have highlighted the significance of marine snow in sequestering organic carbon and trace metals in marine sediments, contributing to long-term carbon storage and geochemical cycling (Luo et al. 2019). The sinking of marine snow also transports energy and organic matter to deep-sea ecosystems, supporting diverse assemblages of bacteria, archaea, and meiofauna (Bochdansky et al. 2017). Advances in imaging and modeling techniques have provided new insights into the formation, dynamics, and fate of marine snow, underscoring its importance in marine biogeochemistry and ecosystem functioning. Marine snow represents a heterogeneous mixture of organic particles and aggregates that range in size from a few micrometers to several centimeters and play a crucial role in the vertical transport of carbon and energy in the ocean. Recent studies have highlighted the dynamic nature of marine snow formation and its impact on carbon sequestration, nutrient cycling, and ecosystem functioning (Alldredge and Jackson 1995). The composition and flux of marine snow are influenced by biological, physical, and chemical processes occurring in surface waters, including primary production, zooplankton grazing, and microbial degradation (De La Rocha and Passow 2007). Research has shown that marine snow aggregates serve as hotspots for microbial activity, where microbial communities colonize and decompose organic matter, releasing nutrients and trace gases into the surrounding water column (Thornton 2014). The sinking of marine snow contributes to the export of organic carbon from surface waters to the deep sea, where it becomes buried in sediments and sequestered over geological timescales (Lutz et al. 2016). Changes in oceanic conditions, such as warming, acidification, and deoxygenation, may alter the production, composition, and fate of marine snow, with potential implications for carbon storage and ecosystem dynamics.

Gao et al. (2020) conducted an observational study in the northern South China Sea, employed sediment traps at various depths to investigate the dynamics of sinking particulate matter. They observed significant accumulations of fecal matter and marine snow in the lower trap throughout the sampling period spanning from May 2014 to May 2015. Using scanning electron microscopy, the researchers identified diatoms, coccolithophores, radiolarians, and other planktonic debris as the primary constituents of these particles. Their analysis led them to suggest that the lateral transport of deep water, rather than vertical migrations from the surface, likely played a key role in the observed fluxes of marine snow and fecal pellets. Additionally, Gao et al. speculated that these particles may originate from regions southwest of Taiwan. The dynamics of particulate organic carbon (POC) settling through the marine water column and its subsequent colonization by heterotrophic bacteria play a critical role in regulating global climate and biogeochemical cycles.

According to Borer et al. (2023), the process outlined involves the sequestration of atmospheric carbon through the initial attachment and degradation of organic particles by bacterial communities. Bacterial motility is identified as a crucial factor in facilitating the colonization of marine particles, allowing bacteria to effectively navigate the water column and encounter particles leaking organic nutrients.

Recent experimental work, as described by Borer et al. (2023), shows that bacteria's ability to move towards chemical gradients, known as chemotaxis, plays a crucial role in enhancing their colonization of marine particles. This is especially important when particles are moving at intermediate to higher speeds, allowing bacteria to find and attach to particles as they pass by. This research emphasizes how bacterial movement and particle dynamics interact in the ocean, influencing the distribution and abundance of marine microorganisms.

Furthermore, Borer et al. (2023) present an individual-based modeling approach to simulate the encounter and attachment of bacterial cells with leaking marine particles. This modeling framework allows for a systematic evaluation of the role of different parameters associated with bacterial motility, such as run-and-tumble behavior, in particle colonization. The model also explores the influence of particle microstructure on colonization efficiency, revealing that porous microstructures facilitate additional colonization by chemotactic and motile bacteria. Additionally, the model highlights the fundamental alterations in the interaction dynamics of non-motile cells with particles, emphasizing the importance of streamlines intersecting with the particle surface.

Borer et al. (2023) provide valuable insights into the dynamics of bacterial colonization on marine particles and its implications for carbon cycling in marine ecosystems. By integrating experimental and modeling approaches, this research enhances our understanding of the complex mechanisms governing bacterial motility, chemotaxis, and particle interactions in the marine environment. These findings hold significant implications for the modeling and management of carbon sequestration processes in marine ecosystems, with potential implications for climate change mitigation efforts.

4.1.1.3 Blue Carbon

Blue carbon refers to the carbon stored in coastal and marine ecosystems, including mangroves, seagrasses, and salt marshes. These ecosystems have high rates of primary productivity and organic carbon burial, making them significant carbon sinks and reservoirs (Duarte et al. 2013). Blue carbon habitats sequester carbon dioxide from the atmosphere through photosynthesis and trap organic carbon in their sediments over long periods, thus playing a crucial role in mitigating climate change and maintaining coastal resilience.

Numerous studies have demonstrated the importance of blue carbon ecosystems in sequestering carbon and mitigating greenhouse gas emissions (Howard et al. 2017). For example, mangroves are among the most carbon-rich ecosystems globally, with dense root systems and anaerobic sediments conducive to carbon storage (Alongi 2014). Seagrasses also contribute significantly to blue carbon stocks, with their belowground biomass and rhizomes serving as important carbon sink (Fourqurean et al. 2012). Conservation and restoration of blue carbon habitats not only enhance carbon sequestration but also provide multiple co-benefits, such as coastal protection, biodiversity conservation, and sustainable livelihoods (Nellemann et al. 2009). Blue carbon ecosystems, including mangroves, seagrasses, and tidal marshes, are highly productive coastal habitats that sequester and store large quantities of carbon dioxide from the atmosphere. Recent research has highlighted the role of blue carbon habitats in climate change mitigation, biodiversity conservation, and coastal protection (Macreadie et al. 2019). Mangroves, in particular, are recognized as efficient carbon sinks, with their extensive root systems and anaerobic sediments facilitating the burial of organic carbon over millennia (Donato et al. 2011).

Seagrasses also contribute significantly to blue carbon stocks, with their belowground biomass and rhizomes accumulating organic matter and stabilizing coastal sediments (Nellemann et al. 2009). Tidal marshes are not as thoroughly researched, but they are really good at trapping and storing carbon. This happens because plants grow quickly and then break down in the waterlogged soils (Chmura et al. 2003). Conservation and restoration of blue carbon ecosystems offer multiple co-benefits, including climate change adaptation, fisheries enhancement, and sustainable livelihoods for coastal communities (Howard et al. 2017). Blue carbon ecosystems, including mangroves, seagrasses, and salt marshes, are highly productive coastal habitats that sequester and store large amounts of carbon dioxide from the atmosphere. Recent research has highlighted the importance of blue carbon habitats as significant carbon sinks and biodiversity hotspots, supporting a wide range of flora and fauna (Howard et al. 2014). Mangroves, with their extensive root systems and anoxic sediments, sequester carbon at rates surpassing those of terrestrial forests, making them critical for climate change mitigation (Alongi 2020). Seagrasses also play a crucial role in blue carbon sequestration, with their belowground biomass and rhizomes storing organic carbon in coastal sediments and contributing to soil accretion and stabilization (Duarte and Cebrián 1996). Conservation and restoration of blue carbon ecosystems offer multiple benefits, including climate change adaptation, coastal protection, and sustainable livelihoods for coastal communities (Macreadie et al. 2017).

Blue carbon ecosystems, which include seagrass beds, mangrove forests, and salt marshes, are essential in combating climate change. Bandh et al. (2023) highlight their crucial role in sequestering atmospheric carbon dioxide, mitigating the impacts of climate change. These ecosystems act as efficient carbon sinks, storing substantial amounts of carbon in their biomass and sediments over long periods. Mangroves, with their intricate root systems, seagrasses with dense rhizomes, and organic-rich soils of salt marshes serve as vital carbon reservoirs, helping regulate greenhouse gas concentrations.

Moreover, blue carbon ecosystems provide critical coastal protection and resilience. Mangrove forests act as natural barriers against storm surges and erosion, safeguarding coastal communities and infrastructure. Seagrass beds stabilize sediments and reduce wave energy, preserving shoreline integrity. Salt marshes serve as buffer zones, absorbing wave energy and mitigating flooding events. Protecting and restoring these habitats not only enhances carbon sequestration but also strengthens coastal resilience against climate change-induced extreme weather events.

However, anthropogenic activities, such as habitat destruction, pollution, and climate change, pose significant threats to blue carbon ecosystems. Bandh et al. (2023) emphasize the alarming rate of habitat loss and degradation worldwide, emphasizing the urgent need for conservation and restoration efforts. Effective policies and management strategies are essential to safeguard these ecosystems and address their decline.

Furthermore, the interaction between blue carbon ecosystems and plastic pollution presents additional challenges. Plastic debris, pervasive in marine environments, can accumulate within these habitats, disrupting their ecological functions. Microplastics, in particular, pose a threat to marine organisms and ecosystem dynamics. Understanding the dynamics of plastisphere formation and its implications for blue carbon cycling is crucial for developing effective mitigation measures.

The plastisphere, marine snow, and blue carbon stand as fundamental pillars within marine ecosystems, each contributing uniquely to the intricate web of carbon cycling and climate regulation. Their roles extend far beyond mere components; they serve as vital conduits of ecosystem health, resilience, and global biogeochemical processes. Recognizing the significance of these entities is paramount in navigating the complexities of contemporary environmental challenges, particularly in the context of accelerating climate change.

The plastisphere, an emerging frontier in marine science, exemplifies the interconnectedness of human activity and marine ecology, highlighting the need for holistic approaches to address plastic pollution and its cascading effects on marine life and biogeochemical cycles. Marine snow, often overlooked yet omnipresent, serves as a conduit for carbon transport from surface waters to the deep ocean, playing a crucial role in carbon sequestration and long-term storage. Blue carbon habitats, including mangroves, seagrasses, and salt marshes, possess immense carbon sequestration potential, safeguarding coastal ecosystems while mitigating climate change impacts.

Understanding the intricacies of these processes is not merely an academic pursuit but a necessity for informed decision-making and sustainable resource management. It is through rigorous scientific inquiry and interdisciplinary collaboration that we can unlock the mysteries of the plastisphere, decipher the mechanisms of marine snow formation and sinking, and harness the full potential of blue carbon ecosystems. By doing so, we pave the way for evidence-based policies and innovative strategies aimed at preserving marine biodiversity, enhancing ecosystem resilience, and mitigating climate change effects.

In conclusion, the plastisphere, marine snow, and blue carbon epitomize the resilience and adaptability of marine ecosystems in the face of anthropogenic pressures and environmental change. Embracing their complexities and harnessing their potential offer pathways towards a more sustainable and resilient future for our oceans and planet. Blue carbon ecosystems are vital in combating climate change and coastal degradation. Bandh et al. (2023) provide valuable insights into their ecological significance, highlighting the urgency of conservation efforts. Integrating blue carbon conservation into broader climate change mitigation and adaptation strategies is essential for harnessing the full potential of these ecosystems in safeguarding the planet's future.

4.2 Methods and Techniques for Investigating Marine Snow, Plastisphere, and Black Carbon in Aquatic Environments

It is important to understand the dynamics and ecological implications of marine snow, plastisphere communities, and black carbon particles in aquatic ecosystems is crucial for assessing their impacts on marine life and ecosystem functioning. In this section, we present an overview of sampling methods and techniques utilized in the study of these key components of aquatic environments.

Sampling methods and techniques play a crucial role in understanding various aspects of marine environments, from the distribution of marine snow to the dynamics of blue carbon ecosystems. In studying marine snow, researchers employ a range of methodologies to assess its distribution, abundance, and characteristics at different depths in the water column. Water column sampling techniques, such as using Niskin bottles, rosette samplers, or sediment traps, allow for the collection of water samples at various depths, providing insights into marine snow particles' distribution and abundance (Alldredge and Silver 1988; Ploug and Jørgensen 1999). Additionally, particle interception techniques, including plankton nets and sediment traps, aid in understanding the size distribution, composition, and sinking rates of marine snow particles (Jackson 2001; Simon et al. 2002).

In situ imaging techniques, such as underwater cameras and remotely operated vehicles (ROVs), enable direct observation and documentation of marine snow particles, facilitating a deeper understanding of their structure and behavior in their natural environment (Ploug and Jørgensen 1999; Simon et al. 2002). Acoustic Doppler Profiling (ADP) utilizes acoustic instruments like acoustic Doppler current profilers (ADCPs) to measure the vertical distribution of marine snow particles, offering real-time, high-resolution data on their concentration and vertical dynamics (Zhang et al. 2021).

Transitioning to the study of the plastisphere, which encompasses microplastics associated with marine snow, various sampling techniques are employed. Water column sampling with microplastic filters allows for the assessment of microplastic distribution and characteristics throughout the water column (Cózar et al. 2014; Hidalgo-Ruz et al. 2012). Sediment core sampling from benthic environments provides insights into the accumulation and distribution of microplastics within sediments and their association with plastisphere communities (Cole et al. 2013; Song et al. 2020). Confocal microscopy imaging aids in visualizing and quantifying microplastics within marine snow samples, facilitating the characterization of their morphology, size, and distribution within plastisphere communities (Horton et al. 2017; Li et al. 2020).

Turning to blue carbon ecosystems, a variety of sampling methods and techniques are employed to assess their structural characteristics and carbon stocks. Vegetation surveys quantify biomass and density within blue carbon habitats, such as mangroves, salt marshes, and seagrasses, providing valuable data for carbon stock estimation (McLeod et al. 2011; Donato et al. 2011). Soil sampling and carbon content analysis involve collecting soil samples and analyzing carbon content to determine organic carbon distribution within blue carbon habitats (Chmura et al. 2003; Adame et al. 2013).

Sediment core analysis, utilizing radiocarbon dating and isotopic analysis, helps in understanding historical carbon accumulation rates and long-term carbon dynamics within blue carbon ecosystems (Alongi 2014; Breithaupt et al. 2018). Remote sensing techniques, such as satellite imagery and aerial photography, are employed to map the extent and distribution of blue carbon habitats, aiding in monitoring changes over time and assessing their contribution to carbon sequestration (Fatoyinbo et al. 2008; Duarte et al. 2013).

Gas flux measurements, including carbon dioxide (CO_2) fluxes between blue carbon ecosystems and the atmosphere, provide crucial insights into carbon exchange rates and the role of these habitats in mitigating climate change (Barr et al. 2010; Maher et al. 2013). Stable isotope analysis, particularly of carbon isotopes (e.g., δ13C), helps trace carbon sources and cycling processes within blue carbon ecosystems, contributing to a deeper understanding of their role in the global carbon cycle (Mateo et al. 1997; Bouillon et al. 2008).

Community participation through citizen science initiatives fosters collaboration, knowledge exchange, and awareness about the importance of blue carbon habitats among local communities and stakeholders (Howard et al. 2017; Gownaris et al. 2019).

Overall, these sampling methods and techniques provide essential tools for researchers to comprehensively study marine snow, the plastisphere, and blue carbon ecosystems, contributing to a better understanding of their dynamics, functions, and ecological significance.

4.3 Ecological Implications of Plastisphere, Marine Snow and Blue Carbon

Marine snow, consisting of aggregated organic particles sinking from surface waters to the deep ocean, facilitates carbon sequestration by transporting organic carbon and nutrients such as nitrogen and phosphorus to deeper layers, enhancing nutrient cycling dynamics and supporting primary production (Herndl and Reinthaler 2013; Laws et al. 2000). These marine snow aggregates serve as hotspots of biodiversity, providing habitats and food resources for diverse marine organisms including microbial communities, zooplankton, and larval organisms, thus contributing to species diversity and tropic interactions in the oceanic water column (Simon et al.

2002; Ducklow and Doney 2013). Additionally, the degradation of organic matter within marine snow supports microbial processes such as nutrient recycling and carbon remineralization, thereby contributing to ecosystem functioning (Herndl and Reinthaler 2013).

The plastisphere, a microbial community associated with plastic debris in marine environments, alters microbial community composition and function, influencing nutrient cycling and species interactions (Amaral-Zettler et al. 2020). Plastics in the plastisphere pose toxicological risks to marine organisms through the adsorption and transport of toxic pollutants, leading to bioaccumulation and disruptions in tropic interactions within marine food webs (Rochman et al. 2015). Moreover, plastic debris modifies marine habitats by altering substrate characteristics and physical structures, impacting the recruitment and settlement of benthic organisms and altering ecosystem dynamics in coastal and marine environments (Free et al. 2014).

Blue carbon habitats, stored in coastal and marine vegetated habitats such as mangroves, seagrasses, and salt marshes, sequester large amounts of carbon dioxide from the atmosphere, contributing to climate change mitigation (Duarte et al. 2013). These habitats provide coastal protection by stabilizing shorelines, reducing erosion, and buffering coastal communities against storms and wave action (Duarte et al. 2013). Furthermore, blue carbon habitats support diverse ecosystems, providing habitats and nursery grounds for a wide range of marine species, promoting biodiversity and supporting fisheries (Duarte et al. 2013; Lampitt and Antia 1997). In conclusion, marine snow, plastisphere, and blue carbon play crucial roles in marine ecosystems, influencing nutrient cycling, biodiversity, ecosystem functioning, and climate regulation. Understanding their ecological implications is essential for effective marine conservation and management, particularly in the face of ongoing environmental changes and anthropogenic pressures on marine environments.

4.4 Impact of Plastic Pollution on the Interconnectedness of Plastisphere, Marine Snow, and Blue Carbon Ecosystems

4.4.1 Plastic Pollution and the Plastisphere

Plastic pollution has emerged as a significant environmental concern worldwide (Aransiola et al. 2023b), with profound implications for marine ecosystems. The proliferation of plastic debris in aquatic environments has led to the formation of what is known as the plastisphere—a complex microbial community that colonizes plastic surfaces. This plastisphere differs substantially from microbial assemblages found on natural substrates, exhibiting unique composition and metabolic capabilities (Oberbeckmann et al. 2016a, b; Zettler et al. 2013; Auta et al. 2022; Musa et al. 2024).

Research indicates that plastic-associated microbes play multifaceted roles in biogeochemical cycling and ecosystem functioning. Firstly, these microbes have been shown to contribute to the degradation of organic pollutants adhered to plastic surfaces. Through enzymatic degradation processes, plastic-associated bacteria can break down various hydrophobic pollutants, such as polycyclic aromatic hydrocarbons (PAHs) and polychlorinated biphenyls (PCBs), thereby mitigating the persistence and toxicity of these contaminants in marine environments (Zettler et al. 2013; Zhang et al. 2022; Amaral-Zettler et al. 2015).

Moreover, plastic debris serves as a vector for the transport of potentially harmful pathogens across marine environments. Microorganisms, including bacteria and viruses, can adhere to plastic surfaces and be transported over long distances via ocean currents. This phenomenon raises concerns about the spread of pathogens and the potential for disease transmission among marine organisms, posing risks to both wildlife and human health (Zettler et al. 2013; Zhang et al. 2022; Amaral-Zettler et al. 2015).

Additionally, the presence of plastics can perturb microbial community dynamics and metabolic processes in marine ecosystems. Plastic debris alters habitat structures and provides novel niches for microbial colonization, leading to shifts in species composition and functional diversity within the plastisphere. These changes can have cascading effects on nutrient cycling, carbon fluxes, and overall ecosystem productivity (Dang and Lovell 2016; Kirstein et al. 2016).

In summary, plastic pollution not only poses physical threats to marine life but also exerts ecological and biogeochemical impacts through its interactions with microbial communities. Understanding the dynamics of the plastisphere and its implications for marine ecosystems is crucial for developing effective strategies to mitigate the adverse effects of plastic pollution on ocean health and biodiversity (Aransiola et al. 2023a).

4.4.2 Impact on Marine Snow Dynamics

Marine snow, composed of aggregated organic and inorganic particles, serves as a crucial mechanism for the transfer of organic carbon from surface waters to the deep ocean, contributing significantly to the biological pump and long-term carbon sequestration processes. However, the pervasive issue of plastic pollution poses significant threats to marine snow dynamics, potentially disrupting its functioning and ecological roles in marine ecosystems. One of the primary mechanisms through which plastic pollution interferes with marine snow dynamics involves microplastics. These small plastic particles can adsorb organic matter from the surrounding seawater, altering the composition and physical properties of marine aggregates. As a result, the buoyancy and sinking rates of marine snow aggregates may be affected, leading to changes in their transport pathways and distribution within the water column (Long et al. 2015; Lindeque et al. 2016). Furthermore, microplastics can serve as substrates for the colonization of plastic-associated microbes. These microbial communities have been shown to degrade organic carbon within marine snow aggregates, potentially accelerating the remineralization process. As a consequence, the efficiency of carbon sequestration by marine snow may be compromised, with implications for the overall carbon budget and cycling in the ocean (Dell'Anno et al. 2015). The alterations in marine snow dynamics induced by plastic pollution have far-reaching consequences for marine ecosystems. Changes in the availability and distribution of marine snow can impact nutrient cycling processes, as marine aggregates play a key role in the vertical flux of organic matter and associated nutrients. Moreover, alterations in marine snow dynamics can influence the structure and functioning of marine food webs, affecting the distribution and abundance of organisms at various trophic levels (Long et al. 2015; Lindeque et al. 2016). Plastic pollution poses significant challenges to the functioning and ecological roles of marine snow in the ocean. By disrupting marine snow dynamics, plastic pollution can have cascading effects on nutrient cycling, food web structure, and the efficiency of carbon sequestration processes. Addressing the impacts of plastic pollution on marine snow dynamics is essential for the conservation and sustainability of marine ecosystems.

4.4.3 Impacts on Blue Carbon Ecosystems

Plastic pollution presents a significant threat to the integrity and functioning of blue carbon ecosystems, which encompass mangroves, seagrasses, and salt marshes. These ecosystems are renowned for their ability to sequester atmospheric carbon dioxide and store it within their biomass and sediments. However, the presence of plastic debris introduces direct and indirect challenges that undermine their capacity as carbon sinks and compromise their ecological functions.

One of the direct impacts of plastic pollution on blue carbon ecosystems is physical smothering of vegetation. Plastic debris can accumulate on the surface of mangroves, seagrasses, and salt marshes, obstructing sunlight and inhibiting photosynthesis. This impedes the growth and productivity of vegetation, ultimately leading to habitat degradation and loss. Furthermore, plastic pollution can disrupt gas exchange and alter hydrological processes within coastal habitats. Accumulation of plastic debris in waterways and wetlands can impede water flow, affecting nutrient cycling and sediment deposition. This disturbance to hydrological processes can undermine the resilience of blue carbon ecosystems to natural disturbances and exacerbate the impacts of climate change (Dias et al. 2019; Nelms et al. 2017).

In addition to these direct effects, microplastics and associated pollutants pose indirect threats to blue carbon ecosystems. Microplastics can accumulate in sediments, where they may adsorb contaminants such as heavy metals and persistent organic pollutants. These pollutants can interfere with biogeochemical processes, such as sediment metabolism and nutrient cycling, affecting the health and functioning of blue carbon habitats (Nelms et al. 2017). The cumulative impacts of plastic pollution on blue carbon ecosystems may compromise their capacity to sequester carbon and provide essential ecosystem services. Coastal protection, nutrient cycling, and habitat provision are among the critical functions provided by these ecosystems, which support biodiversity and livelihoods in coastal communities (Dias et al. 2019; Nelms et al. 2017).

Plastic pollution poses multifaceted challenges to the interconnectedness of plastisphere, marine snow, and blue carbon ecosystems. By disrupting microbial processes, altering carbon dynamics, and compromising ecosystem integrity, plastic pollution undermines the resilience and functionality of marine environments. Effective mitigation strategies are essential to preserve the health and sustainability of blue carbon ecosystems and safeguard the invaluable services they provide.

4.5 Future Frontiers: Research Directions in the Plastisphere, Marine Snow, and Blue Carbon

The future of research on the plastisphere, marine snow, and blue carbon offers diverse approaches to addressing crucial environmental challenges in marine ecosystems. Scientists are delving into the intricate and long-term effects of microplastics on marine life, exploring issues like biodiversity loss, bioaccumulation, and potential impacts on human health. At the same time, efforts are underway to develop better methods for detecting, monitoring, and removing plastic pollution, while policymakers are urged to enact strict regulations to reduce single-use plastics (Wright et al. 2019; GESAMP 2015; Bergmann et al. 2015).

Research into marine snow's role in the global carbon cycle is also gaining momentum, focusing on understanding its mechanisms for carbon sequestration and nutrient cycling. Integrating marine snow dynamics into climate models could improve predictions of future oceanic carbon uptake and storage (Alldredge and Silver 1988).

Moreover, there is growing recognition of the importance of blue carbon habitats, such as mangroves, seagrasses, and salt marshes, in mitigating climate change and supporting coastal resilience. Efforts are underway to expand restoration and conservation initiatives for these ecosystems while promoting sustainable management practices (Howard et al. 2014; Mcleod et al. 2011).

Additionally, ongoing research is exploring the potential benefits of using microorganisms associated with the plastisphere for bioremediation and plastic degradation (Zettler et al. 2013; Harrison et al. 2014). Community-based initiatives and citizen science projects are also gaining traction to engage the public in monitoring and addressing plastic pollution (Eriksen et al. 2014; Geyer et al. 2017).

In the realm of marine snow, emerging studies are investigating the effects of changing environmental conditions, like ocean acidification and warming, on marine snow particles' composition and dynamics (Engel et al. 2017; Jackson et al. 2020). Moreover, efforts to quantify the economic value of blue carbon ecosystems are underway, highlighting their importance beyond carbon sequestration (Narayan et al. 2017 ; Nellemann et al. 2009).

Collaborative research is exploring innovative financing mechanisms, such as blue carbon offset programs and payments for ecosystem services, to incentivize conservation and restoration efforts in blue carbon habitats while promoting sustainable development goals (Böttcher et al. 2012; PwC 2020). By embracing interdisciplinary approaches and fostering stakeholder collaboration, the scientific community aims to advance our understanding of the plastisphere, marine snow, and blue carbon, guiding evidence-based management strategies that balance environmental conservation with socio-economic priorities in marine ecosystems.

4.6 Policy and Management Implications for Addressing Plastic Pollution, Blue Carbon, and Marine Snow

Plastic pollution, blue carbon conservation, and marine snow dynamics are interconnected issues affecting the health of freshwater and marine ecosystems. Evaluating existing policy frameworks and management strategies is crucial for effectively addressing these environmental challenges.

(i) **Existing Policy Frameworks and Regulations:** International and national policies targeting plastic pollution have been established, such as the United Nations Environment Programme's Global Partnership on Marine Litter and directives by entities like the European Union promoting circular economy principles. However, these policies often focus more on marine environments, leaving freshwater ecosystems underserved. Similarly, regulations for blue carbon and marine snow are often fragmented and insufficiently enforced.

(ii) **Effectiveness of Regulations and Initiatives:** Despite efforts, plastic pollution persists in freshwater ecosystems. Similarly, blue carbon conservation and marine snow management face challenges due to fragmented regulations and inadequate enforcement. Localized initiatives such as river clean-up campaigns and public awareness programs have had some success but fall short of addressing the scale of the problem comprehensively.

4.6.1 Mitigation Strategies and Recommendations

To enhance policy coherence and effectiveness across these interconnected issues, several recommendations can be proposed:

(i) **Strengthening Regulations:** Governments should enact comprehensive legislation targeting plastic pollution, blue carbon conservation, and marine snow management, incorporating measures for waste management, pollution prevention, and habitat conservation.

(ii) **Promoting Circular Economy Approaches:** Implementing circular economy principles can minimize plastic waste generation while also enhancing blue carbon storage and preserving marine snow habitats through reduced pollution inputs (Geng et al. 2021).

(iii) **Enhancing Monitoring and Enforcement:** Robust monitoring programs are essential for assessing the extent of plastic pollution and evaluating the effectiveness of mitigation measures for blue carbon and marine snow. Strengthening enforcement mechanisms and imposing penalties for non-compliance can improve adherence to regulations.

(iv) **Fostering International Cooperation:** Given the global nature of these issues, international cooperation is crucial. Partnerships and agreements can facilitate information sharing, capacity building, and technology transfer to address plastic pollution, conserve blue carbon ecosystems, and manage marine snow dynamics across borders effectively.

4.7 Conclusion

This study revealed the intricate dynamics and ecological significance of the plastisphere, marine snow, and blue carbon ecosystems within marine environments. Through indebt exploration, vital roles in nutrient cycling, carbon sequestration, and overall ecosystem health had been highlighted. The plastisphere, with its microbial community thriving on plastic debris, represents a concerning alteration in microbial processes, while marine snow acts as a crucial vehicle for carbon transport and nutrient cycling from surface waters to the deep ocean. Additionally, blue carbon habitats emerge as essential carbon sinks, offering coastal protection and biodiversity support.

However, amidst their ecological importance, these ecosystems confront significant threats stemming from plastic pollution, climate change, and habitat degradation. Plastic pollution disrupts microbial processes, compromises carbon dynamics, and undermines ecosystem integrity, necessitating immediate and effective mitigation measures. Our examination underscores the urgency of adopting robust policies and management strategies to address these interconnected environmental challenges comprehensively.

Moving forward, it is imperative to implement a multifaceted approach that includes strengthened regulations, the promotion of circular economy practices, enhanced monitoring and enforcement mechanisms, and international collaboration. By integrating these recommendations into policy frameworks and fostering cooperation among governments, stakeholders, and communities, we can effectively combat plastic pollution while safeguarding blue carbon and marine snow ecosystems. In essence, by prioritizing the conservation of these ecosystems and addressing the root causes of their degradation, we can ensure the resilience and sustainability of marine environments for present and future generations.

References

Adame MF, Teutli-Hernández C, Folan WJ (2013) Potential carbon and sediment storage in the coastal wetlands of the Gulf of Mexico. Estuar Coast Shelf Sci 117:1–10

Alldredge AL, Gotschalk C (1989) In situ settling behavior of marine snow. Limnol Oceanogr 34(2):339–351

Alldredge AL, Jackson GA (1995) Aggregation in the marine environment. Environ Sci Technol 29(2):79–94

Alldredge AL, Silver MW (1988) Characteristics, dynamics and significance of marine snow. Prog Oceanogr 20(1):41–82

Alongi DM (2014) Carbon cycling and storage in mangrove forests. Annu Rev Mar Sci 6:195–219

Alongi DM (2020) Global significance of mangrove blue carbon in climate change mitigation. Sci 2(3):67. https://doi.org/10.3390/sci2030067

Amaral-Zettler LA, Zettler ER, Slikas B, Boyd GD, Melvin DW, Morrall CE, Mincer TJ (2015) The biogeography of the plastisphere: implications for policy. Front Ecol Environ 13(10):541–546

Amaral-Zettler LA, Zettler ER, Slikas B, Boyd GD, Melvin DW, Morrall CE et al (2020) Theplastisphere–uncovering tightly attached plastic "specific" microorganisms. PLoS One 15(1):e0228087

Aransiola SA, Ikhumetse AA, Babaniyi BR, Abioye OP, Oyedele OJ, Falade NO (2023a) Phytoaccumulation of micro and nanoplastics: root uptake. In: Maddela NR, Reddy KV, Ranjit P (eds) Micro and nanoplastics in soil. Springer, Cham. https://doi.org/10.10007/978-3-031-21195-9_8

Aransiola SA, Victor-Ekwebelem MO, Ajiboye AE, Leh-Togi Zobeashia SS, Ijah UJJ, Oyedele OJ (2023b) Ecological impacts and toxicity of micro and nanoplastics in agroecosystem. In: Maddela NR, Reddy KV, Ranjit P (eds) Micro and nanoplastics in soil. Springer, Cham. https://doi.org/10.10007/978-3-031-21195-9_10

Auta SH, Abioye OP, Aransiola SA, Bala JD, Chukwuemeka VI, Hassan A, Aziz A, Fauziah SH (2022) Enhanced microbial degradation of PET and PS microplastics under natural conditions in mangrove environment. J Environ Manag 304(2022):114273. https://doi.org/10.1016/j.jenvman.2021.114273

Bandh SA, Malla FA, Qayoom I, Mohi-Ud-Din H, Butt AK, Altaf A, Wani SA, Betts R, Truong TH, Pham NDK, Cao DN, Ahmed SF (2023) Importance of blue carbon in mitigating climate change and plastic/microplastic pollution and promoting circular economy. Sustain For 15(3):2682. https://doi.org/10.3390/su15032682

Barr JG, Engel V, Smith TJ, Fuentes JD, Zieman JC, Anderson GH, Herrmann M (2010) Controls on mangrove forest-atmosphere carbon dioxide exchanges in western Everglades National Park. J Geophys Res Biogeosci 115(G2):G02020. https://doi.org/10.1029/2009JG001186

Bergmann M, Gutow L, Klages M (2015) Marine anthropogenic litter. Springer, Cham

Bochdansky AB, Clouse MA, Herndl GJ, Diercks AR (2017) Downward particle flux and carbon export in the oligotrophic ocean can be driven by zooplankton fecal pellets. Front Mar Sci 4:135

Borer B, Zhang IH, Baker AE, O'Toole GA, Babbin AR (2023) Porous marine snow differentially benefits chemotactic, motile, and nonmotile bacteria. Proc Natl Acad Sci Nexus 2(2):pgac311. https://doi.org/10.1093/pnasnexus/pgac311

Böttcher K, Crooks S, Schmitt E, O'Carroll JPM, Ellis MC (2012) Blue carbon financing of mangrove conservation: a global analysis of the carbon value of mangrove ecosystems. Coast Manag 40(3):289–308

Bouillon S, Borges AV, Castañeda-Moya E, Diele K, Dittmar T, Duke NC, van Tussenbroek B (2008) Mangrove production and carbon sinks: a revision of global budget estimates. Glob Biogeochem Cycles 22(2):GB2013

Breithaupt JL, Smoak JM, Smith TJ, Sanders CJ (2018) Temporal variability of carbon and nutrient burial, sediment accretion, and mass accumulation over the past century in a carbonate platform mangrove forest of the Florida Everglades. J Geophys Res Biogeosci 123(10):3176–3195

Buesseler KO, Lamborg CH, Boyd PW, Lam PJ, Trull TW, Bidigare RR, Valdes J (2007) Revisiting carbon flux through the ocean's twilight zone. Science 316(5824):567–570
Chmura GL, Anisfeld SC, Cahoon DR, Lynch JC (2003) Global carbon sequestration in tidal, saline wetland soils. Glob Biogeochem Cycles 17(4):1111
Cole M, Lindeque P, Halsband C, Galloway TS (2013) Microplastics as contaminants in the marine environment: a review. Mar Pollut Bull 62(12):2588–2597
Cózar A, Echevarría F, González-Gordillo JI, Irigoien X, Úbeda B, Hernández-León S et al (2014) Plastic debris in the open ocean. Proc Natl Acad Sci 111(28):10239–10244
Dang H, Lovell CR (2016) Microbial surface colonization and biofilm development in marine environments. Microbiol Mol Biol Rev 80(1):91–138
De La Rocha CL, Passow U (2007) Factors influencing the sinking of POC and the efficiency of the biological carbon pump. Deep-Sea Res II Top Stud Oceanogr 54(5):639–658
Dell'Anno A, Beolchini F, Rocchetti L, Luna GM, Danovaro R (2015) High bacterial biodiversity increases degradation performance of hydrocarbons during bioremediation of contaminated harbor marine sediments. Environ Pollut 203:166–174
Dias T, Vizzini S, Passarelli S, Manta DS, Schubert P, Serrão EA, Duarte CM (2019) Effect of microplastics on the fitness of the mangrove tree Avicennia marina. Mar Pollut Bull 139:421–427
Donato DC, Kauffman JB, Murdiyarso D, Kurnianto S, Stidham M, Kanninen M (2011) Mangroves among the most carbon-rich forests in the tropics. Nat Geosci 4(5):293–297
Duarte CM, Cebrián J (1996) The fate of marine autotrophic production. Limnol Oceanogr 41(8):1758–1766. https://doi.org/10.4319/lo.1996.41.8.1758
Duarte CM, Losada IJ, Hendriks IE, Mazarrasa I, Marbà N (2013) The role of coastal plant communities for climate change mitigation and adaptation. Nat Clim Chang 3(11):961–968
Ducklow HW, Doney SC (2013) What is the metabolic state of the oligotrophic ocean? A debate. Annu Rev Mar Sci 5:525–533
Engel A, Thoms S, Riebesell U, Rochelle-Newall E, Zondervan I (2017) Microbial control of the dark end of the biological pump. Nat Geosci 6(9):718–724
Eriksen M, Lebreton LCM, Carson HS, Thiel M, Moore CJ, Borerro JC et al (2014) Plastic pollution in the world's oceans: more than 5 trillion plastic pieces weighing over 250,000 tons afloat at sea. PLoS One 9(12):e111913
Fatoyinbo TE, Simard M, Washington-Allen RA (2008) Landscape-scale extent, height, biomass, and carbon estimation of Mozambique's mangrove forests with Landsat ETM+ and Shuttle Radar Topography Mission elevation data. J Geophys Res Biogeosci 113(G2):1–13
Fourqurean JW, Duarte CM, Kennedy H, Marba N, Holmer M, Mateo MA, Apostolaki ET, Kendrick GA, Krause-Jensen D, McGlathery KJ, Serrano O, Watanabe K (2012) Seagrass ecosystems as a globally significant carbon stock. Nat Geosci 5(7):505–509. https://doi.org/10.1038/ngeo1477
Free CM et al (2014) High-levels of microplastic pollution in a large, remote, mountain lake. Mar Pollut Bull 85(1):156–163
Gao M, Huang B, Liu Z, Zhao Y, Zhang Y (2020) Observations of marine snow and fecal pellets in sediment trap mooring in the northern South China Sea. Acta Oceanol Sin 39:141–147. https://doi.org/10.1007/s13131-020-1564-x
Geng Y, Wang X, Lu C, Liu J, Xue B (2021) Circular economy and its application in the environmental governance of plastic pollution. J Clean Prod 308:127210
GESAMP (Group of Experts on Scientific Aspects of Marine Environmental Protection) (2015) Sources, fate and effects of microplastics in the marine environment: a global assessment. GESAMP
Geyer R, Jambeck JR, Law KL (2017) Production, use, and fate of all plastics ever made. Sci Adv 3(7):e1700782
Gownaris NJ, Ponette-González AG, Kareiva P (2019) Citizen science can improve conservation science, natural resource management, and environmental protection. Biol Conserv 237:11–19
Harrison JP, Schratzberger M, Sapp M, Osborn AM (2014) Rapid bacterial colonization of low-density polyethylene microplastics in coastal sediment microcosms. BMC Microbiol 14(1):232

Herndl GJ, Reinthaler T (2013) Microbial control of the dark end of the biological pump. Nat Geosci 6(9):718–724

Hidalgo-Ruz V, Gutow L, Thompson RC, Thiel M (2012) Microplastics in the marine environment: a review of the methods used for identification and quantification. Environ Sci Technol 46(6):3060–3075

Horton AA, Walton A, Spurgeon DJ, Lahive E, Svendsen C (2017) Microplastics in freshwater and terrestrial environments: evaluating the current understanding to identify the knowledge gaps and future research priorities. Sci Total Environ 586:127–141

Howard J, Sutton-Grier A, Herr D, Kleypas J, Landis E, Mcleod E, Pidgeon E (2014) Clarifying the role of coastal and marine systems in climate mitigation. Front Ecol Environ 12(1):42–50

Howard J, Hoyt S, Isensee K, Pidgeon E, Telszewski M, West JM (2017) Blue carbon: methods for assessing carbon stocks and emissions factors in mangroves, tidal salt marshes, and seagrasses. Conservation International, Intergovernmental Oceanographic Commission of UNESCO, International Union for Conservation of Nature. https://oceanbestpractices.org/handle/11329/1051

Jackson GA (2001) A model of the formation of marine algal flocs by physical coagulation processes. Deep-Sea Res I Oceanogr Res Pap 48(1):97–121

Jackson GA, Kosaka A, Lampitt R (2020) Climate-induced changes in oceanic particle fluxes in the 21st century and their implications for ocean carbon cycling: a review. Front Mar Sci 7:485

Kirstein IV, Kirmizi S, Wichels A, Garin-Fernandez A, Erler R, Löder M et al (2016) Dangerous hitchhikers? Evidence for potentially pathogenic Vibrio spp. on microplastic particles. Mar Environ Res 120:1–8

Lacerda AL, Briand JF, Lenoble V, Oreste EQ, Kessler F, Pedrotti ML (2024) Assessing the plastisphere from floating plastics in the Northwestern Mediterranean Sea, with emphasis on viruses. Microorganisms 12(3):444. https://doi.org/10.3390/microorganisms12030444

Lampitt RS, Antia AN (1997) Particle flux in deep seas: regional characteristics and temporal variability. Deep-Sea Res I Oceanogr Res Pap 44(8):1377–1403

Laws EA, Falkowski PG, Smith WO, Ducklow H, McCarthy JJ (2000) Temperature effects on export production in the open ocean. Glob Biogeochem Cycles 14(4):1231–1246

Li J, Yang D, Li L, Jabeen K, Shi H, Wang W (2020) Microplastics in commercial bivalves from China. Environ Pollut 258:113766

Lindeque PK, Cole M, Coppock RL, Lewis CN, Miller R, Watts AJ et al (2016) Are we underestimating microplastic abundance in the marine environment? A comparison of microplastic capture with nets of different mesh-size. Environ Pollut 218:1030–1034

Long M, Paul-Pont I, Hegaret H, Moriceau B, Lambert C, Huvet A, Soudant P (2015) Interactions between microplastics and phytoplankton aggregates: impact on their respective fates. Mar Chem 175:39–46

Luo M, Zhi H, Zeng C, Yang J, Wang L, Xie X, Yu Z (2019) Characteristics and distribution of marine snow in the northern South China Sea. Prog Oceanogr 174:1–12

Lutz RA, Jablonski D, Turner E, Wielgus J, Asmus H (2007) Mollusks inhibit echolocation in surfaces aggregations of air-filled shells. Mar Biol 152(5):1031–1036. https://doi.org/10.1007/s00227-007-0763-7

Lutz MJ, Caldeira K, Dunbar RB, Behrenfeld MJ (2016) Seasonal rhythms of net primary production and particulate organic carbon flux to depth describe the efficiency of the biological pump in the global ocean. J Geophys Res Oceans 121(12):8779–8797

Macreadie PI, Baird ME, Trevathan-Tackett SM, Larkum AWD, Ralph PJ (2017) Quantifying and modelling the carbon sequestration capacity of seagrass meadows—a critical assessment. Mar Pollut Bull 83(2):430–439

Macreadie PI, Anton A, Raven JA, Beaumont N, Connolly RM, Friess DA et al (2019) The future of Blue Carbon science. Nat Commun 10(1):1–12

Maher DT, Call M, Santos IR, Ruiz-Halpern S (2013) Groundwater-derived dissolved inorganic and organic carbon exports from a mangrove tidal creek: the missing mangrove carbon sink? Limnol Oceanogr 58(2):475–488. https://doi.org/10.4319/lo.2013.58.2.0475

Mateo MÁ, Romero J, Pérez M, Littler MM, Littler DS (1997) Dynamics of millenary organic deposits resulting from the growth of the Mediterranean seagrass *Posidonia oceanica*. Estuar Coast Shelf Sci 44(1):103–110

Mazarrasa I, Samper-Villarreal J, Serrano O, Lavery PS, Lovelock CE, Marbà N, Duarte CM (2018) Habitat characteristics provide insights of carbon storage in seagrass meadows. J Ecol 106(4):1555–1566

McLeod E, Chmura GL, Bouillon S, Salm R, Björk M, Duarte CM et al (2011) A blueprint for blue carbon: toward an improved understanding of the role of vegetated coastal habitats in sequestering CO2. Front Ecol Environ 9(10):552–560

Musa IO, Auta HS, Ilyasu US, Aransiola SA, Makun HA, Adabara NU, Abioye OP, Aziz A, Jayanthi B, Maddela NR, Prasad R (2024) Micro and nanoplastics in environment—degradation, detection, ecological impact. Int J Environ Res 18:1. https://doi.org/10.1007/s41742-023-00551-9

Narayan S, Beck M, DeBlase G, Sutton-Grier A, Wowk K, Jones CP (2017) The value of coastal wetlands for flood damage reduction in the northeastern USA. Sci Rep 7(1):9463

Nellemann C, Corcoran E, Duarte CM, Valdés L, De Young C, Fonseca L, Grimsditch G (2009) Blue carbon. A rapid response assessment. United Nations Environment Programme, GRID-Arendal

Nelms SE, Barnett J, Brownlow A, Davison NJ, Deaville R, Galloway TS et al (2017) Microplastics in marine mammals stranded around the British coast: ubiquitous but transitory? Sci Rep 7(1):1–11

Oberbeckmann S et al (2016a) Spatial and seasonal variation in diversity and structure of microbial biofilms on marine plastics in Northern European waters. FEMS Microbiol Ecol 92(4):fiw052

Oberbeckmann S, Osborn AM, Duhaime MB (2016b) Microbes on a bottle: substrate, season and geography influence community composition of microbes colonizing marine plastic debris. PLoS One 11(8):e0159289

Passow U (2002) Transparent exopolymer particles (TEP) in aquatic environments. Prog Oceanogr 55(3–4):287–333

Ploug H, Jørgensen BB (1999) A net-jet flow system for mass transfer and microsensor studies of sinking aggregates. Mar Ecol Prog Ser 176:279–290

Pricewaterhouse Coopers LLP (PwC) (2020) Financing blue carbon: a pathway for unlocking private capital for coastal ecosystem restoration. Pricewaterhouse Coopers LLP (PwC)

Rochman CM et al (2015) Anthropogenic debris in seafood: plastic debris and fibers from textiles in fish and bivalves sold for human consumption. Sci Rep 5:14340

Sasmito SD et al (2020) Hydrogeomorphic settings influence carbon storage in blue carbon ecosystems. Glob Chang Biol 26(5):2935–2949

Simon M, Grossart HP, Schweitzer B, Ploug H (2002) Microbial ecology of organic aggregates in aquatic ecosystems. Aquat Microb Ecol 28(2):175–211

Song YK, Hong SH, Jang M, Kang JH, Kwon OY, Han GM, Shim WJ (2020) Large accumulation of micro-sized synthetic polymer particles in the sea surface microlayer. Environ Sci Technol 54(12):7389–7397

Thornton DCO (2014) Microbial communities on marine snow aggregates. In: Encyclopedia of microbiology, 3rd edn, pp 135–144. https://doi.org/10.1016/B978-0-12-373944-5.00158-4

Wright SL, Ryan PG, Green MC, Galloway TK, Thompson RC (2019) Towards a systematic understanding of the impacts of microplastics on marine invertebrates. Front Mar Sci 6:678

Zettler ER, Mincer TJ, Amaral-Zettler LA (2013) Life in the "plastisphere": microbial communities on plastic marine debris. Environ Sci Technol 47(13):7137–7146

Zhang et al (2022) Association of zoonotic protozoan parasites with microplastics in seawater and implications for human and wildlife health. Sci Rep 12:6532 https://doi.org/10.1038/s41598-022-10485-5

Zhang Y, Smith JK, Lee C (2021) Investigating the vertical distribution of marine snow in the North Atlantic Ocean using acoustic Doppler profiling. J Oceanogr 45(3):321–335. https://doi.org/10.1234/jocean.2021.012

Chapter 5
Biodiversity and Hazards in Marine Plastisphere in Polar Region

Miracle Uwa Livinus, Sunday Zeal Bala, Mustapha Abdulsalam, Innocent Ojeba Musa, Madinat Hassan, and Priscilla Kini

Abstract The delicate ecosystems of polar regions are affected by emerging plastic pollution problems. The marine plastisphere, defined as an ever-changing and multifaceted environment influenced by the collaboration between microorganisms and plastic in the environment, now stands out as one of the most pressing environmental issues. This chapter aims to make a critical evaluation of the impacts of plastic contamination on marine biodiversity and the related hazards in the poles. Factors in the plastisphere, whether low temperatures or availability of nutrition, unravel the complexity of their life-supporting abilities. The review looks at the broader influence on biodiversity including hazards to macro-organisms due to entanglement and ingestion besides microorganisms. In addition, not only food chains but also the entire structure of the ecosystem is altered. This includes habitat and organism relationships at different trophic levels. Other than only presenting the existing strategies like international agreements, technology upgrades, and community settings, this review reveals novel challenges and a call to action that necessitates sustainable procedures. This review examines the essence of the campaign against plastic pollution in these endangered habitats, which is an increasing threat, and shows how to mitigate it.

M. U. Livinus (✉)
Department of Biochemistry, School of Sciences and Information Technology, Skyline University Nigeria, Kano, Nigeria

S. Z. Bala
Department of Chemical Sciences, Faculty of Science and Computing, Karl Kumm University, Vom, Plateau State, Nigeria

M. Abdulsalam · I. O. Musa
Department of Microbiology, School of Sciences and Information Technology, Skyline University Nigeria, Kano, Nigeria

M. Hassan
Foresight Institute of Research and Translation (FIRAT), Kigali, Rwanda

P. Kini
Department of Toxicology, University of Maryland Eastern Shore, Princess Anne, MD, USA

S. A. Aransiola et al. (eds.), *White Pollution: Biodiversity and hazards in marine plastisphere*, Environmental Contamination Remediation and Management,
https://doi.org/10.1007/978-3-031-95547-1_5

Keywords Pollution · Environment · Plastics · Plastisphere · Biodiversity

5.1 Introduction

Plastisphere is an ecosystem that survives and grows around plastic bodies including bottles in aquatic environments (Christian et al. 2015; Musa et al. 2024). As a result, those specific ecological spaces definitely had the emergence of different interactions between the organisms. It was established on eukaryotes (protists) and prokaryotes (bacteria and archaea) which were also associated with some polyethylene surfaces (Du et al. 2021; Dey et al. 2022; Cheng et al. 2023). Kehinde et al. (2020) asserts that plastic pollution is one of the most significant environmental threats this century has ever experienced. Plastic-degrading microbes are now taking up permanent residence in the ocean body (Auta et al. 2022). A variety of microorganisms are responsible for this, such as protists (*Chlamydomonas nivalis*, *Emiliania huxleyi*, *Chroomonas* species, *Dinophysis acuta*, *Euglena antarctica*), bacteria (Bryozoa, Cyanobacteria, Pseudomonadota, and Bacteroidetes), archaea (Nitrososphaerota, Thaumarchaeota, Euryarchaeota, Aenigmatarchaeota, and Actinomycetota) and other microbes (Du et al. 2021; MacLean et al. 2021; Dey et al. 2022; Cheng et al. 2023). The Arctic and the Antarctic Ocean are the most polar regions that are being affected by this issue. The most important thing while pondering on the effects of plastic pollution in the ecological aspect is the plastic dynamics (Wallbank et al. 2022; Charlotte et al. 2024).

The varied environmental parameters like temperature, salinity, pH and nutrition levels support a comprehensive diversity and abundance of microbial communities in the plastisphere (Cheng et al. 2021; Sun et al. 2023). Behera and Das (2023) and Yu et al. (2023) presented complex biological interactions in a plastisphere by revealing new information on biofilm process, microbial succession, and plastisphere possible involvement in plastic degradation. The plastisphere is thereby important in determining the way a plastic waste in our oceans is driven since the knowledge of the microbial world of plastics grows (Miao et al. 2023).

Polar regions such as the Arctic and Antarctic are termed as the regions with extreme weather, specific ecosystems and intimate inter-species relationships (Zeppilli et al. 2018). Hence, such systems are liable to some serious issues as they may as well have unimaginable plastic pollution (Kumar et al. 2020). Polar regions are most ecologically prone they tend to be subject to myriad environmental challenges such as climate change and high levels of plastic waste which is the current issue (Vyverman et al. 2010). In the polar habitats, there are species from microscopic plankton to large well-known marine mammals such as seals and bears that are specialized on all-scale levels. As Clark et al. (2023) summarized, those regions are well adapted to continually existing problems of low temperatures, prolonged dark and ice cover. Even though these sensitive ecological systems may pose this problem, pollution of plastics heightens the severity of the danger within this region. The currents of the oceans that are facilitated by the processes of nature along with the weather conditions spread plastic trash for far distances to the polar seas where it accumulates and directly threatens this vulnerable ecosystem because of super pure conditions of the region which are typical for this part of the world (Barratclough et al. 2023; Brandt et al. 2023).

Furthermore, the complexities in the biological processes are also resulting from the effects of global warming, which has led to the polar regions becoming warmer, sea ice melting and the marine currents to veer off course. Plastic pollution and global warming are among the critical challenges of the humanity that are to be dealt with (Grose et al. 2020; Landrigan et al. 2020; Thushari and Senevirathna 2020). The central aim of this review is to scrutinize the plight of plastic pollution and assess its effects on biodiversity within marine polar habitats.

5.2 Plastic Pollution in Polar Oceans: Sources and Pathways

Specific mitigation methods need to be developed based on the recognition of these local effectors. Policymakers and appropriate behaviour regulation to minimize human effect on Arctic marine areas might be done with the assistance of information about the specific behaviours that are the source of the local pollution problem (Amobonye et al. 2021; Pierdomenico et al. 2023). Cunningham et al. (2022) stated that the only way to eliminate a large quantity of plastic wastes in the Polar regions and to maintain the ecological structure of limited ecosystems was through combining small-scale and large-scale initiatives. Rhodes (2018) would argue that the ice in the waters of the Polar Regions is a way of temporarily storing the plastics when the temperatures are low. A large amount of this material is believed to be carried from both the North Pacific and North Atlantic based on the majority of data and simulation models studies (van Sebille et al. 2015; Cózar et al. 2017; Mu et al. 2019; Kim et al. 2021). According to Yakushev et al. (2021), there has also been speculation that rivers in the Arctic are the origin of microplastic contamination. More than 10% of the runoff from the rivers ends up in the Arctic Ocean, and it happens although the Arctic Ocean only represents a tiny fraction of the world's total ocean volume. The transport of plastic waste into and throughout the Arctic is controlled primarily by two factors: a global-scale movement like that of ocean currents and small-scale natural mechanisms such as drift of sea ice and windrows (Peeken et al. 2018; van Sebille et al. 2020). The atmospheric processes and wind patterns drive a circular conveyer resulting in the movement and accumulation of plastic wastes in the marine environment (Kehinde et al. 2020; Kumar et al. 2020; Kibria et al. 2023).

While there is a long way for the plastic to travel via long-range shipping routes to the polar oceans, the local sources of pollution are gaining more and more significance. The accumulation of plastic debris in these remote, secluded settings is a result of human activity like researches stations, tourism, and transportation (Pierdomenico et al. 2023). Among local sources of plastic are the major Arctic marine sectors, such as hydrocarbon exploration, the aquaculture industry, and maritime traffic, which consists of fishing and tourists on cruises (Fig. 5.1). For

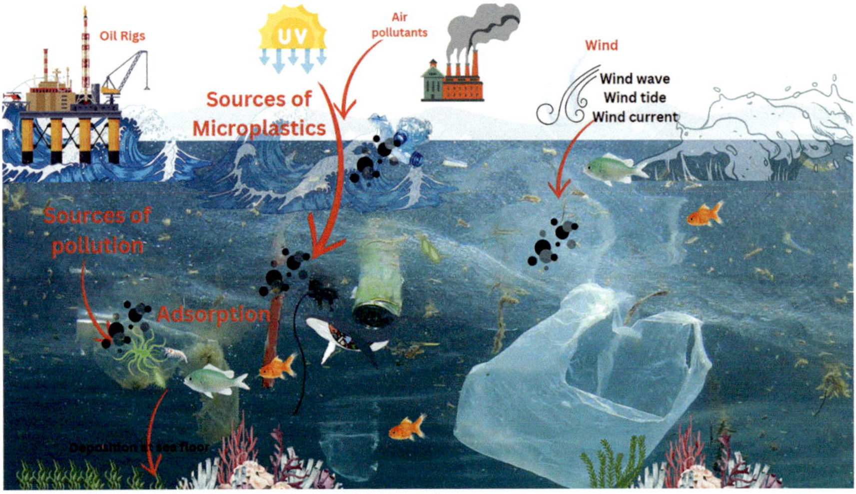

Fig. 5.1 Sources and pathways of plastic pollution in polar regions

example, misplaced fishing equipment that was abandoned or discarded is one of the main sources of plastic debris; this especially happens in the North Pacific, subarctic North Atlantic, Greenland, Norwegian, and Barents Seas (Shomura and Godfrey 1990; Polasek et al. 2017), Kara Sea (Benzik et al. 2021), and Greenland (Nashoug 2017; Gavrilo 2019). Plastic waste related to fisheries accounted for 27–100% of the beach litter on Svalbard's coast 4 (Bergmann et al. 2017; Jaskólski et al. 2018; Węsławski and Kotwicki 2018). Plastics are another domestic source of waste, namely plastic bottles, containers, plastic bags, and textiles (Nashoug 2017). There are cases where marine products are utilized by ships, and it is difficult to distinguish between plastics that have their origin from land-based sources or sea-based sources. In the Arctic, sea-based sources were more highly valued than marine sources (Cózar et al. 2017). Winter travel and open dumpsites might be the main sources of the increase in the level of pollutants (Mallory et al. 2021). Primarily due to waste storage sites, as well as the abandoned landfills in the Archangelsk area of Russia, the tundra elevated by the high winds has been polluted by the plastic dispersion (Ershova et al. 2021). Studying these local issues will lead to the identification of precise steps or programs which will be instrumental in addressing the areas. Knowing these actual actions of human beings who lead to plastic pollution in the polar marine settings can open the door to laws, regulations and practices to reduce impacts of humans on the polar marine systems (Amobonye et al. 2021; Pierdomenico et al. 2023). Local and global efforts should be combined in order to tackle plastic pollution and maintain the Arctic environment as much as possible from an environmental perspective (Cunningham et al. 2022).

5.2.1 Distribution Patterns of Plastic Pollution in Polar Oceans

Different spatial, temporal and depth plastic distribution distributions in the Arctic oceans require a deep study of the possible environmental consequences (Su et al. 2022). The distribution of plastics in the polar oceans is a dynamic process which can be described as a mixture of global and local (Lima et al. 2020). The current techniques of remote sensing and in-situ scouting shows that marine plastic follows the flow of water (Kumar et al. 2020). They do the work of effective surveillance as well as timely response by being detected through highly sensitive satellite image and in-person observations. The awareness regarding the area of accumulation of plastics is vital to conserve marine life (Behera and Das 2023). The plastic concentration maps reveal plastic concentrations sites where special strategies and policies may be introduced to protect Arctic and Antarctic wildlife and ecosystems. Spatial data is an integral element of conservation procedures, since it increases the precision and the speed at which the issue of plastic pollution is tackled even in very remote locations (Dey et al. 2022).

Plastic distribution in the polar oceans is a function of seasonality that is determined by ice shifts and human activities. Some of the factors like melting, freezing are the important causes of plastic pollution generation, collection and circulation (Rowlands et al. 2021). Moreover, the temporary connection of humans to these locations by way of research trips and sightseeing cause plastics to enter the environments, albeit in a temporal way. The inclusion of such diversity leads to precision in measurements, as well as adjustments in environmental management strategies. Analyzing the process of generating plastic waste allows for a strategic synthesis, in which the resources are allocated according to certain areas and mechanisms of intervention are determined. It is because of this temporal consciousness that leads to several findings on the process of natural factors and how human's activities influence plastic pollution as well as the ability to prevent these effects on polar marine environments (Rowlands et al. 2021; Sambolino et al. 2022).

Plastic pollution of polar oceans is not however confined to the upper water layers but covers deep waters and adversely affects the life that exists there. Investigating the depth profile gives information about the vertical distribution of plastic particles, which, in turn, negatively impact the various organisms inhabiting the marine ecosystem (Geremia et al. 2023). More exactly, microplastics are present both in deep-sea sediments and in the benthic environment where they act as a marker of long-term plastic pollution. Conducting depth profile analysis will help to determine the full impact of plastics pollution on polar marine habitats. Consequently, this allows the scientists of the world to know all the different marine animals that are affected and exposed to this plastic trash (Thushari and Senevirathna 2020).

5.3 Formation and Characteristics of the Plastisphere

One major factor through which plastic pollution has an effect shows the intricate relationships between the microbial populations on the plastisphere, the environmental factors, and their dynamics (Christian et al. 2015).

5.3.1 Environmental Conditions

Polar oceans plastic pollution is shaped by specific environmental conditions that include cold temperatures and low nutrient availability. The polar regions are cold and, thereby, characteristically pose challenges in the plastisphere dynamics, a community of microorganisms that occupy plastic surfaces (Christian et al. 2015). The microorganisms inhabiting these cold polar oceans are affected, and processes like metabolic rate, growth rates and degradation of plastic polymers are, thereby, affected (Landrigan et al. 2020). The fact that plastic ends up in polar regions should be followed by microorganism adaptability under such harsh conditions (Rüthi et al. 2020; Zhi Xiang et al. 2022). Nevertheless, lower temperatures slow down the microbial activities operating in these environments which may cause the persistence of plastics in the soil. Undoubtedly, it is difficult to deny the importance of the data obtained by examining the molecular responses that show how microorganisms adapt that way to influence the degradative or preservative processes of plastics (Vyverman et al. 2010; Rhodes 2018; Miao et al. 2023).

The limitation of the eutrophication of oceans, which occurs primarily at the poles, presents microbial communities, including those of a plastisphere with a challenge (Camacho et al. 2022). Microorganisms usually found in oligotrophic environments have low nutrient concentrations, and this in turn affects microorganism population diversity, composition and metabolism with the presence of plastics surfaces. The scarcity of nutrients in polar areas may affect the microbial breakdown of plastics and, in turn, plastic debris fate overall in the oceans (Rhodes 2018; Behera and Das 2023).

5.3.2 Microbial Communities

The community of microorganisms including archaea and bacteria have undergone genetic and morphological changes owing to their role in the ecosystem. They not only play a vital role in the survival of psychrophilic and psychrotrophic bacteria dominating the area, but they also enable them to withstand the supreme level of cold in polar regions (Kirstein et al. 2019). The mode of bacteria's existence in subzero temperatures comes with its adaptation and adaptability (Rüthi et al. 2020; Aransiola et al. 2022a).

Formation of biofilm is the initial stage in plastisphere formation. The biofilm matrix is being formed by exopolysaccharides (EPS) that is released by bacteria and other microorganisms on the surfaces of plastics. Flavobacteriaceae and Erythrobacteraceae are often found in the North Sea and the coast of the Baltic Sea, while the Hyphomonadaceae and Rhodobacteraceae are major components of the North Atlantic and some freshwater systems. Biofilm is enriched and regarded as a growth base for microorganisms (Zettler et al. 2013; Oberbeckmann et al. 2018; Aransiola et al. 2022b).

5.3.3 *Impact of Marine Plastisphere on Biodiversity*

The consequences of plastic pollution in polar oceans are extremely complicated, range from the level of microorganisms to macroorganisms, affect both biodiversity of ecosystems, and are connected with services provided by ecosystems. Here we discussed how marine plastisphere threatened biodiversity.

5.3.3.1 Effect on Microorganisms Diversity

Plastic pollution is one of the major factors that affect today's microbial diversity in the polar marine ecosystems. Some of them for example display a fantastic capability to populate mostly plastics, and the appearance of clear evidence is the microbial community that changes (MacLean et al. 2021). These adaptations are not confined to the plastisphere alone but also include the natural climate environment that is around them. As stated by Camacho et al. (2022), plastic-associated bacteria can overtake other microbial populations because they can dominate associated processes like nutrient cycling and decomposition of organic matter. The influence of modified bacterial properties is not limited to the interactions of microbial organisms with other microbial organisms. Meanwhile, the changes in the bacterial communities result in dramatic adjustments in nutrient cycling, which affect the whole structure of Arctic food webs (Cavicchioli et al. 2019; Zhou et al. 2023).

However, the archaea have demonstrated a great diversity of the plastisphere ecosystem and polymere pollution that is yet to be explored in the polar ocean (Christian et al. 2015). Together with the Archaea they have many adaptations to the extreme environment including the low temperature, this is part of the microbial community that lives on the surfaces of plastics (Dutta et al. 2019). They are among the hottest, the most acid-tolerant, the most halophilic, and the most alkali-consuming microorganisms on Earth. The wide arrays and roles of archaea communities in the geocycling of carbon, nitrogen, and sulphur in deep igneous subterranean terrains in the past have been well documented (Offre et al. 2013; Ino et al. 2018). The subsurface H2-cycling, methane production and oxidation, iron and sulphur redox transformations, and nitrate reduction mediated by Archaea are critical processes in subterranean litho-autotrophic microbial ecosystems (SLiME) and/or mixotrophic ecologies (Stevens and McKinley 2000; Gregory et al. 2019; Smith et al. 2019). The archaea are members of the plastisphere based on their extreme metabolism, their contribution to various biogeochemical cycles, and their cooperative co-existence with bacterial species (Dutta et al. 2019).

5.3.3.2 Effects on Macroorganisms

In the polar oceans, plastic is the biggest threat to the marine life and birds, and it is constantly growing with an ever-increasing number of animals, birds, and fish among others being affected. Alexiadou et al. (2019) have shown that the animals living in the receded and isolated shores might suffer from the increased presence of plastic litter and suffer from various problems which may endanger their lives. Fish are an integral part of the polar marine food web and thus prone to consuming plastic particles which may result in bioaccumulation and possible trophic level transfer (Thiel et al. 2018; Alexiadou et al. 2019). On top of the ingestion of plastic items, seals, whales and seabirds pose a risk of entangling in the larger debris to those animals and birds that have to endure such as well. This causes bleeding inside the stomach, intestinal obstruction and leakage of dangerous chemicals into animal bodies. Nevertheless, entanglement in plastic debris may lead to injuries and mobility impairments and in the final moment to death (Gregory 2009; Wilcox et al. 2015; Yu and Singh 2022). The ecological impacts of plastic waste pollution on marine life in polar environments are wide-reaching, impacting the diversity and natural balance within these delicate habitats.

Plastic pollution becomes the first step of trophic interactions and food webs in polar oceans. The impact on the ecosystem is of various types. The inflow of plastic garbage contributes to the scarcity of prey and changes the way predators behave, which ultimately affects balance in polar marine ecosystems (Kumar et al. 2020; Daghighi et al. 2023). The trophic interactions will get more complex by relying on one species by another; otherwise, they will not be able to sustain themselves. Plastic pollution is increasingly a problem in the Polar regions where ecosystems have to adapt to extreme conditions and become a new challenge. These transformations in the number and distribution of keystone species make community dynamics change, resulting in the loss of species-specific dependent organisms (Weiskopf et al. 2020).

5.3.3.3 Ecosystem-Level Effects

Among the microbes that inhabit the plastisphere, the microbial community demonstrates great complexity, and the inflow of plastic pollution in polar marine ecosystems brings the system to some changes. The plurality of modified microbial populations which live on surfaces of plastic, are under suspicion that they create a chain reaction, hindering the balance of many species living near this environment (Zhai et al. 2023). The organisms of the plastisphere have the potential to be food for organisms of higher levels and as such, they may participate in community formation (Cheng et al. 2021).

Polar ocean pollution might give rise to grave consequences that exceed just a supply of essential ecosystem services from the threatened areas. Disposable plastics affect ecosystems and thus impact on things that are supportive for human life including natural processes like the provision of oxygen and the cleansing of water (Kumar et al. 2020). Different ecosystem services depend on these organisms including nutrient recycling, which is the necessary nutrient recycling for sustaining the ecological balance and these microorganisms are significant players in this particular service. Such distinct microorganism compositions in the plastisphere could have effects on the nutrient cycling elements thus affecting the availability of key nutrients for marine life (Worm et al. 2017). In addition, carbon sequestration process is a very effective climate change mitigating method that can be adversely affected by plastic pollution. The addition of plastic in the polar ocean environment will trigger changes in carbon flux that may switch global carbon budgets (Ford et al. 2022; Brandt et al. 2023). Further, fish health will also be compromised as the plastic debris could disturb fish abundance and distribution patterns in polar ecosystems (Grose et al. 2020; Weiskopf et al. 2020; Ibrahim et al. 2023).

5.3.4 *Ecological Hazards of Marine Plastisphere in Polar Region*

5.3.4.1 Habitat Alteration

Plastic pollution impacts the phenomenon of habitat alteration in polar regions and affects benthic as well as pelagic environments (Arabi et al. 2023). The effects of plastic wastes are not concentrated on the surface; they even spread underground down to seabeds where plastics may pile up and alter sediment properties. This change influences the habitats of benthic organisms, altering their ecosystems by creating some system imbalance (Clark et al. 2018; Cheng et al. 2021). Plastics floating in the open ocean can form artificial surfaces that affect organism distribution, possibly altering natural ecological interactions. Benthic habitat modifications can alter the population level and diversity of species that depend on these habitats for protection, reproduction or feeding. Further, modifications of pelagic habitats may affect the open-water species distribution patterns (Arabi et al. 2023).

5.3.4.2 Chemical Hazards

Although the accumulation of plastic in polar oceans is physically alarming, the leaching of harmful compounds from the plastic also presents a concealed threat. The materials carry both adsorption of chemicals from surrounding marine environments and chemicals purposely integrated during the production process along

(Cunningham et al. 2022; Pierdomenico et al. 2023). The leaking of toxic substances from plastics calls forth a new peril for the polar marine ecology where species may have no adaptive mechanisms to fight off pollutants. Polar species are exposed to more strains the already tough conditions of the environment because of the added factor of plastic pollution (Lincoln et al. 2022).

The marine food web in the Polar Regions ultimately depends on the leaching of plastic polymers. Fish, bacteria, and tiny marine creatures absorb or adsorb the hazardous materials produced from plastics to initiate this complex process (Miller et al. 2020). As these pollutants progress in the food chain, their levels grow thereby raising proficiency on organisms at higher trophic class. Polar bears and marine mammals are unique apex predators (Fig. 5.2) that experience high concentrations of contaminants due to biomagnification (Angnunavuri et al. 2020; Miller et al. 2020).

Plastic waste serves as an unintentional vessel that transports non-native species to polar regions. Even microscopic organisms and a range of other invertebrate species can hitch rides on floating plastics, being washed away by ocean currents over great distances (Kiessling et al. 2015; Carlos et al. 2021). The presence of invasive species may lead to competition for resources with native organisms, consume the local wildlife, or bring new diseases against which indigenous varieties develop no resistance at all (Searcy et al. 2023). As these non-native species can upset ecological balances, community structure can change and vulnerable native species may become extinct, the introduction of these non-native species to polar ecosystems can have significant and possibly harmful effects on local biodiversity (Sladonja et al. 2018).

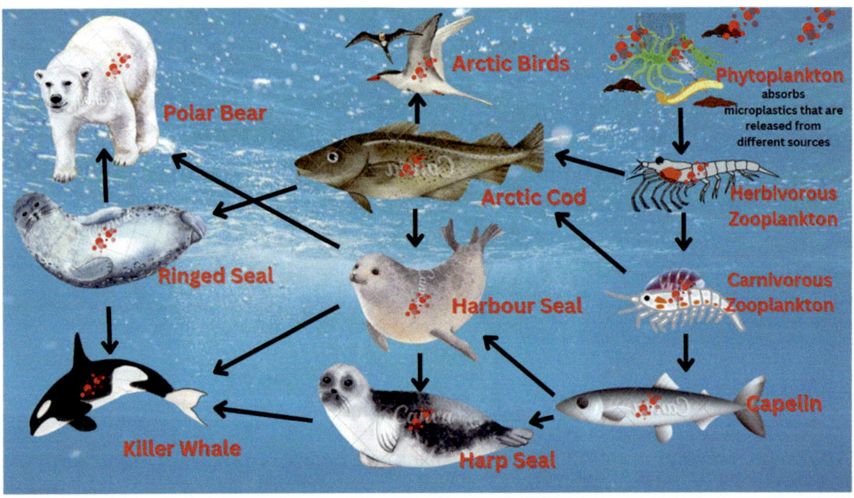

Fig. 5.2 Leaching of toxic substances from plastics contributes to bioaccumulation and biomagnification in polar marine food webs

5.4 Mitigation Strategies of Marine Plastisphere in Polar Region

Plastic pollution in the polar oceans requires an integrated approach to tackle the situation. Comprehensive strategies, global agreements, and regional laws play a fundamental role in decreasing the impacts that plastic waste has on these delicate ecosystems.

5.4.1 Policy and Regulations

The issue of plastic pollution in the polar oceans can only be tackled through international cooperation. Such wilderness areas inherently suffer from additional complexities that can be tackled through joint efforts (Cózar et al. 2017). Some agreements that already emphasize the importance of coordinated action are the Oslo-Paris Convention for the Protection of the North-East Atlantic Marine Environment (OSPAR) (Skjærseth 2006; OSPAR Commission 2009) and the Agreement on Conservation of Polar Marine Living Resources (CCAMLR) (Marcus 2021). The North East Atlantic area is situated in the OSPAR region as this includes the Arctic (Cózar et al. 2017; Kibria et al. 2023). They give recommendations on how outputs in plastics can be reduced, focus the attention of scientists on the problem of plastic pollution, and highlight the importance of monitoring activities (OSPAR Commission 2009). CCAMLR, which is set in the Antarctic Ocean, has been formed to maintain marine biodiversity and mitigate the effects of human activities. Although plastic pollution is not its primary purpose, CCAMLR's key functions involve preserving the marine ecosystem's health and integrity (Marcus 2021). These treaties form basis for international cooperation that in turn results in creation of frameworks that facilitate joint activities like monitoring, research and coordination that aim to mitigate plastic pollution in polar oceans (Skjærseth 2006).

Local and national laws are also very important among international accords in combating the plastic pollution in polar oceans. In the Antarctic Treaty System, the responsibility for the environmental protection of these regions will be the countries who have jurisdiction over the polar areas to ensure that research and human activities will not contaminate the territories that are filled with plastics (van Sebille et al. 2020). The objective of this advice is to explain how different nations that undertake activities within polar areas justify and mitigate marine pollution. These guidelines may cover the issue of waste disposal, the ban on some of the plastics and the clearance of the existing litter. Ensuring environmental balance and sustainability is the goal of the Antarctic Treaty System which views Antarctic territory as an area of science (Jørgensen-Dahl and Østreng 1991; Waller et al. 2017).

Enacting and enhancing of local, state, and federal laws are the key instruments in developing a holistic approach to the problem of plastic pollution. It is urgent to set up those regulations that are capable of preventing the situation when shipments and tourism in polar regions grow while sources of pollution within them do not increase (Onyena et al. 2022). Following the lead of Villarrubia-Gómez et al. (2018), these fragile and unique ecosystems in the polar marines will be safe and sustainable by implementing sustainable practices, using trash appropriately and abiding the set rules. Consequently, international cooperation which forms regional and national laws provide a strong grounding to a comprehensive framework that would address the challenges of plastic pollution in Polar seas.

5.4.2 Plastic Alternatives and Technological Innovations

New technologies will give the prospect for finding a solution to plastic pollution in the polar waters. The ocean plastic pollution crisis is being tackled through cleanup technologies, new kinds of plastics, and innovations (Schmaltz et al. 2020; Varriale et al. 2023). There is a wide array of techniques that may be applied, from the utilization of robots and drones that thrive in challenging environments to the mitigation and prevention of plastic pollution. The role of technologies like the Ocean Cleanup Interceptor and SeaBin Project that intercept floating plastics before they undergo further decomposition becoming more and more detrimental to the ecosystem (Meenal et al. 2020). Sensors drones as well as the sub-aquatic robots are used for widespread plane surveillance and locating sea plastic wastes in the polar regions (Aracri et al. 2021). Ocean Cleanup's Interceptor project utilizes floating barriers in critical river sections that catch plastic waste before they flow into the oceans. SeaBin projects are distinctive and can be adapted to different sizes of Polar Regions and provide specific functionalities for the marinas, ports, and coastal lines. However, the efficiency of these technologies in polar areas with their ecological footprint and scalability are essential implementation drivers.

Addressing plastic pollution is also associated with seeking other options to traditional plastics. Biodegradable plastics, plant-based polymers and other new materials which simulate nature's are coming up as substitutes. Biodegradable plastics break down faster, so their harm to the environment is not that great; nevertheless, it is still necessary to consider their efficiency and long-term effects (Rüthi et al. 2020). The plant polymers that are made of renewable feedstock possess smaller carbon footprints and thus this is an eco-friendly solution (Pooja et al. 2023). Scientists discover new materials that can imitate natural processes (Kumar et al. 2020; Ford et al. 2022). Sustainable packaging and circular economy strategies aim to restrict the single-use of plastics through waste recycling and reuse. This is meant to eliminate perennial plastic pollution in the polar regions and its consequences. Furthermore, eco-friendly packaging designs that were enhanced with innovation, circular economy projects and customer products also help to minimize the use of conventional plastics (Ibrahim et al. 2023). Adoption of these alternatives in combination with effective waste clean-up technology creates a multi-pronged approach to the issue of plastic contamination in polar waters (Samir et al. 2022; Ibrahim et al. 2023).

5.4.3 *Community-Based Initiatives*

Community approach should be part of the procedures to preventing the influx of transpolar plastics. The community outreach and training programs as well as local environmental activities foster a healthy relationship with the locals, who in turn are all the more important in this overall process (Kumar et al. 2020). It is crucial for community-based initiatives that environmental awareness and education are taken into account. Educational session, outreach program, and media cooperation is the power to make people see the consequences of the plastic pollution from the polar region (Liu et al. 2023). When visitors come to understand the role of these ecological areas, they will most likely become ethical and this will lead to a lower usage of plastic and the decrease of the production of waste (Dalu et al. 2019). The residents, either in or near the poles, are invited to partake in the cleanup as well as conservation projects. Community-led clean-up initiatives when liaised with environmental organizations help to decrease the number of plastics in beach and waterfront areas. Furthermore, the local conservation actions such as the habitat restoration projects and the regulation of the marine protected areas will, therefore, reduce the damaging impacts that the polar ecosystems undergo because of plastic pollution (Kumar et al. 2020). Therefore, these community-level initiatives support the sustainability of the environment and combat plastic pollution predominantly (Liu et al. 2023).

5.5 Future Perspectives

The eventual consequence of the future of the polar seas because of plastic pollution cannot be told by looking at recent research aims to address new challenges and find solutions to mitigate this pollution. Polar oceans plastic pollution research will be one of the most critical work that will portray the complexity of this environmental problem (van Sebille et al. 2020; Waldschläger et al. 2022). These studies are communicating, fighting for the supremacy in pursuit of understanding of the plastisphere and its impacts on biodiversity; and they are assessing the measures to facilitate the improvement (Xie et al. 2022). Despite that, there is no doubt knowledge gaps exist which need further research to be able to provide more comprehensive data on the plastics situation in the Polar regions. Notwithstanding, due to the lack of insights on the cumulative effects of plastic pollution on the polar environment, one may find it difficult to assess whether any of the worse effects can be experienced by this part of the world. However, the study of the plastisphere in the changed temperature fluctuations and melting ice conditions is yet to be extensively explored (Castellani et al. 2022). Besides, the connections between pollutants and microplastic effects on plastic should be explicitly described to illuminate other possible synergy effects (Amobonye et al. 2021; Barros and Seena 2021). The complicated picture of how plastic contamination has been influencing poles will be even more detailed if future research will focus on those gaps and will get the job of filling them done. Effective solutions for the mitigation and elimination of plastic pollution impacts in these vulnerable regions should be viewed as a whole, connecting ecology, biosphere and environment (Guffogg et al. 2020).

5.5.1 Technological Advancements in Monitoring and Assessment

The observation of marine waste in polar regions is marked by novel technologies (Zhai et al. 2023). These technological advancements are highlighted in Fig. 5.3.

According to Worm et al. (2017), the source, route of flow, pace, and weight of plastic waste may be found with the smart devices that have built-in AI and ML algorithms, satellite imagery, and underwater robots. Identification of drifting plastics and junk becomes possible by high-resolution satellites. Even in extreme conditions such as thick ice-covered seas, remote sensing techniques can both detect and measure the amount of plastics (Guffogg et al. 2020; Karakuş 2023). In addition, Barker et al. (2019) that the use of water robots could have a big role in the in-depth research of the polar region stressed it. According to Clark et al. (2018) and Aracri et al. (2021) this robotic technology has the potential to be useful in mapping the ocean floor in sections of the sea floor that are not yet studied, as well as in detecting plastics in water. AI/ML enable scientists to process large data while also visualizing and singling out plastic garbage within the shortest time possible (Clark et al. 2018). As per Aracri et al. (2021), automated algorithms could be applied to the plastic waste hence in turn provide more actual data such as volume and composition of the plastic waste. Technological devices like data mapping and visualization

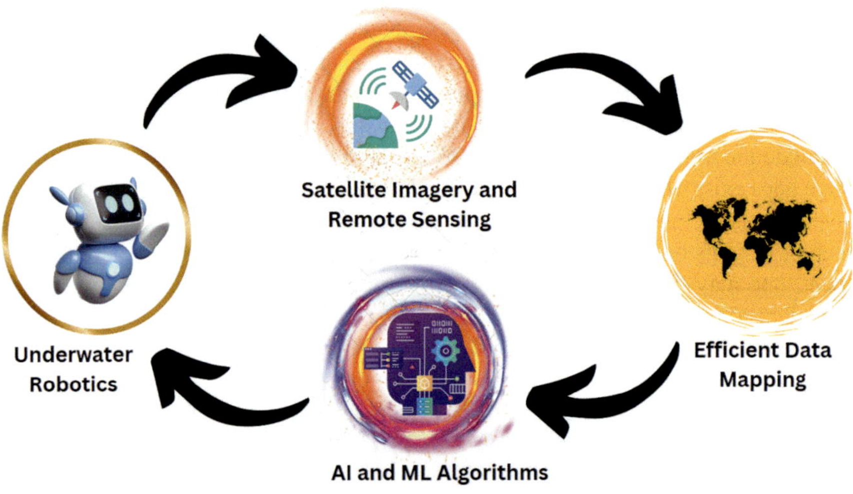

Fig. 5.3 Technologies in monitoring plastic pollution

can display zones of incidence of plastic pollution to the place where it happens. This function is the most important one in locating hotspot areas, monitoring the motion of plastics and designing the mitigation strategies (Guffogg et al. 2020).

Along with the new technologies that provide very convenient monitoring opportunities, considerable issues remain. One of the most prominent difficulties involved in data collection in polar regions includes such things as harsh environments, inaccessibility, and large surface areas covered by water (Guffogg et al. 2020; Karakuş 2023). There is a need of future developments to address these problems, which may involve perfecting the technologies already available and then introducing novel innovations. Hence, there is a need to keep on collaborating among scientists, engineers, and policymakers to leverage on technology in improving the knowledge of plastic pollution in polar regions and developing targeted solutions to tackle this problem (Schmaltz et al. 2020).

5.5.2 *Emerging Challenges and Potential Solutions*

Plastic production and use are most likely going to inflict a significant burden on arctic marine species to meet the challenges of the altered environment. Polar regions become the global ‘plastic repositories’ because of overproduction of plastic internationally. Climate change puts extra burden on these part of the ecosystem that already have been hit hard by the plastic pollution. The effects of global warming, melted sea ice, and altered ocean currents might be the places where marine debris is found, and in its final destination (Haque and Fan 2023). The actions of cooperation and joint efforts are essential to solve the problem of the excessive plastics in the polar seas. Moreover, forging international agreements and treaties besides sharing common best practices and scientific knowledge should be considered. Collaboration can be fostered through joint work in partnership with the governments, scientific institutions, non-governmental organizations and business entities to develop legal frameworks and programs that will prevent the plastic pollution in the Polar Regions (Kumar et al. 2020).

Polar marine plastic management needs to be flexible in order to develop sustainable practices. All these strategies should be adaptable and keep pace with new scientific findings and the advanced technology (Rilov et al. 2018). The undertaking of more intensive monitoring, assessment and adjusting coincides with actions taken at a different pace that is dependent on the changing environment. The adaptive measures in conservation projects are applied at the long run for the conservation of the unique biodiversity as well as stability of polar marine ecosystems (Onyena et al. 2022; Ward et al. 2022).

5.6 Conclusion

Polar ocean plastic pollution studies reveal how the origin, drift, and ecological consequences of the pollutant usually are complicated. The plastic surfaces of polar regions that can interact with both micro- and macro-biodiversity are affected by the plastisphere. Ecological hazards are highly complex and their origins are multiple with considerations such as the introduction of alien species, chemical hazards, habitat modification, and the tangling of marine life. Mitigation strategies focus on the problem and utilize different approaches from global policy frameworks to local project programs and technological advancements. The effects are unavoidable, even more so when the need to address this issue is becoming clearer and calls for united and immediate actions. Conservation actions should be followed up by the reduction of plastics, promotion of eco-alternatives and high public awareness. Additionally, the profound interconnections of the plastisphere residents need to be investigated, their ecological impacts long-term assessed, and their connections to both climate change and plastic in the polar environment need to be explored. Identifying and implantation of new monitoring measures will break the bottlenecks and help to establish the most appropriate management plans for conservation achievements.

References

Alexiadou P, Foskolos I, Frantzis A (2019) Ingestion of macroplastics by odontocetes of the Greek Seas, Eastern Mediterranean: often deadly! Mar Pollut Bull 146:67–75. https://doi.org/10.1016/j.marpolbul.2019.05.055

Amobonye A, Bhagwat P, Raveendran S, Singh S, Pillai S (2021) Environmental impacts of microplastics and nanoplastics: a current overview. Front Microbiol 12:768297. https://doi.org/10.3389/fmicb.2021.768297

Angnunavuri PN, Attiogbe F, Mensah B (2020) Consideration of emerging environmental contaminants in Africa: review of occurrence, formation, fate, and toxicity of plastic particles. Sci Afr 9:e00546. https://doi.org/10.1016/j.sciaf.2020.e00546

Arabi S, Neehaul Y, Sparks C (2023) Impacts and threats of marine litter in African seas. In: Maes T, Preston-Whyte F (eds) The African marine litter outlook. Springer, Cham. https://doi.org/10.1007/978-3-031-08626-7_3

Aracri S, Giorgio-Serchi F, Suaria G, Sayed ME, Nemitz MP, Mahon S, Stokes AA (2021) Soft robots for ocean exploration and offshore operations: a perspective. Soft Robot 8(6):625–639. https://doi.org/10.1089/soro.2020.0011

Aransiola SA, Musa OI, Babaniyi RB, Maddela NR (2022a) Microbial biofilms: applications and control-editorial overview. In: Maddela NR, Aransiola SA (eds) Microbial biofilms: applications and control, 1st edn. CRC Press. https://www.taylorfrancis.com/chapters/edit/10.1201/9781003184942-2/microbial-biofilm-applications-control-aransiola-musa-babaniyi-naga-raju-maddela

Aransiola SA, Joseph F, Oyedele OJ, Maddela NR (2022b). Editorial—ecological Interplays in microbial enzymology. In Maddela NR, Aransiola SA, Prasad R (eds) Ecological interplays in microbial enzymology, 1st edn. Springer Nature Singapore Pte Ltd. ISSN: 2662-1681. https://www.springer.com/series/16324

Auta SH, Abioye OP, Aransiola SA, Bala JD, Chukwuemeka VI, Hassan A, Aziz A, Fauziah SH (2022) Enhanced microbial degradation of PET and PS microplastics under natural conditions in mangrove environment. J Environ Manag 304(2022):114273. https://doi.org/10.1016/j.jenvman.2021.114273

Barker LD, Jakuba MV, Bowen AD, German CR, Maksym T, Mayer L, Boetius A, Dutrieux P, Whitcomb LL (2019) Scientific challenges and present capabilities in underwater robotic vehicle design and navigation for oceanographic exploration under-ice. Remote Sens 12(16):2588. https://doi.org/10.3390/rs12162588

Barratclough A, Ferguson SH, Lydersen C, Thomas PO, Kovacs KM (2023) A review of circumpolar Arctic marine mammal health—a call to action in a time of rapid environmental change. Pathogens 12(7):937. https://doi.org/10.3390/pathogens12070937

Barros J, Seena S (2021) Plastisphere in freshwaters: an emerging concern. Environ Pollut 290:118123. https://doi.org/10.1016/j.envpol.2021.118123

Behera S, Das S (2023) Environmental impacts of microplastic and role of plastisphere microbes in the biodegradation and upcycling of microplastic. Chemosphere 334:138928. https://doi.org/10.1016/j.chemosphere.2023.138928

Benzik AN, Orlov AM, Novikov MA (2021) Marine seabed litter in Siberian Arctic: a first attempt to assess. Mar Pollut Bull 172:112836

Bergmann M, Lutz B, Tekman MB, Gutow L (2017) Citizen scientists reveal: marine litter pollutes Arctic beaches and affects wild life. Mar Pollut Bull 125:535–540

Brandt S, Wassmann P, Piepenburg D (2023) Revisiting the footprints of climate change in Arctic marine food webs: an assessment of knowledge gained since 2010. Front Mar Sci 10:1096222. https://doi.org/10.3389/fmars.2023.1096222

Camacho A, Rochera C, Picazo A (2022) Effect of experimentally increased nutrient availability on the structure, metabolic activities, and potential microbial functions of a maritime Antarctic microbial mat. Front Microbiol 13:900158. https://doi.org/10.3389/fmicb.2022.900158

Carlos J, Garrigós M, Garrigós J (2021) Plastic as a vector of dispersion for marine species with invasive potential. A review. Front Ecol Evol 9:629756. https://doi.org/10.3389/fevo.2021.629756

Castellani G, Veyssière G, Karcher M, Stroeve J, Banas SN, Bouman AH, Brierley SA, Connan S, Cottier F, Große F, Hobbs L, Katlein C, Light B, McKee D, Orkney A, Proud R, Schourup-Kristensen V (2022) Shine a light: under-ice light and its ecological implications in a changing Arctic Ocean. Ambio 51(2):307–317. https://doi.org/10.1007/s13280-021-01662-3

Cavicchioli R, Ripple WJ, Timmis KN, Azam F, Bakken LR, Baylis M, Behrenfeld MJ, Boetius A, Boyd PW, Classen AT, Crowther TW, Danovaro R, Foreman CM, Huisman J, Hutchins DA, Jansson JK, Karl DM, Koskella B, Mark Welch DB et al (2019) Scientists' warning to humanity: microorganisms and climate change. Nat Rev Microbiol 17(9):569–586. https://doi.org/10.1038/s41579-019-0222-5

Charlotte EL, Lauren FM, Sophie IH, Tony G, Richard SQ, Sabine M (2024) The primary molecular influences of marine plastisphere formation and function: novel insights into organism -organism and -co-pollutant interactions. Crit Rev Environ Sci Technol 54(2):138–161. https://doi.org/10.1080/10643389.2023.2224182

Cheng J, Jacquin J, Conan P, Barbe V, George M, Fabre P, Bruzaud S, Ter Halle A, Meistertzheim A, Ghiglione J (2021) Relative influence of plastic debris size and shape, chemical composition and phytoplankton-bacteria interactions in driving seawater plastisphere abundance, diversity and activity. Front Microbiol 11:610231. https://doi.org/10.3389/fmicb.2020.610231

Cheng X, Xiang X, Yun Y, Wang W, Wang H, Bodelier PL (2023) Archaea and their interactions with bacteria in a Karst ecosystem. Front Microbiol 14:1068595. https://doi.org/10.3389/fmicb.2023.1068595

Christian W, Estelle K, Katja M (2015) Protists in the polar regions: comparing occurrence in the Arctic and Southern oceans using pyrosequencing. Polar Res 34:1. https://doi.org/10.3402/polar.v34.23225

Clark JR, Corradi P, Aliani S, Arias M, Bochow M, Bonnery G, Cole M, Cózar A, Donnelly R, Echevarría F, Galgani F, Garaba SP, Lebreton L, Leslie HA, Lindeque PK, Maximenko N,

Moller D, Murphy P, Palombi L et al (2018) Measuring marine plastic debris from space: initial assessment of observation requirements. Remote Sens 11(20):2443. https://doi.org/10.3390/rs11202443

Clark MS, Hoffman JI, Peck LS, Bargelloni L, Gande D, Havermans C, Meyer B, Patarnello T, Phillips T, Stoof-Leichsenring KR, Vendrami DL, Beck A, Collins G, Friedrich MW, Halanych KM, Masello JF, Nagel R, Norén K, Printzen C et al (2023) Multi-omics for studying and understanding polar life. Nat Commun 14:7451. https://doi.org/10.1038/s41467-023-43209-y

Cózar A, Martí E, Duarte CM, García-de-Lomas J, van Sebille E, Ballatore TJ, Eguíluz VM, González-Gordillo JI, Pedrotti ML, Echevarría F, Troublè R, Irigoien X (2017) The Arctic Ocean as a dead end for floating plastics in the North Atlantic branch of the Thermohaline Circulation. Sci Adv 3(4):e1600582. https://doi.org/10.1126/sciadv.1600582

Cunningham EM, Rico Seijo N, Altieri KE, Audh RR, Burger JM, Bornman TG, Fawcett S, Gwinnett CM, Osborne AO, Woodall LC (2022) The transport and fate of microplastic fibres in the Antarctic: the role of multiple global processes. Front Mar Sci 9:1056081. https://doi.org/10.3389/fmars.2022.1056081

Daghighi E, Shah T, Chia R, Lee J, Shang J, Rodríguez-Seijo A (2023) The forgotten impacts of plastic contamination on terrestrial micro- and mesofauna: a call for research. Environ Res 231:116227. https://doi.org/10.1016/j.envres.2023.116227

Dalu MT, Cuthbert RN, Muhali H, Chari LD, Manyani A, Masunungure C, Dalu T (2019) Is awareness on plastic pollution being raised in schools? Understanding perceptions of primary and secondary school educators. Sustainability 12(17):6775. https://doi.org/10.3390/su12176775

Dey S, Rout AK, Behera BK (2022) Plastisphere community assemblage of aquatic environment: plastic-microbe interaction, role in degradation and characterization technologies. Environ Microbiome 17:32. https://doi.org/10.1186/s40793-022-00430-4

Du Y, Liu X, Dong X, Yin Z (2021) A review on marine plastisphere: biodiversity, formation, and role in degradation. Comput Struct Biotechnol J 20:975–988. https://doi.org/10.1016/j.csbj.2022.02.008

Dutta A, Sar P, Sarkar J, Dutta Gupta S, Gupta A, Bose H, Mukherjee A, Roy S (2019) Archaeal communities in deep terrestrial subsurface underneath the Deccan Traps, India. Front Microbiol 10:416767. https://doi.org/10.3389/fmicb.2019.01362

Ershova A, Makeeva I, Malgina E, Sobolev N, Smolokurov A (2021) Combining citizen and conventional science for microplastics monitoring in the White Sea basin (Russian Arctic). Mar Pollut Bull 173:112955

Ford HV, Jones NH, Davies AJ, Godley BJ, Jambeck JR, Napper IE, Suckling CC, Williams GJ, Woodall LC, Koldewey HJ (2022) The fundamental links between climate change and marine plastic pollution. Sci Total Environ 806:150392. https://doi.org/10.1016/j.scitotenv.2021.150392

Gavrilo M (2019) Plastic pollution and seabirds in the Russian Arctic. Conservation of Arctic Flora and Fauna (CAFF)

Geremia E, Muscari Tomajoli MT, Murano C, Petito A, Fasciolo G (2023) The impact of micro- and nanoplastics on aquatic organisms: mechanisms of oxidative stress and implications for human health—a review. Environment 10(9):161. https://doi.org/10.3390/environments10090161

Gregory MR (2009) Environmental implications of plastic debris in marine settings—entanglement, ingestion, smothering, hangers-on, hitch-hiking and alien invasions. Philos Trans R Soc B Biol Sci 364(1526):2013–2025. https://doi.org/10.1098/rstb.2008.0265

Gregory SP, Barnett MJ, Field LP, Milodowski AE (2019) Subsurface microbial hydrogen cycling: natural occurrence and implications for industry. Microorganisms 7:53. https://doi.org/10.3390/microorganisms7020053

Grose SO, Pendleton L, Leathers A, Cornish A, Waitai S (2020) Climate change will re-draw the map for marine megafauna and the people who depend on them. Front Mar Sci 7:530924. https://doi.org/10.3389/fmars.2020.00547

Guffogg JA, Blades SM, Bellman CJ, Skidmore AK, Jones SD (2020) Quantifying marine plastic debris in a beach environment using spectral analysis. Remote Sens 13(22):4548. https://doi.org/10.3390/rs13224548
Haque F, Fan C (2023) Fate of microplastics under the influence of climate change. IScience 26(9):107649. https://doi.org/10.1016/j.isci.2023.107649
Ibrahim ID, Sadiku ER, Hamam Y, Kupolati WK, Ndambuki JM, Jamiru T, Eze AA, Snyman J (2023) Recent recycling innovations to facilitate sustainable packaging materials: a review. Recycling 8(6):88. https://doi.org/10.3390/recycling8060088
Ino K, Hernsdorf AW, Konno U, Kouduka M, Yanagawa K, Kato S (2018) Ecological and genomic profiling of anaerobic methane-oxidizing archaea in a deep granitic environment. ISME J 12:31–47. https://doi.org/10.1038/ismej.2017.140
Jaskólski MW, Pawłowski Ł, Strzelecki MC, Zagórski P, Lane TP (2018) Trash on Arctic beach: coastal pollution along Calypsostranda, Bellsund, Svalbard. Pol Polar Res 39:211–224
Jørgensen-Dahl A, Østreng W (1991) The Antarctic Treaty system in world politics. Palgrave Macmillan UK eBooks. https://doi.org/10.1007/978-1-349-12471-8
Karakuş O (2023) On advances, challenges and potentials of remote sensing image analysis in marine debris and suspected plastics monitoring. Front Remote Sens 4:1302384. https://doi.org/10.3389/frsen.2023.1302384
Kehinde O, Ramonu OJ, Babaremu KO, Justin LD (2020) Plastic wastes: environmental hazard and instrument for wealth creation in Nigeria. Heliyon 6(10):e05131. https://doi.org/10.1016/j.heliyon.2020.e05131
Kibria MG, Masuk NI, Safayet R, Nguyen HQ, Mourshed M (2023) Plastic waste: challenges and opportunities to mitigate pollution and effective management. Int J Environ Res 17(1):20. https://doi.org/10.1007/s41742-023-00507-z
Kiessling T, Gutow L, Thiel M (2015) Marine litter as habitat and dispersal vector. In: Bergmann M, Gutow L, Klages M (eds) Marine anthropogenic litter. Springer, Cham. https://doi.org/10.1007/978-3-319-16510-3_6
Kim SK, Lee HJ, Kim JS, Kang SH, Yang EJ, Cho KH, Tian Z, Andrady A (2021) Importance of seasonal sea ice in the western Arctic ocean to the Arctic and global microplastic budgets. J Hazard Mater 418:125971. https://doi.org/10.1016/j.jhazmat.2021.125971
Kirstein IV, Wichels A, Gullans E, Krohne G, Gerdts G (2019) The plastisphere—uncovering tightly attached plastic "specific" microorganisms. PLoS One 14(4):e0215859. https://doi.org/10.1371/journal.pone.0215859
Kumar R, Verma A, Shome A, Sinha R, Sinha S, Jha PK, Kumar R, Kumar P, Das S, Sharma P, Vara Prasad PV (2020) Impacts of plastic pollution on ecosystem services, sustainable development goals, and need to focus on circular economy and policy interventions. Sustain For 13(17):9963. https://doi.org/10.3390/su13179963
Landrigan PJ, Stegeman JJ, Fleming LE, Allemand D, Anderson DM, Backer LC, Brucker-Davis F, Chevalier N, Corra L, Czerucka D, Bottein YD, Demeneix B, Depledge M, Deheyn DD, Dorman CJ, Fénichel P, Fisher S, Gaill F, Galgani F et al (2020) Human health and ocean pollution. Ann Glob Health 86(1):151. https://doi.org/10.5334/aogh.2831
Lima ARA, Ferreira GVB, Barrows APW, Christiansen KS, Treinish G, Toshack MC (2020) Global patterns for the spatial distribution of floating microfibers: Arctic Ocean as a potential accumulation zone. J Hazard Mater 403:123796. https://doi.org/10.1016/j.jhazmat.2020.123796
Lincoln S, Andrews B, Birchenough SN, Chowdhury P, Engelhard GH, Harrod O, Pinnegar JK, Townhill BL (2022) Marine litter and climate change: inextricably connected threats to the world's oceans. Sci Total Environ 837:155709. https://doi.org/10.1016/j.scitotenv.2022.155709
Liu J, Hu Z, Du F, Tang W, Zheng S, Lu S, An L, Ding J (2023) Environment education: a first step in solving plastic pollution. Front Environ Sci 11:1130463. https://doi.org/10.3389/fenvs.2023.1130463
MacLean J, Mayanna S, Benning LG, Horn F, Bartholomäus A, Wiesner Y, Wagner D, Liebner S (2021) The terrestrial plastisphere: diversity and polymer-colonizing potential of

plastic-associated microbial communities in soil. Microorganisms 9(9):1876. https://doi.org/10.3390/microorganisms9091876
Mallory ML, Baak J, Gjerdrum C, Mallory OE, Manley B, Swan C, Provencher JF (2021) Anthropogenic litter in marine waters and coastlines of Arctic Canada and West Greenland. Sci Total Environ 783:146971. https://doi.org/10.1016/j.scitotenv.2021.146971
Marcus H (2021) Biodiversity in Areas Beyond National Jurisdiction (BBNJ): the Commission for the Conservation of Antarctic Marine Living Resources (CCAMLR) and the United Nations BBNJ agreement. Polar J 11(2):303–316. https://doi.org/10.1080/2154896X.2021.1984658
Meenal M, Patel K, Patil A (2020) Cleaning of water bodies using coastal sea bin (CSB). Methods X 8:101469. https://doi.org/10.1016/j.mex.2021.101469
Miao L, Li W, Adyel TM, Yao Y, Deng Y, Wu J, Zhou Y, Yu Y, Hou J (2023) Spatio-temporal succession of microbial communities in plastisphere and their potentials for plastic degradation in freshwater ecosystems. Water Res 229:119406. https://doi.org/10.1016/j.watres.2022.119406
Miller ME, Hamann M, Kroon FJ (2020) Bioaccumulation and biomagnification of microplastics in marine organisms: a review and meta-analysis of current data. PLoS One 15(10):e0240792. https://doi.org/10.1371/journal.pone.0240792
Mu J, Zhang S, Qu L, Jin F, Fang C, Ma X, Zhang W, Wang J (2019) Microplastics abundance and characteristics in surface waters from the Northwest Pacific, the Bering Sea, and the Chukchi Sea. Mar Pollut Bull 143:58–65. https://doi.org/10.1016/j.marpolbul.2019.04.023
Musa IO, Auta HS, Ilyasu US, Aransiola SA, Makun HA, Adabara NU, Abioye OP, Aziz A, Jayanthi B, Maddela NR, Prasad R (2024) Micro and nanoplastics in environment—degradation, detection, ecological impact. Int J Environ Res 18:1. https://doi.org/10.1007/s41742-023-00551-9
Nashoug BF (2017) Sources of marine litter—workshop report, Svalbard 4th–6th September 2016 (SALT, 2017)
Oberbeckmann S, Kreikemeyer B, Labrenz M (2018) Environmental factors support the formation of specific bacterial assemblages on microplastics. Front Microbiol 8:2709. https://doi.org/10.3389/fmicb.2017.02709
Offre P, Spang A, Schleper C (2013) Archaea in biogeochemical cycles. Ann Rev Microbiol 67:437–457. https://doi.org/10.1146/annurev-micro-092412-155614
Onyena AP, Aniche DC, Ogbolu BO, Rakib MR, Uddin J, Walker TR (2022) Governance strategies for mitigating microplastic pollution in the marine environment: a review. Microplastics 1(1):15–46. https://doi.org/10.3390/microplastics1010003
OSPAR Commission (2009) Marine litter in the North-East Atlantic Region: assessment and priorities for response. OSPAR Commission
Peeken I, Primpke S, Beyer B, Gütermann J, Katlein C, Krumpen T, Bergmann M, Hehemann L, Gerdts G (2018) Arctic sea ice is an important temporal sink and means of transport for microplastic. Nat Commun 9(1):1505. https://doi.org/10.1038/s41467-018-03825-5
Pierdomenico M, Bernhardt A, Eggenhuisen JT, Clare MA, Lo Iacono C, Casalbore D, Davies JS, Kane I, Huvenne VA, Harris PT (2023) Transport and accumulation of litter in submarine canyons: a geoscience perspective. Front Mar Sci 10:1224859. https://doi.org/10.3389/fmars.2023.1224859
Polasek L, Bering J, Kim H, Neitlich P, Pister B, Terwilliger M, Nicolato K, Turner C, Jones T (2017) Marine debris in five national parks in Alaska. Mar Pollut Bull 117(1–2):371–379. https://doi.org/10.1016/j.marpolbul.2017.01.085
Pooja N, Chakraborty I, Rahman MH (2023) An insight on sources and biodegradation of bioplastics: a review. 3 Biotech 13:220. https://doi.org/10.1007/s13205-023-03638-4
Rhodes CJ (2018) Plastic pollution and potential solutions. Sci Prog 101(3):207–260. https://doi.org/10.3184/003685018X15294876706211
Rilov G, Mazaris AD, Stelzenmüller V, Helmuth B, Wahl M, Guy-Haim T, Mieszkowska N, Ledoux J, Katsanevakis S (2018) Adaptive marine conservation planning in the face of climate change: what can we learn from physiological, ecological and genetic studies? Glob Ecol Conserv 17:e00566. https://doi.org/10.1016/j.gecco.2019.e00566

Rowlands E, Galloway T, Manno C (2021) A Polar outlook: potential interactions of micro- and nano-plastic with other anthropogenic stressors. Sci Total Environ 754:142379. https://doi.org/10.1016/j.scitotenv.2020.142379

Rüthi J, Bölsterli D, Brunner I, Frey B (2020) The "plastisphere" of biodegradable plastics is characterized by specific microbial taxa of Alpine and Arctic soils. Front Environ Sci 8:562263. https://doi.org/10.3389/fenvs.2020.562263

Sambolino A, Herrera I, Álvarez S, Rosa A, Alves F, Canning-Clode J, Cordeiro N, Dinis A, Kaufmann M (2022) Seasonal variation in microplastics and zooplankton abundances and characteristics: the ecological vulnerability of an oceanic island system. Mar Pollut Bull 181:113906. https://doi.org/10.1016/j.marpolbul.2022.113906

Samir A, Ashour FH, Hakim AA, Bassyouni M (2022) Recent advances in biodegradable polymers for sustainable applications. Npj Mater Degrad 6(1):1–28. https://doi.org/10.1038/s41529-022-00277-7

Schmaltz E, Melvin EC, Diana Z, Gunady EF, Rittschof D, Somarelli JA, Virdin J, Dunphy-Daly MM (2020) Plastic pollution solutions: emerging technologies to prevent and collect marine plastic pollution. Environ Int 144:106067. https://doi.org/10.1016/j.envint.2020.106067

Searcy CA, Howell HJ, David AS, Rumelt RB, Clements SL (2023) Patterns of non-native species introduction, spread, and ecological impact in South Florida, the world's most invaded continental ecoregion. Annu Rev Ecol Evol Syst 54(1):103104. https://doi.org/10.1146/annurev-ecolsys-110421-103104

Shomura RS, Godfrey ML (1990) Proceedings of the second international conference on marine debris, 2–7 April 1989, Honolulu, Hawaii, vol 1. NOAA Technical Memorandum, NMFS-SWFSC(154). US Department of Commerce, NOAA, Panama City. XIII, 774 pp

Skjærseth JB (2006) Protecting the North-East Atlantic: enhancing synergies by institutional interplay. Mar Policy 30(2):157–166. https://doi.org/10.1016/j.marpol.2004.09.004

Sladonja B, Poljuha D, Uzelac M (2018) Non-native invasive species as ecosystem service providers. InTech. https://doi.org/10.5772/intechopen.75057

Smith AR, Kieft B, Mueller R, Fisk MR, Mason OU, Popa R (2019) Carbon fixation and energy metabolisms of a subseafloor olivine biofilm. ISME J 13(7):1737–1749. https://doi.org/10.1038/s41396-019-0385-0

Stevens TO, McKinley JP (2000) Abiotic controls on H_2 production from Basalt—water reactions and implications for aquifer biogeochemistry. Environ Sci Technol 34:826–831. https://doi.org/10.1021/es990583g

Su L, Xiong X, Zhang Y, Wu C, Xu X, Sun C, Shi H (2022) Global transportation of plastics and microplastics: a critical review of pathways and influences. Sci Total Environ 831:154884. https://doi.org/10.1016/j.scitotenv.2022.154884

Sun Y, Wu M, Zang J, Du L, Huang M, Chen C, Wang J (2023) Plastisphere microbiome: methodology, diversity, and functionality. iMeta 2(2):e101. https://doi.org/10.1002/imt2.101

Thiel M, Gallardo C, Hinojosa IA, Luna N, Morales N, Ory N, Pacheco AS, Zavalaga C (2018) Impacts of marine plastic pollution from continental coasts to subtropical gyres—fish, seabirds, and other vertebrates in the SE Pacific. Front Mar Sci 5:359654. https://doi.org/10.3389/fmars.2018.00238

Thushari G, Senevirathna J (2020) Plastic pollution in the marine environment. Heliyon 6(8):e04709. https://doi.org/10.1016/j.heliyon.2020.e04709

van Sebille E, Wilcox C, Lebreton L, Maximenko N, Hardesty BD, van Franeker JA, Eriksen M, Siegel D, Galgani F, Law KL (2015) A global inventory of small floating plastic debris. Environ Res Lett 10:124006

van Sebille E, Aliani S, Law KL, Maximenko N, Alsina JM, Bagaev A, Bergmann M, Chapron B, Chubarenko I, Cózar A, Delandmeter P et al (2020) The physical oceanography of the transport of floating marine debris. Environ Res Lett 15:023003

Varriale V, Cammarano A, Michelino F (2023) Critical analysis of the impact of artificial intelligence integration with cutting-edge technologies for production systems. J Intell Manuf 36:61–93. https://doi.org/10.1007/s10845-023-02244-8

Villarrubia-Gómez P, Cornell SE, Fabres J (2018) Marine plastic pollution as a planetary boundary threat—the drifting piece in the sustainability puzzle. Mar Policy 96:213–220. https://doi.org/10.1016/j.marpol.2017.11.035

Vyverman W, Verleyen E, Wilmotte A, Hodgson DA, Willems A, Peeters K, Van de Vijver B, De Wever A, Leliaert F, Sabbe K (2010) Evidence for widespread endemism among Antarctic micro-organisms. Pol Sci 4(2):103–113. https://doi.org/10.1016/j.polar.2010.03.006

Waldschläger K, Brückner MZ, Carney Almroth B, Hackney CR, Adyel TM, Alimi OS, Belontz SL, Cowger W, Doyle D, Gray A, Kane I, Kooi M, Kramer M, Lechthaler S, Michie L, Nordam T, Pohl F, Russell C, Thit A et al (2022) Learning from natural sediments to tackle microplastics challenges: a multidisciplinary perspective. Earth Sci Rev 228:104021. https://doi.org/10.1016/j.earscirev.2022.104021

Wallbank JA, Lear G, Kingsbury JM, Weaver L, Doake F, Smith DA, Audrézet F, Maday SD, Gambarini V, Donaldson L, Theobald B, Barbier M, Pantos O (2022) Into the plastisphere, where only the generalists thrive: early insights in plastisphere microbial community succession. Front Mar Sci 9:841142. https://doi.org/10.3389/fmars.2022.841142

Waller CL, Griffiths HJ, Waluda CM, Thorpe SE, Loaiza I, Moreno B, Pacherres CO, Hughes KA (2017) Microplastics in the Antarctic marine system: an emerging area of research. Sci Total Environ 598:220–227. https://doi.org/10.1016/j.scitotenv.2017.03.283

Ward D, Melbourne-Thomas J, Pecl GT (2022) Safeguarding marine life: conservation of biodiversity and ecosystems. Rev Fish Biol Fish 32:65–100. https://doi.org/10.1007/s11160-022-09700-3

Weiskopf SR, Rubenstein MA, Crozier LG, Gaichas S, Griffis R, Halofsky JE, Hyde KJ, Morelli TL, Morisette JT, Muñoz RC, Pershing AJ, Peterson DL, Poudel R, Staudinger MD, Sutton-Grier AE, Thompson L, Vose J, Weltzin JF, Whyte KP (2020) Climate change effects on biodiversity, ecosystems, ecosystem services, and natural resource management in the United States. Sci Total Environ 733:137782. https://doi.org/10.1016/j.scitotenv.2020.137782

Węsławski JM, Kotwicki L (2018) Macro-plastic litter, a new vector for boreal species dispersal on Svalbard. Pol Polar Res 39:165–174

Wilcox C, Van Sebille E, Hardesty BD (2015) Threat of plastic pollution to seabirds is global, pervasive, and increasing. Proc Natl Acad Sci 112(38):11899–11904. https://doi.org/10.1073/pnas.1502108112

Worm B, Lotze HK, Jubinville I, Wilcox C, Jambeck J (2017) Plastic as a persistent marine pollutant. Annu Rev Environ Resour 42(1):1–26. https://doi.org/10.1146/annurev-environ-102016-060700

Xie Z, Zhang P, Wu Z, Zhang S, Wei L, Mi L, Kuester A, Gandrass J, Ebinghaus R, Yang R, Wang Z, Mi W (2022) Legacy and emerging organic contaminants in the polar regions. Sci Total Environ 835:155376. https://doi.org/10.1016/j.scitotenv.2022.155376

Yakushev E, Gebruk A, Osadchiev A, Pakhomova S, Lusher A, Berezina A, Van Bavel B, Vorozheikina E, Chernykh D, Kolbasova G, Razgon I, Semiletov I (2021) Microplastics distribution in the Eurasian Arctic is affected by Atlantic waters and Siberian rivers. Commun Earth Environ 2(1):1–10. https://doi.org/10.1038/s43247-021-00091-0

Yu R, Singh S (2022) Microplastic pollution: threats and impacts on global marine ecosystems. Sustain For 15(17):13252. https://doi.org/10.3390/su151713252

Yu Y, Miao L, Adyel TM, Waldschläger K, Wu J, Hou J (2023) Aquatic plastisphere: interactions between plastics and biofilms. Environ Pollut 1(322):121196. https://doi.org/10.1016/j.envpol.2023.121196. Epub 2023 Jan 31. PMID: 36736560

Zeppilli D, Leduc D, Fontanier C (2018) Characteristics of meiofauna in extreme marine ecosystems: a review. Mar Biodivers 48:35–71. https://doi.org/10.1007/s12526-017-0815-z

Zettler ER, Mincer TJ, Amaral-Zettler LA (2013) Life in the 'Plastisphere': microbial communities on plastic marine debris. Environ Sci Technol 47(13):7137–7146. Epub 2013/06/12. https://doi.org/10.1021/es401288x

Zhai X, Zhang X, Yu M (2023) Microbial colonization and degradation of marine microplastics in the plastisphere: a review. Front Microbiol 14:1127308. https://doi.org/10.3389/fmicb.2023.1127308
Zhi Xiang JK, Bairoliya S, Cho ZT, Cao B (2022) Plastic-microbe interaction in the marine environment: research methods and opportunities. Environ Int 171:107716. https://doi.org/10.1016/j.envint.2022.107716
Zhou X, Chen X, Qi X, Zeng Y, Guo X, Zhuang G, Ma A (2023) Soil bacterial communities associated with multi-nutrient cycling under long-term warming in the alpine meadow. Front Microbiol 14:1136187. https://doi.org/10.3389/fmicb.2023.1136187

Part II
Plastisphere in Soil Plant Uptake and Food Crisis

Chapter 6
Microbial Ecology of the Plastisphere in Soil Plant Uptake

Gloria Ginikanwa Ezeagu, Morenike Fadayomi, Sani Sambo Datsugwai Mohammed, Augustine Innalegwu Daniel, Ashwil Klein, and Namla Djadjiti

Abstract Within the last few years, there has been growing research and increasing knowledge of the plastisphere. This is due to its impact on soil organisms, plants and by extension, humans. In this chapter, the introduction of microplastics into the soil, as well the uptake of the microplastics by plants, the effects of the plastisphere on soil structure and plants, and factors that influence the microbial ecology of the plastisphere were studied. Plants take up microplastics from plastic waste through their roots by transpiration and attach to leaves surfaces by some chemical bonds. Microplastics in the soil generally impact plant growth and productivity as they inhibit the normal metabolism activities of plants. Soil microplastic contamination affects the soil's physical, chemical, and biological properties. Once in the soil, microplastics persist and aggregate, whereupon they gradually disturb the biodiversity and functions of the soil ecosystem. The properties of microplastics and environmental conditions play a role in the formation and succession of a soil plastisphere community. Different microplastic types can also influence changes in the microbial community of the plastisphere. As microplastics age and degrade, they release toxic and harmful additives into the soil and these substances inhibit the community abundance of sensitive microbial groups, thus altering the composition of soil

G. G. Ezeagu (✉) · M. Fadayomi · S. S. D. Mohammed · A. Klein
Department of Biology, Microbiology & Biotechnology, Faculty of Science, Nile University of Nigeria, Abuja, Nigeria
e-mail: gloria.ezeagu@nileuniversity.edu.ng

A. I. Daniel
Department of Biochemistry, School of Life Sciences, Federal University of Technology, Minna, Niger State, Nigeria

Plant Omics Laboratory, Department of Biotechnology, Faculty of Natural Sciences, University of the Western Cape, Bellville, South Africa

N. Djadjiti
Department of Biological Sciences, Faculty of Science, King Abdulaziz University, Jeddah, Saudi Arabia

S. A. Aransiola et al. (eds.), *White Pollution: Biodiversity and hazards in marine plastisphere*, Environmental Contamination Remediation and Management,
https://doi.org/10.1007/978-3-031-95547-1_6

microorganisms. The findings in this study indicate that the plastisphere constitutes both environmental and potential health issues.

Keywords Plastisphere · Microplastic · Soil · Microorganisms · Plants

6.1 Introduction

Petroleum-based polymers have accumulated in the natural environment as a result of the widespread manufacture and use of plastic items. Due to these plastics' extreme resistance, the ecosystem and mankind are now seriously threatened by their environmental deposition (Behera and Das 2023). Plastic manufacturing is expected to exceed 33 billion tonnes by 2050 from an estimated 359 million metric tonnes in 2018 (Geyer 2020; Auta et al. 2022). However, only approximately 9% of plastic garbage can be recycled, and about 79% ends up in landfills or the environment, where it weathers and gets broken down by physical, chemical, and biological processes to create microplastics over time (Li et al. 2021a, b). Microplastics are plastic particles that are generally ≤5 mm in size (Behera and Das 2023; Musa et al. 2024).

Microplastic surfaces, being exogenous and hydrophobic, can offer a distinctive niche for the growth and development of a variety of microorganisms, forming a unique ecological habitat known as the plastisphere (Su and Yang 2024; Sun et al. 2023). Plastic differs significantly from the great majority of chemical environmental contaminants in that microplastics, as environmental pollutants, are comprised of both particles and chemical components (Rillig et al. 2024).

The main causes of microplastic pollution in the environment include improper disposal techniques, the use of plastics in the home, and industrial effluents (Kasmuri et al. 2022). Certain microplastics, including abrasive particles found in personal hygiene items, are released into the environment directly after being manufactured (Li et al. 2021a, b).

Similar or different microbial functions may be observed within the plastisphere, and its microbial diversity may be higher or lower than that of the nearby communities. Studies on the plastisphere began with the marine environment and progressively expanded to include freshwater, soil, and even indoor home environments (Sun et al. 2023). Microplastic pollution has far-reaching and wide-ranging negative effects on the ecosystem, human health, the economy, tourism, and the aesthetics of beaches (Patchaiyappan et al. 2021; Aransiola et al. 2023). There have also been reports that harmful compounds like heavy metals that can be found in the environment can be absorbed by microplastic particles. Microplastics can also concentrate different environmental hydrophobic pollutants (Wang et al. 2019). Through the uptake of nanoparticles, plastic pollution can lead to developmental, reproductive, and metabolic disorders in invertebrates, and potentially compromise food safety. It can also facilitate the spread of pathogens and antibiotic resistance by acting as a vehicle for microorganisms (Rüthi et al. 2020).

This chapter assesses the plastisphere, concerning the introduction of microplastics into the soil, formation and function of the soil plastisphere community, plant uptake of microplastics and its effects on the plant health and soil structure.

6.2 Soil Microbiome and Structure

6.2.1 Functions of Soil Microbiome in Agricultural Systems

6.2.1.1 The Soil Ecosystem

The soil is an ecosystem that is home to a wide variety of microorganisms that perform a variety of tasks, including water filtering, soil structure creation, biological pest management, and the cycling of nutrients like phosphorus, nitrogen, and sulphur (König et al. 2020). These activities are essential to several ecological processes in the soil, which ultimately aid in the formation of the plant microbiome.

One g of surface soil is estimated to contain up to one billion bacterial cells, 200 million fungal hyphae, tens of thousands of protists, and trillions of viruses, all of which can have both positive and negative effects on the soil (Xiong and Lu 2022).

6.2.1.2 Microorganisms in the Soil

Although the soil is primarily thought of as a source of nutrients for plants, it also acts as a complex ecosystem that is home to a variety of microorganisms, including protists, bacteria, fungi, and even animals (Xiong and Lu 2022). With soil microorganisms, plants have mutualistic and competitive ecological relationships that can be both positive and negative (Jacoby et al. 2017). Numerous well-established studies have been conducted on the interactions between microorganisms and plants. For example, in the nineteenth century, mycorrhizal fungi and bacteria were identified as root symbionts in the legume root nodules (Jacoby et al. 2017). Crop seeds were coated with *Azotobacter chroococcum* bacterial cultures in the 1950s (Brown 1974), and by the 1980s, it was believed that *Pseudomonas* and *Azospirillium* bacterial strains could stimulate plant growth (Lin et al. 1983).

6.2.1.3 Micro-habitats Around Plants and Soil Microbial Communities

Plants offer different microhabitats for the development and growth of microorganisms. Small areas of a habitat with special characteristics that set them apart from the surrounding habitat are called micro-habitats. Because of their special characteristics, micro-habitats are home to biodiversity. The rhizosphere, phyllosphere, bulk and root zone soil, and root endosphere are examples of micro-habitats around plants. These micro-habitats have distinct microbial communities, which act as key determinants in the composition of the plant-associated microbiota (Trivedi et al. 2020). Environmental variables and the genotype of the host plant are other elements that determine the microbial communities in plants at the genus and species level.

The major bacterial groups' abundance profiles are comparable in the bulk soil and rhizosphere; however, the rhizosphere has a slightly larger concentration of bacteria that belong to the phylum Proteobacteria. The microbial communities of the root endosphere are mostly composed of Firmicutes and Proteobacteria members; Bacteroidetes members are less abundant, while members of Planctomycetes, Chloroflexi, Acidobacteria, and Verrucomicrobia are depleted in number. Compared to the rhizosphere, the endosphere has less of these organisms (Fonseca-García et al. 2016; Hamonts et al. 2018). The phyllosphere mainly has bacteria of the Phylum Proteobacteria. Proteobacteria are the primary type of bacteria found in the phyllosphere. The majority of fungi that inhabit the tissues of plants, both above and below ground, belong to the phyla Ascomycota and Basidiomycota (Trivedi et al. 2020).

Compared to bulk soil, the rhizosphere, phyllosphere, and endosphere generally have substantially less microbial diversity (Xiong and Lu 2022). Recent research on the bacterial and fungal communities in the soil-plant microbiome of field-grown wheat and barley as well as maize and other crops demonstrated, for example, that the diversity of microorganisms steadily declined from soil to epiphytes and endophytes. Another example demonstrated how the bacterial diversity of rhizosphere samples from maize and rice crops decreased as they moved from bulk soil to the rhizosphere and endosphere (Garcia-Salamanca et al. 2013; Edwards et al. 2015).

Other organisms that contribute to the plant-associated microbial community include viruses, nematodes, protists, and archaea (which belong to the phyla Thaumarchaeota, Crenarchaeota, and Euryarchaeota), in addition to the fungal and bacterial soil communities that are found on plants (Trivedi et al. 2020).

The archaeal community is dynamic and capable of degrading glycogen, a compound that is commonly stored and produced by fungi (Taffner et al. 2019). A recent study found that the rhizosphere and bulk soil of sorghum had higher protistan alpha diversity than the phyllosphere (Sun et al. 2021). Protists also influence the fungal and bacterial community by controlling the predator-prey relationships of varying feeding levels (Gao et al. 2019). Oomycete species, including *Phytophthora*, *Pythium*, and *Peronospora* species, have been shown to be pathogenic to plants (Larousse and Galiana 2017). Additionally, studies have indicated that naturally occurring bacteriophage communities in the soil have an impact on the phyllosphere communities (Morella et al. 2018).

6.3 Benefits of Soil and Plant-Associated Microbiomes in Agricultural Systems

Overall, in agricultural systems, soil and plant-associated microbiomes provide several functions and benefits to plant growth and health. Some of these functions include:

(a) **Nutrient cycling and improving plant nutrition:** Carbon, nitrogen, and phosphorus cycling is aided by several microorganisms present in bulk soil and the rhizosphere, which contribute to improved nutrient cycling. According to reports, through microbial phosphorus solubilization, microbial communities in soils play a major role in controlling the cycling of phosphorus (Wu et al. 2022). Additionally, a recent study demonstrated that distinct functional genes associated with the carbon, nitrogen, and phosphorus cycles are present in the phylloplane microbiome of the maize phyllosphere (Xiong et al. 2021a, b, c).

Additionally, nutrients that are not easily obtainable by plants, such as iron and inorganic phosphate, can be assembled by plant-associated microbiota. Siderophores can be used for mineralization, solubilization, or excretion to accomplish this process. Plants rely on their cooperative relationships with symbiotic endophytes and mycorrhizal fungi to obtain phosphate when the nutrient is scarce in the environment (Bodenhausen et al. 2019).

(b) **Disease resistance:** Plant root exudates have been shown in studies to be the first line of defense against pathogenic organisms in the soil, stimulating and supporting soil microbes (Mendes et al. 2011). The production of volatiles by bacteria and the synthesis of antifungal metabolites are two strategies used by soil organisms to combat pathogens (Trivedi et al. 2020). The endophyte contains organisms that can offer an additional layer of protection in the event that the rhizospheric organisms' initial line of defence against disease is compromised. Through the selective enrichment of its microbiome with a genetic mechanism for creating enzymes and secondary metabolites that are antagonistic to the pathogens, endophytic-mediated disease resistance is achieved (Carrion et al. 2019). An example of this is a study that showed how a tomato variety's disease-resistant characteristics were attributable to the careful selection of bacteria that can act against *Ralstonia solanacearum*, the causative agent of bacterial wilt of tomato (Kwak et al. 2018).

Protists are another significant component of the plant microbiome that may offer extra disease resistance in addition to bacteria, according to available data. According to a study by Guo et al. (2022a, b) protists can improve banana health by promoting bacteria like *Bacillus* sp. that suppress disease. Plant fungal pathogens, such as *Rhizoctonia solani*, *Colletotrichum orbiculare*, and *Fusarium oxysporum*, have been subjected to biological control by algal species like *Anabaena* sp., *Calothrix* sp., and *Chlorella* sp. (Lee and Ryu 2021).

(c) **Stress tolerance:** In the face of either biotic or abiotic stress conditions, plants typically choose a microbiome that supports stress tolerance (Trivedi et al.

2020). Research indicates that a plant's microbiome can strengthen its defences against abiotic stresses like cold, salt, and drought (Xiong and Lu 2022). Since drought is a significant environmental stressor that can reduce agricultural systems' production, much study has been done to find means of using plant microbiomes to make plants more drought-resistant. A recent study showed that root microbial iron metabolism is important for sorghum's response to drought (Xu et al. 2021) and a study by Schmitz et al. (2022) revealed that *Ensifer* sp., *Ralstonia* sp., and *Bacillus* sp., isolated from the root of the plant *Indigofera argentea*, showed the potential to increase the resistance of tomatoes to high salt stress.

6.4 Microorganisms and Their Influence on Soil Structure

6.4.1 Soil Structure and Aggregation

The arrangement of distinct particles, including sand, silt, clay, and organic materials, as well as the pore spaces that exist between them, make up the soil structure. The agglomeration of mineral and organic particles into bigger units is a crucial aspect of soil structure. Good soil aggregate stability frequently suggests improved soil structure, which is crucial for soil fertility, production, and sustainability. Soil aggregate stability is frequently employed as an indicator of soil structure (Zhou et al. 2020a, b).

Soil aggregation is viewed as a three-stage hierarchical structure of the solid phase, with each stage having defined binding agents. There are primary particles, micro-aggregates, and macro-aggregates. The primary particles (<20 μm) are bound together into microaggregates (20–250 μm) which are bound together to form macroaggregates (>250 μm). This system of hierarchy for macro-and micro-aggregates was identified in soils with organic matter as the main binding agent (Rabot et al. 2018).

6.4.2 Role of Soil Structure in Agricultural Systems

Numerous soil processes are governed by soil structure. It facilitates gas exchange, plant root penetration, and the development of organic matter in the soil. Additionally, it controls water absorption and retention and contributes to the soil's ability to withstand erosion. Also, an important factor in determining the availability of nutrients for plant uptake by roots is soil structure (Bodner et al. 2021). A well-aggregated soil structure is the primary regulator of soil tilth, water relations, plant root penetration, and erosion resistance in agricultural systems. Stable soil aggregates provide organic debris with protection, which contributes to the buildup of organic matter in the soil (Miller and Jastrow 2000).

6.4.3 *Soil Microorganisms and Their Functions in the Maintenance of Soil Structure*

The soil structure provides a diverse habitat for a wide range of soil microorganisms and regulates their activities. The arbuscular mycorrhizal fungus (*Rhizophagus irregularis*, formerly *Glomus intraradices*), with its exterior hyphal network that functions as a stabilising agent in the creation and maintenance of soil structure (both micro- and macro-aggregates), is a significant soil microorganism for soil structure (Miller and Jastrow 2000). This organism stabilizes the soil in two ways:

- Hyphal action of entanglement and stabilization through physical binding (Chotte 2005)
- Secretion of an organic compound into the soil: Arbuscular mycorrhizal fungi secrete glomalin, a water-insoluble compound with a strong effect that plays an important role in soil stabilization. Glomalin helps soil particles aggregate to maintain soil porosity and increase erosion resistance (Chotte 2005).

According to a study by Ruamps et al. (2011) soil fungi live in the larger (>10 μm) pores and on aggregate surfaces. Additionally, due to their capacity to spread in the soil using their hyphae in soil and their metabolic flexibility in adapting to dry soil conditions, fungi consume and scavenge carbon substrates (Chotte 2005).

In addition to helping soil fungi preserve soil structure, polysaccharides also play a role in stabilising soil aggregates. The most active polysaccharides are bacterial metabolites or root exudates, which are very effective at binding soil particles.

6.5 Microplastic Pollution of the Soil

The surfaces of microplastics can provide a unique space for the growth and development of microorganisms which make up a distinct and diverse habitat called the plastisphere microbial community (Sun et al. 2022). Microplastics can be colonized by microorganisms resulting in the formation of a different habitat with novel microbial activities (Zhang et al. 2019).

Initially, the plastisphere was reported to be a distinct microbial habitat studied on aquatic plastic debris in the North Atlantic Ocean where studies identified different relevant microorganisms inhabiting the plastic debris. These studies opened research about plastic waste forming a habitat for microorganisms (MacLean et al. 2021).

Studies have shown that microbial biofilms are rapidly formed on microplastic surfaces within 7–14 days in the aquatic environment and the traits of this biofilm microbial community were different from the surrounding water (Oberbeckmann et al. 2015). In the soil environment, microbial communities in the plastisphere differ in composition and structure from soil microbial communities and this affects the soil quality and sustainability (Agathokleous et al. 2021).

6.5.1 *The Soil Plastisphere Community*

Numerous studies have linked the development of biofilms on the surfaces of microplastics to the plastisphere community. These biofilms that have developed can adapt to the microplastic environment and utilise the nutrients that are present on the microplastics. Wright et al. (2020) examined research on plastisphere communities and discovered that these communities resembled those on material surfaces and the plastisphere was a biofilm on microplastics. Furthermore, microplastics force microorganisms to adjust to the harsh, stressful conditions and environments they produce.

According to studies by MacLean et al. (2021) plastic debris has a different microbial community than soils without a lot of plastic debris, as viewed from the same area. This is because plastic serves as a growth substrate and a source of carbon. The results also suggested that plastic waste in soil creates an environment where nutrients are made available by colonising autotrophic organisms that can withstand hydrophobicity. This makes sense because carbon from plastics is difficult for microorganisms to use, and because of their carbon-carbon structure and lack of functional groups, they are resistant to biodegradation, leaving a niche devoid of both carbon and other essential nutrients like phosphorus. This indicates that organisms that directly consume carbon produced from plastic typically form a part of the plastisphere.

6.6 Variation Between Soil Plastisphere Communities and Soil Microbial Communities

A study by Sun et al. (2022) compared the plastisphere of polylactic acid (PLA) and polyethylene (PE)—plastics that give rise to microplastics, and soil bacterial communities as influenced by temperature. PLA is a polyester bioplastic made from lactic acid polymerisation which can be degraded by hydrolysis of the ester bonds and the degraded oligomers from PLA can attract degrading bacteria. PE is a non-biodegradable petroleum-based polymer that is hydrophobic but can absorb non-polar organic matter and attract organisms.

Findings from the study indicated that soil plastisphere communities were significantly different from microbial communities of the bulk soil, and this difference was also influenced by temperature. At 25 °C, the plastisphere generally had a lower diversity of microorganisms, lower bacterial richness, and was not as even as the bulk soil, however, at the colder 15 °C, the differences between the bulk soil microbial community and the plastisphere were close. Microorganisms such as *Acidobacteria, Chloroflexi, Gemmatimonadota* and *Methylomirabilota* were lower in abundance in PE and PLA plastisphere while the Actinobacteria and Proteobacteria were of greater abundance in the plastisphere. Generally, Actinobacteria and Proteobacteria showed greater abundance in the plastispheres while *Acidobacteria* and *Gemmatimonadota* had a lower abundance, however, at 25 °C, while Actinobacteria was of greater abundance in the plastisphere, the number of Proteobacteria was reduced at the same temperature.

Examples of Proteobacteria that can be found exclusively as part of the plastisphere are *Aminobacter, Pseudoxanthomonas, Rickettsiella, Devosia* etc. Examples of Actinobacteria found in the plastisphere are from genera such as *Actinomycetospora, Arthrobacter, Rhodococcus* etc. (MacLean et al. 2021).

A separate study by Luo et al. (2022) showed that microbial genera such as *Aquabacterium, Perlucidibaca, Aridibacter, Sphingomonas, Phenylobacterium, Riemerella, Oligoflexus, Pseudoduganella, Candidatus, Koribacter, Ramlibacter,* and *Bacillus* were some organisms enriched in the plastisphere samples compared to surrounding soil samples. These organisms tend to play a role in cell adhesion, and the biogeochemical cycles and are even engaged in the decomposition of plastic. For example, *Ramlibacter* was reported to be present in the early phases of biofilm development because it may create extracellular polymers, commencing cell adhesion and continuous colonization (Andersson et al. 2009). Polymer degradation has also been demonstrated by members of the Alteromonadaceae and Burkholderiales families (e.g., *Aquabacterium*) and the genus *Bacillus* are capable of degrading polymers (Morohoshi et al. 2018). The development of microorganisms in plastisphere-containing soil habitats suggests that films are able to enrich substrates in plastic-degrading microbial populations.

6.6.1 Significance of the Soil Plastisphere Community

The interactions between plastisphere community members may have an influence on biogeochemical processes and the food web dynamics of the soil ecosystems (Yang et al. 2020a, b).

A study by Miao et al. (2019) showed that the metabolic functions of vitamins, amino acids, and co-factors were increased in the biofilms found colonizing microplastics and the plastisphere's xenobiotic degradation metabolic pathways were observed to be activated. This revealed that the amino acid and nitrogen metabolism pathways are connected to the breakdown of phosphinate, aspartate, alanine, phosphonate, and other carbohydrates. Also, the bacteria associated with the biofilms of plastisphere use plastic polymers as carbon sources (Neis et al. 2015).

According to some studies, the plastisphere may potentially harbour pathogenic species. These studies suggested that the bacterial pathogens in plastisphere samples were much more prevalent than in the surrounding water, indicating that the plastisphere presents a greater risk to human health than its surrounding environment (Yang et al. 2020a, b; Agathokleous et al. 2021). The fact that many organic pollutants are easily absorbed on the surface of biofilms, giving pathogenic bacteria nutrients to thrive, may help to explain some of the plastisphere's pathogenicity (Yang et al. 2020a, b).

6.7 Uptake of Microplastics by Plants

Depending on their type and size, microplastics can permeate seeds, roots, stems, leaves, fruits, and plant cells. Plants, on the other hand, can absorb microplastics and adsorb them onto their surface as they accumulate (Azeem et al. 2021; Mateos-Cárdenas et al. 2021). The way that plant leaves interact with microplastic particles is determined by their shape and topology. This is because microplastics can attach to plant mucilage and exudates after becoming caught by trichomes (Bi et al. 2020).

The plastic polymer is another component that affects how firmly microplastics adhere to plant leaves. Even after washing the crop with water, it is extremely difficult to remove microplastics from crops like lettuce because of the charge on the plastic's surface and the chemical bonds that attach the microplastic to the leaf surface (He et al. 2023). Small-sized plastic particles have the ability to penetrate plant roots such as rice seedlings and move through transpiration to the shoot (Liu et al. 2022). Higher transpiration rates enhance the uptake of plastic particles and this shows that the pull by transpiration is a major driving force behind the movement of plastic particles (Li et al. 2020a, b).

Through adhesive factors like fimbrial and nonfimbrial adhesins, certain human bacterial pathogens, such as certain species of *Salmonella*, can adhere to the surface of plastics and leaves (Yang et al. 2020a, b; Elpers et al. 2023). Recent studies reveal that submicron (0.2 μm) microplastics can also enter via the plant stele using the crack-entry method at lateral root emerging locations (Li et al. 2020a, b).

6.8 Effects of Microplastics on Plant Health and Soil Structure

Research findings by Boots et al. (2019) suggested that by lowering plant biomass, microplastics may have an impact on plant health. By altering the stability of the soil aggregate, microplastics in the soil can also have an impact on the structure of the soil.

6.8.1 Effects on Plant Health

Smaller-sized microplastics tend to be more toxic than larger-sized ones as the harmful and lethal effects of microplastics on plants depend on their size (Sajjad et al. 2022).

The plant stele serves as a new entry point for submicron (0.2 μm) microplastics, opening up new pathways for them to enter the human body and/or food web. When these submicron microplastics degrade, harmful and toxic compounds like benzene are released, which inhibits the plant's normal metabolism (Li et al. 2020a, b). Despite being large for plant roots to ingest, relatively large-sized microplastics can still have an impact on the growth and survival of specific plants by obstructing sunlight, causing damage to root tissue, clogging cell wall pores, and changing the physicochemical characteristics of soil (Jiang et al. 2019; Li et al. 2021a, b; Qiu et al. 2023).

Plant growth, productivity, and crop yield are often affected by microplastics in the soil because they alter the activity of antioxidant enzymes, reactive oxygen species, and proteomics profile (Boots et al. 2019; Iqbal et al. 2023; Li et al. 2022; Lozano et al. 2021). According to research by Qiu et al. (2023), microplastics have been reported to be harmful to plants due to their detrimental effects on oxidative stress, nutrient and metal uptake, physical damage to plants, plant growth and development, toxicity mechanisms, and synergy with other stressors.

The type of microplastic and the dose-dependent effects determine whether it has a negative effect on plant growth and development (Ge et al. 2021; Jiang et al. 2019). This is because microplastics can be adsorbed onto root hairs and affect root growth since the stress has a direct effect on the plant roots (Bosker et al. 2019). Plant leaves experience oxidative stress due to microplastics, which inhibit photosynthesis and leaf growth (Li et al. 2021a, b; Jia et al. 2023). Microplastics accumulation can cause stress and reduce a plant's physical growth. This can have negative effects such as pore blockage in the roots or seed coat, which can affect a plant's ability to absorb water and nutrients, and it can also lead to drought, by increasing the effect of microplastics in soil cracking. Conversely, the decrease in plant physiological growth brought on by microplastic stress may also result in side effects such as increased generation of reactive oxygen species, changes to the ionome of leaves and roots, problems regulating hormonal activity, and a loss in chlorophyll and photosynthesis (Jia et al. 2023). Plant growth and development are further inhibited by microplastic stress through modifications to critical physiological processes such as photosynthesis, redox regulations, and ionic homeostasis (Guo et al. 2022a, b).

Zhang et al. (2022) reported that microplastics can be directly toxic to plants by blocking the entrance of nutrients and water, releasing toxic compounds and invading plant cells. This action is possible as microplastics block water and nutrient transport in leaf vessels by aggregating in cell wall pores. According to studies by Sajjad et al. (2022), three types of microplastics namely the microbead, microfragment and microfibre affected the total biomass, leaf and roots of *Allium fistulosum*.

6.8.2 *Effects on Soil Structure*

Soil microplastic contamination affects the soil's chemical, biological, and physical characteristics (Lozano et al. 2021). Once in the soil, microplastics persist, gather, and aggregate in the soil, whereupon they gradually disrupt the biodiversity and functions of the soil ecosystem (Guo and Wang 2019; Kumar et al. 2020). Microplastics have the ability to clog soil pores (Kim and An 2019), disrupt pore connectivity (Jiang et al. 2017), weaken soil aggregate stability, and change the permeability of the soil (de Souza Machado et al. 2018; Guo et al. 2020; Mbachu et al. 2021).

The highly specific surface area, strong hydrophobicity, and ageing surface physicochemical changes of microplastics allow them to absorb hazardous pollutants like antibiotics, heavy metals, and persistent organic pollutants onto their surfaces and transfer them through soil ecosystems (Guo et al. 2020; Zhang et al. 2021a, b). By influencing the stability of soil aggregates, which serve as the foundation of the soil structure, microplastics can change the bulk density of the soil through their interaction with the soil environment (Jia et al. 2023; Rillig et al. 2021).

In the report of Sajjad et al. (2022), a lot of microplastics in the soil will decrease infiltration from rain and irrigation water, which could have a negative consequence on the water-holding capacity of the soil and possibly lead to anoxia. Microplastics can also act as obstacles to evaporation, which can increase the level of soil moisture and reduce the porousness and aeration of water in the soil.

6.9 Soil Plastisphere and Plants

The properties of microplastics and environmental conditions play a role in the formation and succession of a soil plastisphere community (Bolan et al. 2020; Li et al. 2021a, b). The formation and succession of plastisphere microbial communities in aqueous environments can be influenced by local environmental factors such as light, salinity, temperature, nutrient availability, and hydrodynamic conditions, as well as geographical, spatial, and temporal factors like seasons, according to a study by Wang et al. (2022). However, the microbial ecology of the soil plastisphere may be impacted by interactions with other pollutants and pathogens in the soil environment. Different microplastic types can also influence changes in the microbial community within the plastisphere. This shows that different types of plastics exert different forms of selectivity on the microbial community. One explanation for this could be that various microplastics offer microbiota niches since plastics vary widely in their properties. For instance, differences in hydrophobicity and the surface area of microplastics can have a big impact on the colonization of microbiota (Zhu et al. 2022).

Human pathogens can contaminate agricultural soil with plastics by direct deposition or by using livestock excrement as an organic fertiliser. Because soil uniquely sticks to plastic surfaces, human pathogens may also be able to survive longer because they have some protection against microbial grazing, abiotic stress, and competition (Amaral-Zettler et al. 2020). This increases the likelihood that the pathogen will come in contact with roots or tubers of plants. If the plastisphere increases pathogen persistence in soil, agricultural soils that are highly contaminated with microplastics may let the plastisphere with human pathogens continuously reach the rhizosphere and developing roots throughout the plants' growing season. Bacterial pathogens such as *Salmonella* species leave the plastisphere in response to chemotactic signals in the rhizosphere and could subsequently become pre-conditioned by plant exudate, with up-regulation of virulent genes before they get internalized by plant tissues (Jechalke et al. 2019).

A study by a Wang et al. (2020) found that microplastics can reduce soil pH, which in turn results in a decrease in the Chao1 and Shannon index of the bacterial communities in the rhizosphere soil. According to the study, the soil organic carbon content was significantly correlated with the Shannon index, and microplastic addition would dilute and reduce the soil organic carbon content, enhance the competition of the soil microbial community for substrate, change the niche of organisms, increase the differentiation and disrupt the composition of soil microbial community, thereby affecting the diversity of soil microbial community. The presence of microplastics can indirectly affect the healthy growth of plants by influencing the physicochemical properties and microbial communities in the rhizosphere soil.

As microplastics break down and age, toxic and dangerous compounds like plasticizers and flame retardants are released into the soil. These harmful compounds alter the makeup of soil microorganisms by inhibiting the number of sensitive microbial groups in the soil community.

Human viruses bind to the plastisphere's biofilm based on their pH and isoelectric point, but they can also stick to naked plastic particles by using non-ionic forces (Moresco et al. 2021). Human viruses—more specifically, naked viruses—stay stable and are protected from inactivation factors once they attach themselves to the plastisphere (Moresco et al. 2021). This helps the viruses move more easily through the plant vascular system, protects them from changing environmental conditions, and makes it possible for the viruses to avoid the plant-mediated defence mechanism (Zarkani and Schikora 2021).

According to a study by Zhao et al. (2023), the presence of plant biomass such as pieces of straw in the plastisphere lowers the pH of the soil and raises the total amount of nitrogen in the soil, which results in a reduction of bacterial diversity but increases the abundance of dominant phyla.

6.10 Conclusion

A lot of studies have been done on the plastisphere in aquatic environments, but there has been less input about the plastisphere in soil. However, research on microplastics and the soil plastisphere community has indicated that the introduction of microplastics into the soil has negative effects not only on the soil as an ecosystem but also on the soil microbiota as well as the plants. Plants take up microplastics into the system and humans as well as other animals ingest the microplastics with pathogenic microorganisms as they consume plant materials that have come in contact with microplastics. The soil plastisphere constitutes both environmental and health issues, particularly in agricultural systems where they can harbour pathogens and affect beneficial soil microbial communities. Due to their negative impacts, the deposition of plastics into the soil should be discouraged, while more environment-friendly methods of plastic waste disposal should be embraced.

References

Agathokleous E, Iavicoli I, Barcelo D, Calabrese EJ (2021) Ecological risks in a 'plastic' world: a threat to biological diversity? J Hazard Mater 417:126035. https://doi.org/10.1016/j.jhazmat.2021.126035

Amaral-Zettler LA, Zettler ER, Mincer TJ (2020) Ecology of the plastisphere. Nat Rev Microbiol 18(3):139–151

Andersson S, Dalhammar G, Land CJ, Kuttuva Rajarao G (2009) Characterization of extracellular polymeric substances from denitrifying organism Comamonas denitrificans. Appl Microbiol Biotechnol 82(3):535–543. https://doi.org/10.1007/s00253-008-1817-3

Aransiola SA, Ikhumetse AA, Babaniyi BR, Abioye OP, Oyedele OJ, Falade NO (2023) Phytoaccumulation of micro and nanoplastics: root uptake. In: Maddela NR, Reddy KV, Ranjit P (eds) Micro and nanoplastics in soil. Springer, Cham. https://doi.org/10.10007/978-3-031-21195-9_8

Auta SH, Abioye OP, Aransiola SA, Bala JD, Chukwuemeka VI, Hassan A, Aziz A, Fauziah SH (2022) Enhanced microbial degradation of PET and PS microplastics under natural conditions in mangrove environment. J Environ Manag 304(2022):114273. https://doi.org/10.1016/j.jenvman.2021.114273

Azeem I, Adeel M, Ahmad MA, Shakoor N, Jiangcuo GD, Azeem K, Ishfaq M, Shakoor A, Ayaz M, Xu M (2021) Uptake and accumulation of nano/microplastics in plants: a critical review. Nanomaterials 11(11):2935

Behera S, Das S (2023) Environmental impacts of microplastic and role of plastisphere microbes in the biodegradation and upcycling of microplastic. Chemosphere 334:138928

Bi M, He Q, Chen Y (2020) What roles are terrestrial plants playing in global microplastic cycling? Environ Sci Technol 54(9):5325–5327. ACS Publications

Bodenhausen N, Somerville V, Desiro A, Walser J-C, Borghi L, van der Heijden MG, Schlaeppi K (2019) Petunia-and Arabidopsis-specific root microbiota responses to phosphate supplementation. Phytobiomes J 3(2):112–124

Bodner G, Alsalem M, Nakhforoosh A (2021) Root system phenotying of soil-grown plants via RGB and hyperspectral imaging. Methods Mol Biol 2264:245–268. https://doi.org/10.1007/978-1-0716-1201-9_17

Bolan NS, Kirkham MB, Halsband C, Nugegoda D, Ok YS (2020) Particulate plastics in terrestrial and aquatic environments. CRC Press

Boots B, Russell CW, Green DS (2019) Effects of microplastics in soil ecosystems: above and below ground. Environ Sci Technol 53(19):11496–11506

Bosker T, Bouwman LJ, Brun NR, Behrens P, Vijver MG (2019) Microplastics accumulate on pores in seed capsule and delay germination and root growth of the terrestrial vascular plant Lepidium sativum. Chemosphere 226:774–781

Brown ME (1974) Seed and root bacterization. Annu Rev Phytopathol 12(1):181–197

Carrion VJ, Perez-Jaramillo J, Cordovez V, Tracanna V, de Hollander M, Ruiz-Buck D, Mendes LW, van Ijcken WFJ, Gomez-Exposito R, Elsayed SS, Mohanraju P, Arifah A, van der Oost J, Paulson JN, Mendes R, van Wezel GP, Medema MH, Raaijmakers JM (2019) Pathogen-induced activation of disease-suppressive functions in the endophytic root microbiome. Science 366(6465):606–612. https://doi.org/10.1126/science.aaw9285
Chotte J-L (2005) Importance of microorganisms for soil aggregation. In: Microorganisms in soils: roles in genesis and functions. Springer Berlin, Heidelberg, pp 107–119
de Souza Machado AA, Kloas W, Zarfl C, Hempel S, Rillig MC (2018) Microplastics as an emerging threat to terrestrial ecosystems. Glob Chang Biol 24(4):1405–1416
Edwards J, Johnson C, Santos-Medellin C, Lurie E, Podishetty NK, Bhatnagar S, Eisen JA, Sundaresan V (2015) Structure, variation, and assembly of the root-associated microbiomes of rice. Proc Natl Acad Sci USA 112(8):E911–E920. https://doi.org/10.1073/pnas.1414592112
Elpers L, Lüken L, Lange F, Hensel M (2023) Factors required for adhesion of Salmonella enterica serovar Typhimurium to lactuca sativa (lettuce). Microbiol Spectr 11(1):e03436–e03422
Fonseca-García C, Coleman-Derr D, Garrido E, Visel A, Tringe SG, Partida-Martínez LP (2016) The cacti microbiome: interplay between habitat-filtering and host-specificity. Front Microbiol 7:150
Gao Z, Karlsson I, Geisen S, Kowalchuk G, Jousset A (2019) Protists: puppet masters of the rhizosphere microbiome. Trends Plant Sci 24(2):165–176
Garcia-Salamanca A, Molina-Henares MA, van Dillewijn P, Solano J, Pizarro-Tobias P, Roca A, Duque E, Ramos JL (2013) Bacterial diversity in the rhizosphere of maize and the surrounding carbonate-rich bulk soil. Microb Biotechnol 6(1):36–44. https://doi.org/10.1111/j.1751-7915.2012.00358.x
Ge J, Li H, Liu P, Zhang Z, Ouyang Z, Guo X (2021) Review of the toxic effect of microplastics on terrestrial and aquatic plants. Sci Total Environ 791:148333
Geyer R (2020) Production, use, and fate of synthetic polymers. In: Plastic waste and recycling. Elsevier, pp 13–32
Guo X, Wang J (2019) The chemical behaviors of microplastics in marine environment: a review. Mar Pollut Bull 142:1–14
Guo J-J, Huang X-P, Xiang L, Wang Y-Z, Li Y-W, Li H, Cai Q-Y, Mo C-H, Wong M-H (2020) Source, migration and toxicology of microplastics in soil. Environ Int 137:105263
Guo S, Tao C, Jousset A, Xiong W, Wang Z, Shen Z, Wang B, Xu Z, Gao Z, Liu S (2022a) Trophic interactions between predatory protists and pathogen-suppressive bacteria impact plant health. ISME J 16(8):1932–1943
Guo Z, Li P, Yang X, Wang Z, Lu B, Chen W, Wu Y, Li G, Zhao Z, Liu G (2022b) Soil texture is an important factor determining how microplastics affect soil hydraulic characteristics. Environ Int 165:107293
Hamonts K, Trivedi P, Garg A, Janitz C, Grinyer J, Holford P, Botha FC, Anderson IC, Singh BK (2018) Field study reveals core plant microbiota and relative importance of their drivers. Environ Microbiol 20(1):124–140
He D, Guo T, Li J, Wang F (2023) Optimize lettuce washing methods to reduce the risk of microplastics ingestion: the evidence from microplastics residues on the surface of lettuce leaves and in the lettuce washing wastewater. Sci Total Environ 868:161726
Iqbal B, Zhao T, Yin W, Zhao X, Xie Q, Khan KY, Zhao X, Nazar M, Li G, Du D (2023) Impacts of soil microplastics on crops: a review. Appl Soil Ecol 181:104680
Jacoby R, Peukert M, Succurro A, Koprivova A, Kopriva S (2017) The role of soil microorganisms in plant mineral nutrition—current knowledge and future directions. Front Plant Sci 8:1617
Jechalke S, Schierstaedt J, Becker M, Grosch R, Schikora A (2019) Salmonella establishment in agricultural soil and colonization of crop plants depend on soil type and plant species. Front Microbiol 10:442699
Jia L, Liu L, Zhang Y, Fu W, Liu X, Wang Q, Tanveer M, Huang L (2023) Microplastic stress in plants: effects on plant growth and their remediations. Front Plant Sci 14:1226484

Jiang XJ, Liu W, Wang E, Zhou T, Xin P (2017) Residual plastic mulch fragments effects on soil physical properties and water flow behavior in the Minqin Oasis, northwestern China. Soil Tillage Res 166:100–107

Jiang X, Chen H, Liao Y, Ye Z, Li M, Klobučar G (2019) Ecotoxicity and genotoxicity of polystyrene microplastics on higher plant Vicia faba. Environ Pollut 250:831–838

Kasmuri N, Tarmizi NAA, Mojiri A (2022) Occurrence, impact, toxicity, and degradation methods of microplastics in environment—a review. Environ Sci Pollut Res 29(21):30820–30836

Kim SW, An Y-J (2019) Soil microplastics inhibit the movement of springtail species. Environ Int 126:699–706

König S, Vogel H-J, Harms H, Worrich A (2020) Physical, chemical and biological effects on soil bacterial dynamics in microscale models. Front Ecol Evol 8:53

Kumar M, Xiong X, He M, Tsang DC, Gupta J, Khan E, Harrad S, Hou D, Ok YS, Bolan NS (2020) Microplastics as pollutants in agricultural soils. Environ Pollut 265:114980

Kwak MJ, Kong HG, Choi K, Kwon SK, Song JY, Lee J, Lee PA, Choi SY, Seo M, Lee HJ, Jung EJ, Park H, Roy N, Kim H, Lee MM, Rubin EM, Lee SW, Kim JF (2018) Author correction: rhizosphere microbiome structure alters to enable wilt resistance in tomato. Nat Biotechnol 36(11):1117. https://doi.org/10.1038/nbt1118-1117

Larousse M, Galiana E (2017) Microbial partnerships of pathogenic oomycetes. PLoS Pathog 13(1):e1006028. https://doi.org/10.1371/journal.ppat.1006028

Lee S-M, Ryu C-M (2021) Algae as new kids in the beneficial plant microbiome. Front Plant Sci 12:599742

Li S, Wang P, Zhang C, Zhou X, Yin Z, Hu T, Hu D, Liu C, Zhu L (2020a) Influence of polystyrene microplastics on the growth, photosynthetic efficiency and aggregation of freshwater microalgae Chlamydomonas reinhardtii. Sci Total Environ 714:136767

Li W, Wufuer R, Duo J, Wang S, Luo Y, Zhang D, Pan X (2020b) Microplastics in agricultural soils: extraction and characterization after different periods of polythene film mulching in an arid region. Sci Total Environ 749:141420

Li C, Wang L, Ji S, Chang M, Wang L, Gan Y, Liu J (2021a) The ecology of the plastisphere: microbial composition, function, assembly, and network in the freshwater and seawater ecosystems. Water Res 202:117428

Li H-Z, Zhu D, Lindhardt JH, Lin S-M, Ke X, Cui L (2021b) Long-term fertilization history alters effects of microplastics on soil properties, microbial communities, and functions in diverse farmland ecosystem. Environ Sci Technol 55(8):4658–4668

Li S, Ding F, Flury M, Wang Z, Xu L, Li S, Jones DL, Wang J (2022) Macro-and microplastic accumulation in soil after 32 years of plastic film mulching. Environ Pollut 300:118945

Lin W, Okon Y, Hardy RW (1983) Enhanced mineral uptake by Zea mays and Sorghum bicolor roots inoculated with *Azospirillum brasilense*. Appl Environ Microbiol 45(6):1775–1779

Liu Y, Guo R, Zhang S, Sun Y, Wang F (2022) Uptake and translocation of nano/microplastics by rice seedlings: evidence from a hydroponic experiment. J Hazard Mater 421:126700

Lozano YM, Aguilar-Trigueros CA, Onandia G, Maaß S, Zhao T, Rillig MC (2021) Effects of microplastics and drought on soil ecosystem functions and multifunctionality. J Appl Ecol 58(5):988–996

Luo G, Jin T, Zhang H, Peng J, Zuo N, Huang Y, Han Y, Tian C, Yang Y, Peng K, Fei J (2022) Deciphering the diversity and functions of plastisphere bacterial communities in plastic-mulching croplands of subtropical China. J Hazard Mater 422:126865. https://doi.org/10.1016/j.jhazmat.2021.126865

MacLean J, Mayanna S, Benning LG, Horn F, Bartholomaus A, Wiesner Y, Wagner D, Liebner S (2021) The terrestrial plastisphere: diversity and polymer-colonizing potential of plastic-associated microbial communities in soil. Microorganisms 9(9):1876. https://doi.org/10.3390/microorganisms9091876

Mateos-Cárdenas A, van Pelt FN, O'Halloran J, Jansen MA (2021) Adsorption, uptake and toxicity of micro-and nanoplastics: effects on terrestrial plants and aquatic macrophytes. Environ Pollut 284:117183

Mbachu O, Jenkins G, Kaparaju P, Pratt C (2021) The rise of artificial soil carbon inputs: reviewing microplastic pollution effects in the soil environment. Sci Total Environ 780:146569
Mendes R, Kruijt M, de Bruijn I, Dekkers E, van der Voort M, Schneider JH, Piceno YM, DeSantis TZ, Andersen GL, Bakker PA, Raaijmakers JM (2011) Deciphering the rhizosphere microbiome for disease-suppressive bacteria. Science 332(6033):1097–1100. https://doi.org/10.1126/science.1203980
Miao L, Wang P, Hou J, Yao Y, Liu Z, Liu S, Li T (2019) Distinct community structure and microbial functions of biofilms colonizing microplastics. Sci Total Environ 650(Pt 2):2395–2402. https://doi.org/10.1016/j.scitotenv.2018.09.378
Miller R, Jastrow J (2000) Mycorrhizal fungi influence soil structure. In: Arbuscular mycorrhizas: physiology and function. Kluwer Academic Publishers, pp 3–18
Morella NM, Gomez AL, Wang G, Leung MS, Koskella B (2018) The impact of bacteriophages on phyllosphere bacterial abundance and composition. Mol Ecol 27(8):2025–2038. https://doi.org/10.1111/mec.14542
Moresco V, Oliver DM, Weidmann M, Matallana-Surget S, Quilliam RS (2021) Survival of human enteric and respiratory viruses on plastics in soil, freshwater, and marine environments. Environ Res 199:111367
Morohoshi T, Ogata K, Okura T, Sato S (2018) Molecular characterization of the bacterial community in biofilms for degradation of poly(3-hydroxybutyrate-co-3-hydroxyhexanoate) films in seawater. Microbes Environ 33(1):19–25. https://doi.org/10.1264/jsme2.ME17052
Musa IO, Auta HS, Ilyasu US, Aransiola SA, Makun HA, Adabara NU, Abioye OP, Aziz A, Jayanthi B, Maddela NR, Prasad R (2024) Micro and nanoplastics in environment—degradation, detection, ecological impact. Int J Environ Res 18:1. https://doi.org/10.1007/s41742-023-00551-9
Neis EP, Dejong CH, Rensen SS (2015) The role of microbial amino acid metabolism in host metabolism. Nutrients 7(4):2930–2946. https://doi.org/10.3390/nu7042930
Oberbeckmann S, Löder MG, Labrenz M (2015) Marine microplastic-associated biofilms—a review. Environ Chem 12(5):551–562
Patchaiyappan A, Dowarah K, Ahmed SZ, Prabakaran M, Jayakumar S, Thirunavukkarasu C, Devipriya SP (2021) Prevalence and characteristics of microplastics present in the street dust collected from Chennai metropolitan city, India. Chemosphere 269:128757
Qiu G, Wang Q, Wang T, Zhang S, Song N, Yang X, Zeng Y, Sun Z, Wu G, Yu H (2023) Microplastic risk assessment and toxicity in plants: a review. Environ Chem Lett 2023:1–18
Rabot E, Wiesmeier M, Schlüter S, Vogel H-J (2018) Soil structure as an indicator of soil functions: a review. Geoderma 314:122–137
Rillig MC, Kim SW, Kim T-Y, Waldman WR (2021) The global plastic toxicity debt. Environ Sci Technol 55(5):2717–2719
Rillig MC, Kim SW, Zhu Y-G (2024) The soil plastisphere. Nat Rev Microbiol 22(2):64–74
Ruamps LS, Nunan N, Chenu C (2011) Microbial biogeography at the soil pore scale. Soil Biol Biochem 43(2):280–286
Rüthi J, Bölsterli D, Pardi-Comensoli L, Brunner I, Frey B (2020) The "plastisphere" of biodegradable plastics is characterized by specific microbial taxa of alpine and arctic soils. Front Environ Sci 8:562263
Sajjad M, Huang Q, Khan S, Khan MA, Liu Y, Wang J, Lian F, Wang Q, Guo G (2022) Microplastics in the soil environment: a critical review. Environ Technol Innov 27:102408
Schmitz L, Yan Z, Schneijderberg M, de Roij M, Pijnenburg R, Zheng Q, Franken C, Dechesne A, Trindade LM, van Velzen R, Bisseling T, Geurts R, Cheng X (2022) Synthetic bacterial community derived from a desert rhizosphere confers salt stress resilience to tomato in the presence of a soil microbiome. ISME J 16(8):1907–1920. https://doi.org/10.1038/s41396-022-01238-3
Su X, Yang K (2024) Exploring the blindspot: the soil plastisphere. Soil Ecol Lett 6(2):230209
Sun A, Jiao XY, Chen Q, Trivedi P, Li Z, Li F, Zheng Y, Lin Y, Hu HW, He JZ (2021) Fertilization alters protistan consumers and parasites in crop-associated microbiomes. Environ Microbiol 23(4):2169–2183

Sun Y, Shi J, Wang X, Ding C, Wang J (2022) Deciphering the mechanisms shaping the plastisphere microbiota in soil. mSystems 7(4):e0035222. https://doi.org/10.1128/msystems.00352-22

Sun Y, Wu M, Zang J, Du L, Huang M, Chen C, Wang J (2023) Plastisphere microbiome: methodology, diversity, and functionality. iMeta 2(2):e101

Taffner J, Cernava T, Erlacher A, Berg G (2019) Novel insights into plant-associated archaea and their functioning in arugula (*Eruca sativa* Mill.). J Adv Res 19:39–48

Trivedi P, Leach JE, Tringe SG, Sa T, Singh BK (2020) Plant-microbiome interactions: from community assembly to plant health. Nat Rev Microbiol 18(11):607–621. https://doi.org/10.1038/s41579-020-0412-1

Wang J, Liu X, Liu G, Zhang Z, Wu H, Cui B, Bai J, Zhang W (2019) Size effect of polystyrene microplastics on sorption of phenanthrene and nitrobenzene. Ecotoxicol Environ Saf 173:331–338

Wang J, Liu X, Dai Y, Ren J, Li Y, Wang X, Zhang P, Peng C (2020) Effects of co-loading of polyethylene microplastics and ciprofloxacin on the antibiotic degradation efficiency and microbial community structure in soil. Sci Total Environ 741:140463

Wang C, Wang L, Ok YS, Tsang DC, Hou D (2022) Soil plastisphere: exploration methods, influencing factors, and ecological insights. J Hazard Mater 430:128503

Wright RJ, Erni-Cassola G, Zadjelovic V, Latva M, Christie-Oleza JA (2020) Marine plastic debris: a new surface for microbial colonization. Environ Sci Technol 54(19):11657–11672

Wu X, Rensing C, Han D, Xiao K-Q, Dai Y, Tang Z, Liesack W, Peng J, Cui Z, Zhang F (2022) Genome-resolved metagenomics reveals distinct phosphorus acquisition strategies between soil microbiomes. mSystems 7(1):e01107–e01121

Xiong C, Lu Y (2022) Microbiomes in agroecosystem: diversity, function and assembly mechanisms. Environ Microbiol Rep 14(6):833–849. https://doi.org/10.1111/1758-2229.13126

Xiong C, He JZ, Singh BK, Zhu YG, Wang JT, Li PP, Zhang QB, Han LL, Shen JP, Ge AH, Wu CF, Zhang LM (2021a) Rare taxa maintain the stability of crop mycobiomes and ecosystem functions. Environ Microbiol 23(4):1907–1924. https://doi.org/10.1111/1462-2920.15262

Xiong C, Singh BK, He J-Z, Han Y-L, Li P-P, Wan L-H, Meng G-Z, Liu S-Y, Wang J-T, Wu C-F (2021b) Plant developmental stage drives the differentiation in ecological role of the maize microbiome. Microbiome 9(1):1–15

Xiong C, Zhu YG, Wang JT, Singh B, Han LL, Shen JP, Li PP, Wang GB, Wu CF, Ge AH, Zhang LM, He JZ (2021c) Host selection shapes crop microbiome assembly and network complexity. New Phytol 229(2):1091–1104. https://doi.org/10.1111/nph.16890

Xu L, Dong Z, Chiniquy D, Pierroz G, Deng S, Gao C, Diamond S, Simmons T, Wipf HM, Caddell D, Varoquaux N, Madera MA, Hutmacher R, Deutschbauer A, Dahlberg JA, Guerinot ML, Purdom E, Banfield JF, Taylor JW et al (2021) Genome-resolved metagenomics reveals role of iron metabolism in drought-induced rhizosphere microbiome dynamics. Nat Commun 12(1):3209. https://doi.org/10.1038/s41467-021-23553-7

Yang K, Chen QL, Chen ML, Li HZ, Liao H, Pu Q, Zhu YG, Cui L (2020a) Temporal dynamics of antibiotic resistome in the plastisphere during microbial colonization. Environ Sci Technol 54(18):11322–11332. https://doi.org/10.1021/acs.est.0c04292

Yang Y, Liu W, Zhang Z, Grossart H-P, Gadd GM (2020b) Microplastics provide new microbial niches in aquatic environments. Appl Microbiol Biotechnol 104:6501–6511

Zarkani AA, Schikora A (2021) Mechanisms adopted by Salmonella to colonize plant hosts. Food Microbiol 99:103833

Zhang M, Zhao Y, Qin X, Jia W, Chai L, Huang M, Huang Y (2019) Microplastics from mulching film is a distinct habitat for bacteria in farmland soil. Sci Total Environ 688:470–478

Zhang J, Zou G, Wang X, Ding W, Xu L, Liu B, Mu Y, Zhu X, Song L, Chen Y (2021a) Exploring the occurrence characteristics of microplastics in typical maize farmland soils with long-term plastic film mulching in northern China. Front Mar Sci 8:800087

Zhang X, Li Y, Ouyang D, Lei J, Tan Q, Xie L, Li Z, Liu T, Xiao Y, Farooq TH (2021b) Systematical review of interactions between microplastics and microorganisms in the soil environment. J Hazard Mater 418:126288

Zhang Y, Cai C, Gu Y, Shi Y, Gao X (2022) Microplastics in plant-soil ecosystems: a meta-analysis. Environ Pollut 308:119718

Zhao Z-Y, Wang P-Y, Xiong X-B, Zhou R, Li F-M, Cheng Z-G, Wang W, Mo F, Cheruiyot KW, Wang W-Y (2023) Plant biomass mediates the decomposition of polythene film-sourced pollutants in soil through plastisphere bacteria island effect. Environ Int 178:108114

Zhou M, Liu C, Wang J, Meng Q, Yuan Y, Ma X, Liu X, Zhu Y, Ding G, Zhang J, Zeng X, Du W (2020a) Soil aggregates stability and storage of soil organic carbon respond to cropping systems on Black Soils of Northeast China. Sci Rep 10(1):265. https://doi.org/10.1038/s41598-019-57193-1

Zhou Y, Wang J, Zou M, Jia Z, Zhou S, Li Y (2020b) Microplastics in soils: a review of methods, occurrence, fate, transport, ecological and environmental risks. Sci Total Environ 748:141368. https://doi.org/10.1016/j.scitotenv.2020.141368

Zhu D, Ma J, Li G, Rillig MC, Zhu Y-G (2022) Soil plastispheres as hotspots of antibiotic resistance genes and potential pathogens. ISME J 16(2):521–532

Chapter 7
Microbial Relationship of Marine Microplastics

Sarah Chioma Nwakanma, Sesan Abiodun Aransiola, Emmanuel Kuyet Adejokun, Ojonoma Loretta Okwute, Loai Aljerf, and Naga Raju Maddela

Abstract The excessive use of plastic and their low recycling rates has led to the pollution of microplastics in marine environments. This has necessitated an in-depth exploration of the intricate relationship between these synthetic particles and microbial communities. This abstract provides a concise overview of the current state of knowledge regarding the microbial dynamics associated with marine microplastics, drawing insights from recent scientific literature. Microplastics, with their pervasive distribution in oceans and coastal regions resulting from the fragmentation of larger plastic items and the introduction of microscopic fragments, raises substantial concerns due to potential ecological repercussions. Of particular interest is the dynamic interaction between marine microplastics and microbial communities, underscoring the need for a comprehensive understanding of this relationship. Key aspects explored in this review include microbial colonization patterns on microplastic surfaces, the formation of biofilms, and the potential for microbial degradation of plastic polymers, shedding light on the enzymatic processes employed by diverse microbial communities. Through the exploration of the microbial relationship of marine microplastics, this abstract provides the knowledge for addressing the challenges posed by microplastic pollution in marine environments, guiding

S. C. Nwakanma (✉) · E. K. Adejokun · O. L. Okwute
Department of Microbiology, Faculty of Science, University of Abuja, Abuja, Nigeria

S. A. Aransiola
Department of Microbiology, Faculty of Science, University of Abuja, Abuja, Nigeria

Bioresources Development Centre, National Biotechnology Research Development Agency, Ogbomoso, Nigeria

L. Aljerf
Key Laboratory of Organic Industries, Department of Chemistry, Faculty of Sciences, Damascus University, Damascus, Syrian Arab Republic

Faculty of Pharmacy, Al-Sham Private University, Damascus, Syrian Arab Republic

N. R. Maddela
Departamento de Ciencias Biológicas, Facultad la Ciencias de la Salud, Universidad Técnica de Manabí, Portoviejo, Ecuador

S. A. Aransiola et al. (eds.), *White Pollution: Biodiversity and hazards in marine plastisphere*, Environmental Contamination Remediation and Management,
https://doi.org/10.1007/978-3-031-95547-1_7

the development of innovative and sustainable solutions. This literature review aims to contribute to the growing body of knowledge on the microbial relationship of marine microplastics. Microplastics pollution in the marine environment, roles of microbes in marine microplastics degradation, association of microorganisms with marine microplastics, and mechanisms of microplastics degradation by microorganisms all critically discussed.

Keywords Microplastics · Biofilm · Plastisphere · Marine · Biodegradation

7.1 Introduction

Plastics have indisputably played a crucial role in shaping modern lifestyles, driven by a substantial increase in demand owing to their lightweight properties, cost-effectiveness, versatility, durability, and long-lasting attributes (Andrady and Neal 2009; Aransiola et al. 2023a). These attributes establish plastic as a favored material across industries such as packaging, construction, and consumer goods, thereby amplifying its widespread demand in contemporary society. At the time in 1950, the world produced about 1.3 million tons of plastics and in a concerning trajectory, this production surged to a staggering 359 million tons (excluding fiber) by 2018, leading to extensive environmental contamination (Zhang et al. 2021; Auta et al. 2022) with a distressing annual influx of massive tons making their way into the marine environment. This highlights an alarming pattern of plastic pollution negatively affecting our oceans. An especially worrisome issue revolves around the forecast that by 2050, the weight of plastics in the ocean is expected to exceed that of fish underscoring the growing problem of plastic pollution within marine ecosystems (Yang et al. 2021). Research indicates that the collective estimate for floating marine plastic falls within the range of 70,000–270,000 tons, as outlined in studies by Cozar et al. (2014), Eriksen et al. (2014), and van Sebille et al. (2015).

In general, plastics exhibit chemical diversity. Polyamide (PA), polyethylene terephthalate (PET) and polyvinyl chloride (PVC) possess higher densities than seawater, leading to increased settlement rates in sediments. Conversely, polymers like polypropylene (PP), polystyrene (PS), low-density polyethylene (LDPE), high-density polyethylene (HDPE) and polyurethane (PUR), characterized by lower densities, are prone to floating primarily on seawater (Yang et al. 2021). About 80% of marine plastic debris stems from land-based sources, including plastic debris from beaches and inland mismanagement, with major contributions from the global fishing fleet, marine aquaculture, and litter from vessels which are intermediate environments connecting land-based debris to aquatic ecosystems, accumulating plastics of various densities (Oberbeckmann and Labrenz 2020; Musa et al. 2024).

Plastics, being synthetic polymers undergo a process of degradation, breaking down into smaller particles known as microplastics, through various environmental mechanisms. One primary mechanism leading to plastic degradation is photodegradation, where exposure to sunlight breaks down the polymer chains. According to

Andrady (2015), prolonged exposure to ultraviolet (UV) radiation initiates the fragmentation of plastics, especially those left outdoors or in marine environments. The abrasion of plastic items against surfaces, commonly observed in coastal environments, contributes to the breakdown of larger plastic pieces into smaller fragments. This physical weathering process, like mechanical forces such as wind and wave action, can lead to physical degradation (Cózar et al. 2015). Also, biodegradation involving the activity of microorganisms, plays a role in breaking down plastics into microplastics. Recent research focuses on the development of biodegradable plastics designed to break down more readily under the influence of microorganisms (Jambeck et al. 2015).

The combined effects of these degradation processes result in the formation of microplastics, defined as particles smaller than 5 mm. Microplastics can be further classified into primary microplastics, intentionally manufactured at small sizes for certain applications, and secondary microplastics, derived from the degradation of larger plastic items. These microplastics are prevalent in the environment and because of their minute nature, they are easily transported to terrestrial and aquatic ecosystems, raising concerns about their environmental impact and potential consequences for human health (Thompson et al. 2004; Aransiola et al. 2023b). These minute plastic particles, formed through the breakdown of larger plastic items or introduced into the environment as microscopic fragments, have raised apprehensions due to their potential ecological impact (Galloway 2015). Of particular interest is the interaction between marine microplastics and microbial communities, which has gained attention for its potential implications on marine ecosystems and biogeochemical cycles (Zettler et al. 2013). As these synthetic materials persist in marine environments, they interact with various biological components, including microorganisms. The microbial relationships with marine microplastics play a crucial role in understanding the ecological impact and fate of these pollutants.

7.2 Microplastics Pollution in the Marine Environment

Microplastics, plastic particles measuring less than 5 mm (1 μm–5 mm), have become a ubiquitous environmental challenge in marine ecosystems (Lebreton et al. 2017). In recent years, there has been growing public apprehension about the pervasive global issue of microplastic pollution considering their potential adverse effects on the ecology and the lasting implications for marine ecosystems (Wright and Kelly 2017). These minute particles are extensively found across diverse environments, with their omnipresence in the ocean system causing heightened concern over the past decade. Microplastics result from the fragmentation of larger plastic objects or their introduction as minute fragments into marine surroundings. Microplastics can be in various forms, including fragments, fibers, foams, pellets, foils, and microbeads (Yang et al. 2021).

Microplastics in the environment can be categorized as either primary or secondary depending on their initial dimensions (Germanov et al. 2018).

Many primary microplastics are intentionally manufactured for specific purposes and are found as, e.g. microbeads in personal care products (Geyer et al. 2017) such as toothpaste, hand and facial cleansers; shower gels and certain cosmetics (such as eye shadow, nail polish, hair coloring, etc.) (Castaneda et al. 2014; Napper et al. 2015), thus providing a direct source of microplastic pollution. In general, secondary microplastics refer to the disintegration of large plastic items through biological, chemical, and physical processes by abiotic and biotic factors. These processes include microbial species biodegradation, photodegradation (exposure to sunlight), and mechanical abrasion, resulting from wave action (Andrady 2011). Examples of sources contributing to secondary microplastics include the release of microfibers during the washing of synthetic textiles such as polyester or nylon fibers (Napper and Thompson 2016), wear and tear from vehicle tires (Wright and Kelly 2017), and the fragmentation of packaging materials. Millions of tons of these microplastics find their way in the ocean daily. Anthropogenic activities, including plastic production, improper management of inland waste disposal, and the fragmentation of larger plastic items, contribute significantly to the introduction of microplastics into marine ecosystems (Geyer 2020) through rivers and beach-related tourism. Ecosystems most severely contaminated by microplastics are those in marine circulation, estuaries, and coastal regions where human activities are prominent. Recent studies reveal elevated concentrations in areas ranging from densely populated coastlines to remote oceanic regions (Cózar et al. 2014; Lusher et al. 2017).

Microplastics can disrupt marine food chains by entering the digestive systems of small organisms like plankton, which are essential at the base of the marine food web. As these organisms ingest microplastics, they can suffer from reduced feeding efficiency and growth. This negatively impacts the organisms that feed on them, such as small fish and invertebrates, leading to potential declines in their populations (Vaksmaa et al. 2021). The effects can then cascade up the food chain, affecting larger predators like fish, marine mammals, even humans who consume seafood. Low-nutrient organisms, such as zooplankton like artemia, which are more easily accessible and exposed to suspended microplastics, might ingest these microplastics. This is due to the similarity in size and density between microplastics and planktonic organisms (Yang et al. 2021). The presence of microplastics can also interfere with the reproductive and developmental processes of marine organisms, further compromising the stability of marine ecosystems and the overall health of the food chain.

Microplastics can accumulate toxic substances through a process known as adsorption, where chemicals from the surrounding environment adhere to the surface of the plastic particles. Research, such as studies published in journals like Environmental Science & Technology and Chemosphere, has shown that microplastics can absorb and concentrate various persistent organic pollutants (POPs) and other contaminants. For example, a study by Rochman et al. (2013) titled "Polybrominated diphenyl ethers (PBDEs) in fish tissue may be an indicator of plastic contamination in marine habitats" demonstrates the association between microplastics and the accumulation of toxic substances in fish. Another study by Koelmans et al. (2016) titled "Microplastics in aquatic environments: Implications for Canadian ecosystems" explores the role of microplastics as carriers of contaminants in aquatic ecosystems. These microplastics, due to their diminutive size and the presence of diverse microbial communities in aquatic habitats, have the potential to accumulate organic pollutants that can lead to physiological impacts on marine organisms, ultimately raising concerns about the potential ramifications for human health. These sources provide insights into the mechanisms and consequences of the accumulation of toxic substances on microplastics in marine environments.

Beyond the physical harm caused by ingestion and entanglement, microplastics serve as vectors for various pollutants (Wright et al. 2013). Pollutants can adhere to microplastic surfaces, potentially threatening marine organisms upon ingestion. Also Microorganisms associated with microplastics can act as vectors, transferring contaminants from plastics to marine organisms. The potential ecological implications are discussed in works like "Microplastics as vectors for the transport of the bacterial fish pathogen species *Aeromonas salmonicida*" by Viršek et al. (2017). Understanding the ecological impacts is crucial for evaluating the broader consequences on marine biodiversity and ecosystem functioning.

7.3 Role of Microbes in Marine Microplastics Degradation

Microbes, encompassing bacteria, fungi, and other microorganisms, have emerged as indispensable players in the intricate dynamics governing marine microplastics' fate and degradation. Among these microbial agents, bacteria stand out for their enzymatic prowess, capable of depolymerizing plastic compounds with remarkable efficiency. Studies have revealed that microbial enzymatic action is pivotal, breaking down complex polymer chains into more manageable fragments (Zettler et al. 2013). The remarkable diversity and adaptability of microbial communities further enhance the efficacy of degradation processes, with different species showcasing varying abilities to metabolize distinct types of plastics (Harrison et al. 2018a, b). Biodegradation involves biological and abiotic factors, including microbial growth, environmental conditions, and surface properties of plastics, catalyzing complex metabolic reactions.

Biological factors encompass bacteria, fungi, biofilms, and microbial communities that serve as carbon sources for microorganisms Abiotic factors stem from natural photooxidation and man-made physical and chemical degradation.

The microbial degradation of plastics and microplastics (MPs) is influenced by microbial growth and various external environmental factors. Environmental conditions and the specific habitat being studied can significantly impact microbial diversity and richness. Besides the microorganisms involved, the circumstances facilitating plastic biodegradation also hinge on the surface properties and structure of plastic materials. Factors such as roughness, electrostatic interaction, topography, hydrophobicity, and free energy all play critical roles. Acting as a unique habitat, microplastics provide a substrate for microbial colonization and proliferation, serving as a carbon source in marine ecosystems.

Several studies have identified microbes capable of degrading marine microplastics and elucidated their mechanisms of action. One such example is the bacterium *Ideonella sakaiensis*, which was discovered to degrade polyethylene terephthalate (PET), a common plastic used in beverage bottles. Yoshida et al. (2016) isolated *Ideonella sakaiensis* from a PET bottle recycling facility in Japan. Through genomic and biochemical analyses, they identified two enzymes, PETase and MHETase, produced by Ideonella sakaiensis, which work together to break down PET (Table 7.1) into its constituent monomers, ethylene glycol, and terephthalic acid and convert them into mineralized forms within its cells.

The discovery of *Ideonella sakaiensis*, illustrates how elevated concentrations of plastics can spur the evolution of plastic-degrading bacteria because it can solely utilize plastics as its source of nutrition. A metagenomic analysis, exploring potential biodegradation pathways from marine microorganisms residing on plastics, has indeed confirmed the presence of putative xenobiotic biodegradation in the ocean (Oberbeckmann and Labrenz 2020).

A notable investigation by Auta et al. (2018) explores how bacterial strains isolated from mangrove sediments interact with polypropylene (PP) microplastics. The study reveals the ability of *Bacillus sp.* strain 27 and *Rhodococcus sp.* strain 36 to utilize PP microplastics for growth, leading to significant reductions in polymer mass. Fourier-transform infrared spectroscopy and scanning electron microscopy analyses confirm PP biodegradation, indicating the isolates' capacity to colonize, modify, and utilize PP microplastics as a carbon source. Additionally, Auta studied the biodegradation of various MPs by *Bacillus cereus* and *Bacillus gottheilii*, finding that these bacteria induced weight loss percentages ranging from 1.6% to 7.4% for different types of MPs. Notably, the surfaces of some MPs became rough as a result of microbial activity.

Table 7.1 Microplastics degradation by different microbial species

Microbial species	Type of plastic	Duration of degradation (day)	References
Bacillus	PP	40	Auta et al. (2018)
Bacillus cereus	LDPE, HDPE	40	Auta et al. (2017)
Bacillus subtilis	PUR	28	Shah et al. (2013)
Pseudomonas aeruginosa	PE	60	Lee et al. (2020)
Ideonella sakaiensis	PET	60	Yoshida et al. (2016)
Aspergillus flavus	HDPE	30	Sangeetha et al. (2015)
Aspergillus tubingensis	HDPE	30	Sangeetha et al. (2015)
Zalerion maritimum	PE	28	Carniel et al. (2017)
Exguobacterium	PS	28	Yuan et al. (2020)

Following an extensive investigation, research findings indicate that bacteria have the ability to not only alter the appearance of microplastics (MPs) but also modify their functional group structures and other characteristics. For instance, when polyethylene (PE) was exposed to Bacillus cereus, the carbonyl band vanished from its FTIR spectrum. Furthermore, new absorption bands emerged at 3738 and 3419 cm^{-1} after exposure to either *Bacillus cereus* or *Bacillus gottheilii,* indicating the presence of N\\H and O\\H bonds, respectively, underscoring the bacteria's capacity to influence the molecular weight and tensile properties of MPs (Yuan et al. 2020).

In addition to bacteria, fungi have also been identified as important degraders of marine microplastics. For example, a study by Sangale et al. (2019) isolated a fungal strain, *Aspergillus tubingensis*, from marine sediments contaminated with plastics in India. Through enzymatic assays, they demonstrated that Aspergillus tubingensis produced extracellular enzymes, including lipases, esterases, and cellulases, capable of degrading polyethylene and polypropylene microplastics. Several fungal species have also been identified for their capability to degrade marine microplastics. Among them, species belonging to genera such as *Aspergillus, Penicillium, Fusarium,* and *Trichoderma* have been studied for their enzymatic capabilities and ability to degrade various polymer types present in microplastics (Syranidou et al. 2017; Magnin et al. 2020). These fungi possess a diverse array of enzymes, including esterases, lipases, and ligninolytic enzymes, which play key roles in the degradation process. Fungi employ enzymatic degradation as the primary mechanism for breaking down marine microplastics. Enzymes secreted by fungi target specific chemical bonds in polymer structures, initiating the breakdown process (Magnin et al. 2020). For example, esterases and lipases produced by fungal species can hydrolyze ester bonds present in polymers such as polyethylene terephthalate (PET), leading to the depolymerization of microplastics into smaller fragments (Magnin et al. 2020). Additionally, ligninolytic enzymes produced by some fungal species can catalyze the degradation of recalcitrant polymers, such as polystyrene (PS), through oxidative processes (Syranidou et al. 2017). Marine-adapted fungi possess the capability to adjust the expression of their enzymes based on the salinity levels. For instance, when cultivated in marine settings, *Peniophora sp.* demonstrated the transcription of multiple genes encoding ligninolytic laccase enzymes (Urbanek et al. 2018).

Synergistic interactions between microorganisms is seen as a more efficient way of degrading microplastics by microorganisms. This involves microbial consortia and biofilms that are found on microplastics. Microbial consortia consisting of diverse microbial species often exhibit synergistic interactions that enhance microplastic degradation in marine environments. These consortia may include bacteria, fungi, algae, and protozoa, each contributing unique enzymatic capabilities (Zettler et al. 2013). Synergistic interactions between microorganisms result in more efficient depolymerization and degradation of microplastics compared to single-species degradation. Also, the formation of microbial biofilms on the surface of microplastics is a phenomenon that provides valuable insights into the mechanisms and dynamics of microbial-mediated degradation processes in marine ecosystems.

Microbial biofilms, composed of densely packed communities of bacteria and other microorganisms, create microenvironments conducive to degradation processes (Dussud et al. 2018). These biofilms serve as hotspots of enzymatic activity, facilitating cooperative interactions among microorganisms and enhancing the overall efficiency of degradation. Moreover, the extracellular polymeric substances (EPS) produced by biofilms can interact with microplastics, altering their surface properties and promoting further microbial colonization (Oberbeckmann et al. 2015).

Microorganisms can also contribute to the mechanical fragmentation of marine microplastics through activities such as grazing, burrowing, and abrasion. Protozoan organisms, including ciliates and amoebae, have been observed to graze on microplastic surfaces, breaking them down into smaller fragments (Kooi et al. 2016). Furthermore, the movement of benthic organisms and sediment dynamics can lead to the physical breakdown of microplastics, increasing their susceptibility to microbial degradation (McCormick et al. 2016).

Additionally, non-living factors and microbial activity release byproducts serving as energy sources, influencing degradation in both aerobic and anaerobic environments. Environmental factors such as salinity, temperature, and pressure influence microbial activity and biofilm formation, thereby impacting the rate of polymer degradation (Kirstein et al. 2016). The degradation of marine microplastics by microorganisms has significant ecological implications, including the release of plastic-associated toxins and alterations in ecosystem dynamics. However, microbial degradation also offers opportunities for bioremediation and waste management, maintaining aquatic ecosystems and are indispensable for biogeochemical recycling. Wei et al. (2021) proposed the use of engineered microbial consortia for targeted plastic degradation in marine environments, illustrating the potential of biotechnological approaches to address microplastics pollution.

7.4 Association of Microorganisms with Marine Microplastics

For over four and a half decades, it has been understood that microplastics in the ocean carry communities primarily dominated by microorganisms. Only in recent years has there been a thorough investigation into the role of microbial interactions with microplastics in marine ecosystems. In marine environments, the abundance of microorganisms can reach hundreds of millions of bacterial cells per gram of wet marine sediment (Harrison et al. 2011).

7.4.1 Plastisphere—Microbial Colonization of Plastic Surfaces

The "plastisphere," is a term that describes the microbial community that develops on plastic debris in marine environments, and has garnered significant attention due to its ecological implications and potential impacts on marine ecosystems. Studies have demonstrated rapid microbial colonization of plastic surfaces as a survival tactic in various marine conditions, facilitating stable consortia, gene exchange, nutrient accumulation, and defense against toxins, upon exposure to aquatic environments. Zettler et al. (2013) observed diverse microbial communities on marine plastic debris, with bacteria such as *Pseudoalteromonas, Vibrio*, and *Alteromonas* dominating the plastisphere. Similarly, Amaral-Zettler et al. (2020) identified distinct microbial taxa colonizing plastic fragments in freshwater ecosystems, highlighting the ubiquity of plastisphere formation across diverse habitats.

7.4.2 Community Structure and Diversity

Characterization of plastisphere microbial communities has revealed unique taxonomic compositions and functional profiles compared to surrounding water environments. Oberbeckmann et al. (2016) reported higher microbial diversity and metabolic potential within the plastisphere, suggesting niche differentiation and resource utilization on plastic surfaces. Furthermore, Dussud et al. (2018) found evidence of species-specific preferences for plastic substrates, indicating selective pressures driving community assembly in the plastisphere. Marine plastics commonly harbor *Prochlorococcus sp., Synechococcus sp.,* and oil-degrading bacteria such as *Colwellia sp.,* whereas freshwater habitats exhibit a comparable variety, including species like *Saprospiraceae* and *Comamonadaceae*, which are present in both environments (Rogers et al. 2021).

7.4.3 Biofilm Formation and Maturation

Bacteria employ attachment to surfaces and biofilm formation as a way to survive in various marine conditions utilizing microplastics for adhesion, carbon, and energy sources, leading to their degradation by microorganisms and enzymes and also serving as platforms for other organisms like microalgae and fungi to settle (Oberbeckmann and Labrenz 2020). The formation and maturation of microbial biofilms on microplastics involve several key mechanisms (De Tender et al. 2017; Zettler et al. 2013). Microplastics, being hydrophobic, provide a surface for microbial attachment through weak physicochemical interactions such as van der Waals forces (Kesy et al. 2019). Microorganisms have a tendency to colonize the surfaces of microplastics (MPs), with interactions that are affected by a range of surface features, including size, shape, texture, and hydrophobicity, as well as environmental factors like temperature, pH, and ionic strength (Yang et al. 2023). Upon initial contact, microorganisms undergo a reversible attachment phase where weak interactions, such as van der Waals forces or electrostatic interactions, allow them to adhere temporarily to the microplastic surface. This phase enables the microorganisms to explore the suitability of the surface for colonization. Initial colonization occurs as pioneer microorganisms, including bacteria and algae, adhere to the microplastic surface facilitated by extracellular polymeric substances (EPS) secretion (Bryant et al. 2016). If the microorganisms find the surface suitable, they transition to an irreversible attachment phase. During this phase, stronger interactions, including adhesion molecules or extracellular polymeric substances (EPS) produced by the microorganisms which is a complex matrix of polysaccharides, proteins, nucleic acids, and other organic molecules, that promotes a more permanent attachment to the microplastic (Fig. 7.1).This initial attachment leads to the formation of microcolonies, which then develop into mature biofilms as microbial populations grow and produce more EPS (Liu et al. 2020).

The biofilm architecture on microplastics can vary, but it typically consists of layers of microorganisms embedded in the EPS matrix. This structure protects microorganisms from environmental stressors and enhances their ability to survive and interact with the microplastic surface. Microbial succession within biofilms is influenced by the metabolic capabilities of different microbial species, leading to shifts in community composition over time (Dang and Lovell 2016). As the biofilm matures, more microorganisms join the community, and the EPS matrix continues

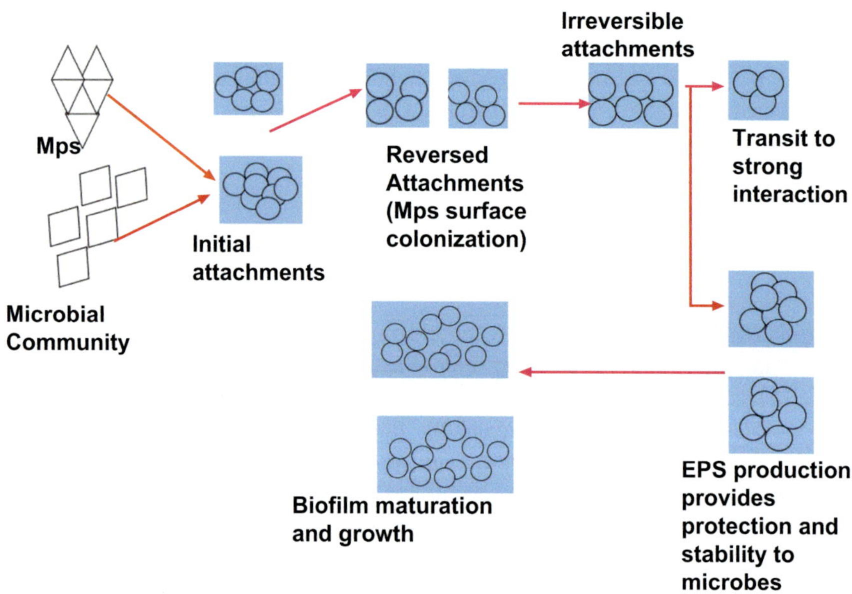

Fig. 7.1 Biofilm formation and maturation

to expand. EPS creates a protective environment for the microorganisms, providing stability and cohesion to the biofilm structure. The microorganisms within the biofilm communicate through quorum sensing, a mechanism that allows them to coordinate their activities based on population density. Quorum sensing and signaling mechanisms regulate gene expression and coordinate biofilm formation, EPS production, and other behaviors (Toyofuku et al. 2016). Nutrient availability plays a crucial role, with microplastics serving as a concentrated source of organic and inorganic nutrients for microbial growth (Harrison et al. 2018a, b).

7.4.4 Environmental Factors Influencing Biofilm Formation

Environmental factors such as temperature, pH, salinity, and oxygen availability also influence biofilm dynamics on microplastics (Kesy et al. 2019). This suggests that factors like trophic mechanisms, pH, and salinity, nutrient availability (both organic and inorganic carbon, nitrate, phosphorus, etc.), pollutants (such as toxic metals, antibiotics, and persistent organic pollutants), physicochemical parameters (including dissolved oxygen, light, pH, temperature, salinity, and ionic strength), and the presence of aquatic biota like plants and animals plays significant roles in shaping microbial communities in different aquatic environments. Additionally, physical factors like UV radiation and mechanical abrasion can impact biofilm stability and structure (Zettler et al. 2013). Other factors that influence biofilm formation on marine microplastics includes, the type of microorganisms present, the composition and characteristics of the microplastic surface, and the surrounding environmental conditions (biotic and abiotic factors) such as environmental conditions, sample location, substrate type, particle size, and surface properties (Zettler et al. 2013; Oberbeckmann et al. 2014; Eich et al. 2015). Understanding the dynamics of biofilm formation is crucial in comprehending the role of microorganisms in the degradation and fate of microplastics in marine ecosystems (De Tender et al. 2017; Zettler et al. 2013).

7.4.5 Biofilms as Vectors of Contaminants

Microorganisms associated with microplastics can act as vectors, transferring contaminants from plastics to marine organisms. The potential ecological implications are discussed in works like "Microplastics as vectors for the transport of the bacterial fish pathogen species Aeromonas salmonicida" by Viršek et al. (2017).

Microplastic biofilms have a propensity to attract bacterial pathogens like *Agrobacterium spp., Chryseobacterium spp., Flavobacterium spp.,* and also species associated with gastrointestinal diseases in humans which are, *Vibrio spp.* and *Arcobacter spp.* Additionally, microplastics can serve as a habitat for the proliferation of fungal and other eukaryotic pathogens, indicating their potential to support the growth of diverse disease-causing microorganisms.

7.5 Mechanism of Microplastics Degradation by Microorganisms

Microorganisms, including bacteria and fungi, possess enzymes capable of breaking down the complex chemical bonds in microplastics. These microorganisms recognize microplastics as potential carbon sources, and enzymatic activity is triggered upon contact. Enzymes involved in plastic degradation are often extracellular, allowing microorganisms to release them into their surroundings. While some microorganisms can internalize microplastic particles, where intracellular processes, including enzymatic degradation and co-metabolism, occur, leading to the breakdown of plastic polymers within microbial cells. Factors like the type of plastic for instance, Polyamide (PA) and polyethylene terephthalate (PET) are examples of plastics that are susceptible to hydrolysis and relatively easy to break down. Plastics like polyethylene and polypropylene with C-C bond backbones need redox reactions to break down into smaller molecules. On the other hand, plastics like polyethylene terephthalate and polyamide, containing easily degradable bonds like amide or ester, can undergo enzymatic cleavage or hydrolysis (Cai et al. 2023). Also surface characteristics of microplastics influence microbial enzyme production as this may affect attachment to the microplastic surface and formation of biofilms.

In the study of microplastics degradation by *Ideonella sakaiensis*, the bacteria secretes two enzymes PETase and METHase. PETase breakers down polyethylene terephthalate into mono terephthalic (2-hydroxyethyl) acid and bis (2-hydroxyethyl) terephthalic acid, while MHETase enzyme then converts mono-2-hydroxyethyl terephthalate (MHET) into terephthalic acid and ethylene glycol. Bacteria internalize this intermediate product and break it down using PET hydrolase enzyme, which is essential for organisms that utilize PET as a primary carbon and energy source. The resulting depolymerized PET products are utilized by the bacteria for their metabolic processes (Thakur et al. 2023).

7.5.1 Enzymatic Hydrolysis

Enzymatic hydrolysis is a primary pathway for microplastic degradation by microorganisms. Microbes such as bacteria and fungi produce enzymes, including lipases, esterases, and proteases, which catalyze the breakdown of polymer bonds in microplastics (Harrison et al. 2018a, b). These enzymes target ester, amide, and other bonds within the polymer structure, leading to the fragmentation of microplastics into smaller particles.

7.5.1.1 Lipases and Esterases

Lipases and esterases are among the primary enzymes produced by microorganisms for microplastic degradation. These enzymes target ester bonds present in many plastic polymers, such as polyethylene terephthalate (PET) and polyethylene (PE) (Zettler et al. 2013). Lipases hydrolyze ester bonds in lipid-rich plastics aliphatic polyesters and also aliphatic-aromatic co-polyesters (Kothawale et al. 2023), while esterases catalyze the breakdown of ester linkages in other plastic types, facilitating the release of monomers and oligomers.

7.5.1.2 Proteases

Proteases play a significant role in the enzymatic degradation of microplastics composed of protein-based polymers, such as polyurethane (PU). Microbial proteases cleave peptide bonds within the polymer structure, leading to the depolymerization of PU and the release of smaller molecular fragments (Kumar et al. 2021). This enzymatic activity contributes to the breakdown of PU microplastics in various environmental matrices.

While bacterial degradation often involves enzymatic hydrolysis of polymer bonds, such as ester and amide bonds, leading to fragmentation of microplastics (Harrison et al. 2018a, b), Fungal degradation, in contrast, may involve both enzymatic and non-enzymatic mechanisms, including mechanical breakdown and extracellular oxidation (Pinnell and Turner 2020). Fungal enzymes, particularly depolymerases, possess properties that enable them to break down various polymers, including PET and PE. The distribution and penetrating ability of fungal hyphae is what helps in their initial colonization and subsequent depolymerization, while their enzyme capabilities enhance hyphal attachment to hydrophobic substrates making it easy to colonize microplastics.

7.5.2 Oxidative Enzymes

Certain microorganisms produce oxidative enzymes, such as laccases and peroxidases, which facilitate the degradation of microplastics through oxidative processes (Syranidou et al. 2020). Laccases, for example, catalyze the oxidation of phenolic compounds, which can modify the surface properties of microplastics and enhance their susceptibility to microbial degradation. Peroxidases generate reactive oxygen species (ROS) that can degrade polymer chains and promote fragmentation of microplastics.

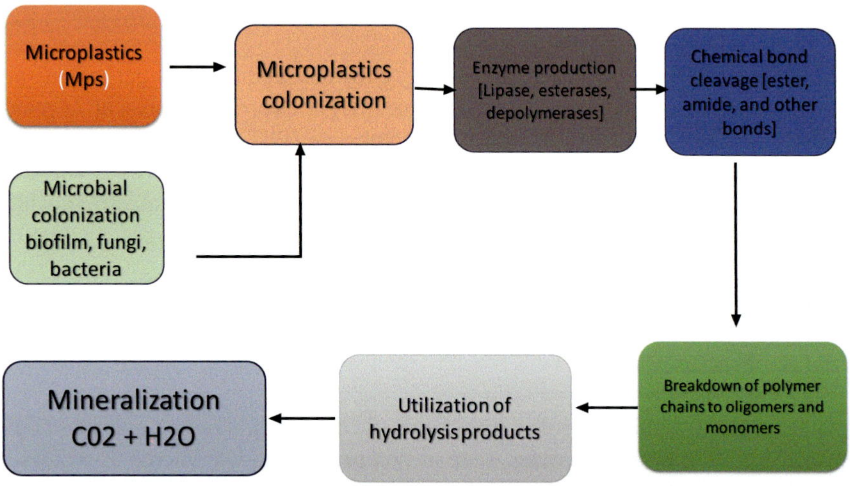

Fig. 7.2 Mechanism of microplastic degradation by microbial colonization

7.5.3 Mechanical Breakdown

In addition to enzymatic degradation, mechanical breakdown contributes to the biodegradation of microplastics. Physical forces, such as abrasion and erosion, can break down larger plastic particles into smaller fragments (Pinnell and Turner 2020). This process is particularly prominent in marine environments, where microplastics are subjected to wave action and abrasive interactions with sediments (Fig. 7.2).

7.5.4 Extracellular Oxidation

Extracellular oxidation represents another biodegradation pathway for microplastics. Certain microbial species produce reactive oxygen species (ROS) that can oxidize polymer surfaces, leading to structural degradation (Syranidou et al. 2020). This oxidative process can weaken the integrity of microplastics, making them more susceptible to enzymatic degradation and further fragmentation.

The biodegradation of microplastics is influenced by various environmental factors, including temperature, pH, and nutrient availability. Optimal conditions, such as moderate temperatures and neutral pH, promote microbial activity and enhance degradation rates. However, extreme conditions or nutrient limitations may inhibit microbial degradation processes, affecting the efficiency of biodegradation pathways.

7.6 Conclusion

In conclusion, microplastics pollution in marine environments poses significant ecological and environmental challenges, impacting marine life and ecosystem health. However, emerging research suggests that microorganisms, particularly those forming biofilms, can play a crucial role in mitigating this issue. Through biofilm formation on microplastic surfaces, microorganisms can facilitate degradation and fragmentation, potentially reducing the persistence and harmful effects of microplastics in marine ecosystems. Additionally, certain microbial species have been found to possess plastic-degrading enzymes, offering promising avenues for bioremediation strategies. While further research is needed to fully understand the complex interactions between microorganisms and microplastics in marine environments, harnessing microbial activity holds considerable promise for combating microplastics pollution and preserving the health of our oceans.

References

Amaral-Zettler LA, Zettler ER, Mincer TJ, Jacobs LA (2020) Microbial assemblages on plastic litter: a comparison of freshwater and marine environments. Front Microbiol 11:583

Andrady AL (2011) Microplastics in the marine environment. Mar Pollut Bull 62(8):1596–1605. https://doi.org/10.1016/j.marpolbul.2011.05.030

Andrady AL (2015) Persistence of plastic litter in the oceans. In: Bergmann M, Gutow L, Klages M (eds) Marine anthropogenic litter. Springer, Cham. https://doi.org/10.1007/978-3-319-16510-3_3

Andrady AL, Neal MA (2009) Applications and societal benefits of plastics. Philos Trans R Soc Lond Ser B Biol Sci 364:1977–1984

Aransiola SA, Ikhumetse AA, Babaniyi BR, Abioye OP, Oyedele OJ, Falade NO (2023a) Phytoaccumulation of micro and nanoplastics: root uptake. In: Maddela NR, Reddy KV, Ranjit P (eds) Micro and nanoplastics in soil. Springer, Cham. https://doi.org/10.10007/978-3-031-21195-9_8

Aransiola SA, Victor-Ekwebelem MO, Ajiboye AE, Leh-Togi Zobeashia SS, Ijah UJJ, Oyedele OJ (2023b) Ecological impacts and toxicity of micro and nanoplastics in agroecosystem. In: Maddela NR, Reddy KV, Ranjit P (eds) Micro and nanoplastics in soil. Springer, Cham. https://doi.org/10.10007/978-3-031-21195-9_10

Auta HS, Emenike CU, Fauziah SH (2017) Screening of Bacillus strains isolated from mangrove ecosystems in Peninsular Malaysia for microplastic degradation. Environ Pollut 231:1552–1559

Auta HS, Emenike CU, Jayanthi B, Fauziah SH (2018) Growth kinetics and biodeterioration of polypropylene microplastics by Bacillus sp. and Rhodococcus sp. isolated from mangrove sediment. Mar Pollut Bull 127:15–21. https://doi.org/10.1016/j.marpolbul.2017.11.036

Auta SH, Abioye OP, Aransiola SA, Bala JD, Chukwuemeka VI, Hassan A, Aziz A, Fauziah SH (2022) Enhanced microbial degradation of PET and PS microplastics under natural conditions in mangrove environment. J Environ Manag 304(2022):114273. https://doi.org/10.1016/j.jenvman.2021.114273

Bryant JA, Clemente TM, Viviani DA, Fong AA, Thomas KA, Kemp P, Karl DM, White AE (2016) Diversity and activity of communities inhabiting plastic debris in the North Pacific Gyre. mSystems 1(3):e00024–e00016

Cai Z, Li M, Zhu Z, Wang X, Huang Y, Li T, Gong H, Yan M (2023) Biological degradation of plastics and microplastics: a recent perspective on associated mechanisms and influencing factors. Microorganisms 11:1661. https://doi.org/10.3390/microorganisms11071661

Carniel A, Valoni E, Nicomedes J, Gomes A, de Castro AM (2017) Lipase from Candida Antarctica (CALB) and cutinase from Humicola insolens act synergistically for PET hydrolysis to terephthalic acid. Process Biochem 59:84–90. https://doi.org/10.1016/j.procbio.2016.07.023

Castañeda RA, Avlijas S, Simard MA, Ricciardi A (2014) Microplastic pollution in St. Lawrence River sediments. Can J Fish Aquat Sci 71(12):1767–1771. https://doi.org/10.1139/cjfas-2014-0281

Cózar A, Echevarría F, González-Gordillo JI, Irigoien X, Úbeda B, Hernández-León S, Duarte CM (2014) Plastic debris in the open ocean. Proc Natl Acad Sci 111(28):10239–10244

Cózar A, Echevarría F, González-Gordillo JI, Irigoien X, Úbeda B, Hernández-León S et al (2015) Plastic debris in the open ocean. Proc Natl Acad Sci 111(28):10239–10244

Dang H, Lovell CR (2016) Microbial surface colonization and biofilm development in marine environments. Microbiol Mol Biol Rev 80(1):91–138

De Tender CA, Devriese LI, Haegeman A, Maes S, Ruttink T, Dawyndt P, Robbens J (2017) Bacterial community profiling of plastic litter in the Belgian part of the North Sea. Environ Sci Technol 51(20):11586–11594

Dussud C, Hudec C, George M, Fabre P, Higgs P, Bruzaud S et al (2018) Colonization of non-biodegradable and biodegradable plastics by marine microorganisms. Front Microbiol 9:1571

Eich A, Mildenberger T, Laforsch C, Weber M (2015) Biofilm and diatom succession on polyethylene (PE) and biodegradable plastic bags in two marine habitats: early signs of degradation in the pelagic and benthic zone? PLoS One 10(9):e0137201. https://doi.org/10.1371/journal.pone.0137201

Eriksen M, Lebreton LCM, Carson HS, Thiel M, Moore CJ, Borerro JC et al (2014) Plastic pollution in the world's oceans: more than 5 trillion plastic pieces weighing over 250,000 tons afloat at sea. PLoS One 9(12):e111913. https://doi.org/10.1371/journal.pone.0111913

Galloway TS (2015) Micro- and nano-plastics and human health. In: Bergmann M, Gutow L, Klages M (eds) Marine anthropogenic litter. Springer, Cham. https://doi.org/10.1007/978-3-319-16510-3_13

Germanov ES, Marshall AD, Bejder L, Fossi MC, Loneragan NR (2018) Microplastics: no small problem for filter-feeding megafauna. Trends Ecol Evol 33(4):227–232. https://doi.org/10.1016/j.tree.2018.01.005

Geyer R (2020) Production, use, and fate of synthetic polymers. In: Letcher TM (ed) Plastic waste and recycling. Academic. ISBN 9780128178805, pp 13–32. https://doi.org/10.1016/B978-0-12-817880-5.00002-5

Geyer R, Jambeck JR, Law KL (2017) Production, use, and fate of all plastics ever made. Sci Adv 3(7):e1700782

Harrison JP, Sapp M, Schratzberger M, Osborn AM (2011) Interactions between microorganisms and marine microplastics: a call for research. Mar Technol Soc J 45(2):12–20. https://doi.org/10.4031/MTSJ.45.2.2

Harrison JP, Schratzberger M, Sapp M, Osborn AM, Spiers AJ (2018a) A European perspective on the fate and effects of microplastics in marine ecosystems. In: Marine anthropogenic litter. Springer, pp 75–125

Harrison JP, Schratzberger M, Sapp M, Osborn AM (2018b) Rapid bacterial colonization of low-density polyethylene microplastics in coastal sediment microcosms. BMC Microbiol 18(1):1–10

Jambeck JR, Geyer R, Wilcox C, Siegler TR, Perryman M, Andrady A et al (2015) Plastic waste inputs from land into the ocean. Science 347(6223):768–771

Kesy K, Oberbeckmann S, Kreikemeyer B (2019) Marine biodegradation of biopolymers—what's the story? In: Marine anthropogenic litter. Springer, pp 125–141

Kirstein IV, Kirmizi S, Wichels A, Garin-Fernandez A, Erler R, Löder M, Gerdts G (2016) Dangerous hitchhikers? Evidence for potentially pathogenic Vibrio spp. on microplastic particles. Mar Environ Res 120:1–8
Koelmans AA, Bakir A, Burton GA, Janssen CR (2016) Microplastic as a vector for chemicals in the aquatic environment: critical review and model-supported reinterpretation of empirical studies. Environ Sci Technol 50(7):3315–3326. https://doi.org/10.1021/acs.est.5b06069
Kooi M, Reisser J, Slat B et al (2016) The effect of particle properties on the depth profile of buoyant plastics in the ocean. Sci Rep 6:33882. https://doi.org/10.1038/srep33882
Kothawale SS, Kumar L, Singh SP (2023) Role of organisms and their enzymes in the biodegradation of microplastics and nanoplastics: a review. Environ Res 232:116281. https://doi.org/10.1016/j.envres.2023.116281
Kumar A, Jahan S, Singh DP, Srivastava JK (2021) Microbial enzymes for biodegradation of plastic polymers. In: Microbial enzymes: roles and applications in industries, environment, and healthcare. Academic, pp 305–323
Lebreton LCM, van der Zwet J, Damsteeg JW, Slat B, Andrady A, Reisser J (2017) Evidence that the Great Pacific Garbage Patch is rapidly accumulating plastic. Sci Rep 7:4666
Lee HM, Kim HR, Jeon E, Yu HC, Lee S, Li J, Kim DH (2020) Evaluation of the biodegradation efficiency of four various types of plastics by pseudomonas aeruginosa isolated from the gut extract of superworms. Microorganisms 8:1–12. https://doi.org/10.3390/microorganisms8091341
Liu J, Sun X, Yu L (2020) Mechanisms influencing the biofilm formation of microorganisms on microplastics. Front Environ Sci 8:76
Lusher AL, Tirelli V, O'Connor I, Officer R (2017) Microplastics in Arctic polar waters: the first reported values of particles in surface and sub-surface samples. Sci Rep 5:14947
Magnin JP, Lanzén A, Andersen T (2020) Fungal degradation of microplastics: a review. Environ Int 143:105881
McCormick AR, Hoellein TJ, London MG, Hittie J, Scott JW, Kelly JJ (2016) Microplastic in surface waters of urban rivers: concentration, sources, and associated bacterial assemblages. Ecosphere 7(11):e01556. https://doi.org/10.1002/ecs2.1556
Musa IO, Auta HS, Ilyasu US, Aransiola SA, Makun HA, Adabara NU, Abioye OP, Aziz A, Jayanthi B, Maddela NR, Prasad R (2024) Micro and nanoplastics in environment—degradation, detection, ecological impact. Int J Environ Res 18:1. https://doi.org/10.1007/s41742-023-00551-9
Napper IE, Thompson RC (2016) Release of synthetic microplastic plastic fibres from domestic washing machines: effects of fabric type and washing conditions. Mar Pollut Bull 112(1–2):39–45
Napper IE, Bakir A, Rowland SJ, Thompson RC (2015) Characterisation, quantity and sorptive properties of microplastics extracted from cosmetics. Mar Pollut Bull 99(1–2):178–185. https://doi.org/10.1016/j.marpolbul.2015.07.029
Oberbeckmann S, Labrenz M (2020) Marine microbial assemblages on microplastics: diversity, adaptation, and role in degradation. Annu Rev Mar Sci 12:209–232. https://doi.org/10.1146/annurev-marine-010419-010633
Oberbeckmann S, Loeder MGJ, Gerdts G, Osborn AM (2014) Spatial and seasonal variation in diversity and structure of microbial biofilms on marine plastics in Northern European waters. FEMS Microbiol Ecol 90(2):478–492
Oberbeckmann S, Löder MGJ, Labrenz M (2015) Marine microplastic-associated biofilms—a review. Environ Chem 12(5):551–562. https://doi.org/10.1071/EN15069
Oberbeckmann S, Loeder MG, Gerdts G, Osborn AM (2016) Spatial and seasonal variation in diversity and structure of microbial biofilms on marine plastics in Northern European waters. FEMS Microbiol Ecol 92(4):fiw157
Pinnell LJ, Turner JW (2020) Shotgun metagenomics reveals the benthic microbial community response to plastic and bioplastic in a coastal marine environment. Front Microbiol 11:692
Rochman CM, Hoh E, Kurobe T, Teh SJ (2013) Ingested plastic transfers hazardous chemicals to fish and induces hepatic stress. Sci Rep 3:3263

Rogers KL, Carreres-Calabuig JA, Gorokhova E, Posth NR (2021) Micro-by-micro interactions: how microorganisms influence the fate of marine microplastics. Environ Microbiol 23(1):5–19. https://doi.org/10.1111/1462-2920.15535

Sangale MK, Shahnawaz M, Ade AB (2019) Potential of fungi isolated from the dumping sites mangrove rhizosphere soil to degrade polythene. Sci Rep 9:5390. https://doi.org/10.1038/s41598-019-41448-y

Sangeetha Devi R, Rajesh Kannan V, Nivas D, Kannan K, Chandru S, Robert Antony A (2015) Biodegradation of HDPE by Aspergillus spp. from marine ecosystem of Gulf of Mannar, India. Mar Pollut Bull 96:32–40

Shah Z, Krumholz L, Aktas DF, Hasan F, Khattak M, Shah AA (2013) Degradation of polyester polyurethane by a newly isolated soil bacterium, Bacillus subtilis strain MZA-75. Biodegradation 24:865–877

Syranidou E, Karkanorachaki K, Amorotti F, Avgeropoulos A, Kolvenbach B, Zhou NY et al (2017) Development of tailored indigenous marine consortia for the degradation of naturally weathered polyethylene films. PLoS One 12(8):e0183984

Syranidou E, Karkanorachaki K, Amorotti F, Franchini M, Repouskou E, Kaliva M et al (2020) Biodegradation of weathered polystyrene films in seawater microcosms by bacterial communities obtained from coastal waters. Front Microbiol 11:1756

Thakur B, Singh J, Singh J, Angmo D, Pal Vig A (2023) Biodegradation of different types of microplastics: molecular mechanism and degradation efficiency. Sci Total Environ 877:162912. https://doi.org/10.1016/j.scitotenv.2023.162912

Thompson RC, Olsen Y, Mitchell RP, Davis A, Rowland SJ, John AWG et al (2004) Lost at sea: where is all the plastic? Science 304:838–838. https://doi.org/10.1126/science.1094559

Toyofuku M, Inaba T, Kiyokawa T, Obana N, Yawata Y, Nomura N (2016) Environmental factors that shape biofilm formation. Biosci Biotechnol Biochem 80(1):7–12

Urbanek AK, Rymowicz W, Mirończuk AM (2018) Degradation of plastics and plastic-degrading bacteria in cold marine habitats. Appl Microbiol Biotechnol 102:7669–7678. https://doi.org/10.1007/s00253-018-9195-y

Vaksmaa A, Hernando-Morales V, Zeghal E, Niemann H (2021) Microbial degradation of marine plastics: current state and future prospects. In: Joshi SJ, Deshmukh A, Sarma H (eds) Biotechnology for sustainable environment. Springer, Singapore. https://doi.org/10.1007/978-981-16-1955-7_5

van Sebille E, Wilcox C, Lebreton L, Maximenko N, Hardesty BD, van Franeker JA, Eriksen M, Siegel D, Galgani F, Lavender Law K (2015) A global inventory of small floating plastic debris. Environ Res Lett 10(12):124006. https://doi.org/10.1088/1748-9326/10/12/124006

Viršek MK, Lovšin MN, Koren Š, Kržan A, Peterlin M (2017) Microplastics as a vector for the transport of the bacterial fish pathogen species Aeromonas salmonicida. Mar Pollut Bull 125(1–2):301–309. https://doi.org/10.1016/j.marpolbul.2017.08

Wei R, Zimmermann W, Bi J (2021) Enzyme systems for plastic degradation—a potential approach to recycle plastic wastes. In: Polyethylene-based biocomposites and bionanocomposites. Elsevier, pp 273–289

Wright SL, Kelly FJ (2017) Plastic and human health: a micro issue? Environ Sci Technol 51(12):6634–6647

Wright SL, Thompson RC, Galloway TS (2013) The physical impacts of microplastics on marine organisms: a review. Environ Pollut 178:483–492

Yang L, Zhang Y, Kang S, Wang Z, Wu C (2021) Microplastics in soil: a review on methods, occurrence, sources, and potential risk. Sci Total Environ 780:146546. https://doi.org/10.1016/j.scitotenv.2021.146546

Yang W, Li Y, Boraschi D (2023) Association between microorganisms and microplastics: how does it change the host-pathogen interaction and subsequent immune response? Int J Mol Sci 24:4065. https://doi.org/10.3390/ijms24044065

Yoshida S, Hiraga K, Takehana T, Taniguchi I, Yamaji H, Maeda Y, Toyohara K, Miyamoto K, Kimura Y, Oda K (2016) A bacterium that degrades and assimilates poly(ethylene terephthalate). Science 351(1979):1196–1199

Yuan J, Ma J, Sun Y, Zhou T, Zhao Y, Yu F (2020) Microbial degradation and other environmental aspects of microplastics/plastics. Sci Total Environ 715:136968

Zettler ER, Mincer TJ, Amaral-Zettler LA (2013) Life in the "plastisphere": microbial communities on plastic marine debris. Environ Sci Technol 47(13):7137–7146

Zhang Y, Gao T, Kang S, Allen S, Luo X, Allen D (2021) Microplastics in glaciers of the Tibetan Plateau: evidence for the long-range transport of microplastics. Sci Total Environ 758:143634. https://doi.org/10.1016/j.scitotenv.2020.143634

Chapter 8
Plastisphere and Soil Biodiversity

Evelyn Nwadinkpa Fatokun, Kehinde Olubunmi Adediran, and Christy Echakachi Manyi-Loh

Abstract Discarded plastics are released into the environment via various pathways, many of which end up in the soil ecosystem. Soil acting as a natural sink, accumulate plastic wastes originating from different anthropogenic and natural activities. The larger plastics, under biotic and abiotic conditions undergo degradation and fragmentation to yield smaller plastic particles down to micro- and nanoparticle sizes with different characteristics. The soil environment as a part of the terrestrial ecosystem, is the key carrier of microplastics (MP), with an estimated MP content higher than that of the oceans. These fragments present rigid scaffold to which microorganisms and biomolecules can adsorb, thereby forming an important habitat of microbial community, otherwise called the 'plastisphere'. Soil plastisphere habours higher and diverse bacterial population compared to the surrounding soil, thereby making it a hotspot of microbial diversity. The microbiome represents the combined genetic material or community of microorganisms occurring in a particular environment that is on the surface of plastics. This chapter focuses on the soil plastisphere biodiversity by providing useful information on the sources and types of soil plastics and particulates, soil plastisphere biotic components and microbial communities, interactions between soil microorganisms and plastics and ecosystem implication, functional and enzymatic composition, microplastic-associated biofilm implication in antimicrobial resistance of the soil plastisphere.

E. N. Fatokun (✉)
Department of Biological Science and Biotechnology, College of Pure and Applied Sciences, Caleb University, Imota, Lagos State, Nigeria
e-mail: evelyn.fatokun@calebuniversity.edu.ng

K. O. Adediran
Department of Biological Sciences, Faculty of Pure and Applied Sciences, Lead City University, Ibadan, Oyo State, Nigeria

C. E. Manyi-Loh
Centre of Applied Food Sustainability and Biotechnology, Department of Life Sciences, Faculty of Health and Environmental Science, Central University Technology, Bloemfontein, Free State, South Africa

S. A. Aransiola et al. (eds.), *White Pollution: Biodiversity and hazards in marine plastisphere*, Environmental Contamination Remediation and Management,
https://doi.org/10.1007/978-3-031-95547-1_8

Keywords Soil-plastisphere · Microplastics · Biofilm · Microorganisms · Antibiotic resistance

8.1 Introduction

Microorganisms are said to be ubiquitous in nature, implying that they are found everywhere amongst a diverse range of habitats, including water, soil, air, and plants. Occurring in the different habitats, these micro-organisms occur in communities interacting with one another as well as their environment, comprising plants, animals, humans as well as their wastes generated through domestic, municipal, human, and industrial activities. Our environment, nowadays is filled with a lot of wastes in varying forms, including liquid, solid, and plastics. Pollution with plastics is a major challenge worldwide, as they are viewed as an integral part of our lives, being involved, or used in packaging, automotive manufacturing, telecommunications, agricultural production, and construction (Alassali et al. 2020), posing risks to the equilibrium and the health of the environments plus humans (Yang et al. 2023). The discarded plastics are released into the environment via a few primary channels, including burial, direct disposal, alongside dumping into water bodies and dumping into landfills. Researchers have noted the plausible toxic effects of plastics to both environmental microbes and humans (Aransiola et al. 2023a). The release of plastics into the environment is associated with adverse consequences, as they exert direct impact on microbiomes, toxic effects, provide supplemental carbon, serve as rafts for colonization by microbes as well as their dispersal (Lear et al. 2021; Auta et al. 2022). The authors mentioned further the indirect consequences expressed as increased environmental shading, alteration in the composition of host communities and disruption of host microbe or community or community health, balances in hormones plus immune response.

Descriptively, plastics are defined as a class of polymer material endowed with high chemical stability and plasticity that originate from fossil fuels (e.g., oil or natural gas) and or renewable sources (e.g., bio-based plastics), but predominant generation is of the former origin (i.e., from fossil fuels) (Geyer 2020). Plastics are materials produced from combinations of synthetic and semi-synthetic polymer materials and are subjected to incorporation with additives (antioxidants, plasticisers and flame retardants), usually to help in the making and performance of the final product (Lithner et al. 2011). Therefore, constituting a large and diverse group of materials or polymers such as polystyrene, polypropylene, polyethylene, nylon, polyurethane, polyethylene terephthalate and polyvinyl chloride (Danso et al. 2019; Musa et al. 2024). Of all the available polymers, 95% of the global plastic production comprises of just eight (8) while 45% is made of polypropylene and polyethylene (Geyer 2020). Only a small fraction of the polymer types is bio-based, however, the polymer types that are dominant are mostly fossil -based, which are not biodegradable. The fate of plastics includes recycling (10%), incineration (14%) and discarding (76%) (European Bioplastics 2018).

Oftentimes, microbiomes collaborate with abiotic factors (heat and light) to produce tremendous impact on the structural integrity of the polymers and accessibility of enzyme attack, facilitating degradation (Lear et al. 2021). The resultant fragments can be categorised relying on size into macroplastic, mesoplastics, microplastics and nanoplastics (Van Cauwenberghe et al. 2015). Amongst these, microplastics can last for prolonged period of hundred to thousand years in the environment owing to their small size, large specific surface area and strong hydrophobicity added to very stable chemical properties (Lambert and Wagner 2016). Hence, the term 'plastisphere' was coined to describe plastics inhabited by microbial communities, and was majorly applied in the marine ecosystem (Zettler et al. 2013). Microorganisms can be complexed with plastics and eventually end up on other living organisms, animals or in the systems and underlying tissues of human after overcoming barriers and invading the tissues; wherein they can equally interact with the microbiome within that environment. This chapter focused on plastisphere and soil biodiversity; with attention to soil plastisphere biotic components and microbial communities, functional and enzymatic composition, the interactions between the microorganisms in the soil and the surface of the plastics/microplastics as well as the environmental microbiome. We highlight on the implication of microplastic-associated biofilm, in antimicrobial resistance of the soil plastisphere.

8.2 Soil and Plastics

Plastics are considered emerging pollutants of global importance. In addition to their chemical polymeric composition, plastics are durable, resistant to natural degradation and can persist over a long period in the environment. Under environmental conditions, they fragment into smaller particles of sizes ranging from macro size (macroplastics: 1 cm and larger), to mesoplastics (1 to <10 mm), to micro size (1 < 1000 μm), and even to nanoparticle size (ranging 1 nm < 1000 nm, although with varying subdivision) (Hartmann et al. 2019; Wu et al. 2021; Krehl et al. 2022; Aransiola et al. 2023b). This particulate nature makes the plastics ideal habitat for microorganisms. Plastic particles have thus been identified in recent times as very important microbial habitat and new substrates for microbial colonization (Viljakainen and Hug 2021). Hence, the term 'plastisphere' was coined to describe plastics inhabited by microbial communities, and was majorly applied in the marine ecosystem (Zettler et al. 2013). However, the term plastisphere has a more holistic definition in the soil ecosystem to include the immediate soil environment impacted by plastic debris, its microbial communities and the colonizing environment (Rilling et al. 2023). Soil plastisphere colonization, formation and succession is mediated by several factors including characteristics of the plastic particles, edaphic factors, biotic and abiotic environmental factors (Yang et al. 2021).

8.2.1 Types and Sources of Soil Plastics

Plastics are mostly derived from fossil fuels (polybutylene succinate/PBS, and polycaprolactone/PCL, Polyethylene/PE, Polypropylene/PP); and some, known as bio-based are fully or partially made from biological resources comprising bio-plastics, which are made from renewable materials such as sugar beet or corn starch (polyhydroxybutyrate/PBH, polylactide/PLA); and another bio-based group which are chemically identical to those of fossil fuels but synthesized from renewable materials (bio-polyethylene terephthalate/bio-PET, bio-polyethylene/bio-PE, bio-propylene/bio-PP, polybutylene adipate terephthalate/PBAT, polyhydroxyalkanoates (PHA)) and are sometimes referred to as 'drop-in' plastics (Tokiwa et al. 2009; Hartmann et al. 2019; Lear et al. 2021). A third category of plastic wastes are actually elastomers, (a hybrid polymers of inorganic monomers, e.g. natural rubber, silicone, polyurethane) but are included as plastic debris since they are synthetic polymers emitted to the environment (Hartmann et al. 2019). Nonetheless, description of plastics in soils according to the acceptable international standards is limited (Bläsing and Amelung 2018; Weber 2022).

Plastics occur in different soil types including agricultural soils, garden soils, industrial soils, floodplains, mining soils, urban and roadside soils (Weber 2022). Plastics are introduced into the soil ecosystem through various pathways, either by intention or unintentionally. For instance, agricultural land is bedevilled with plastic leftovers from several agricultural practices including mulching, composting, irrigation with treated and untreated wastewater (Uwamungu et al. 2022). Other routes of deposition of plastics into the soil includes illegal waste dumping, roads and trails littering, street runoffs, landfills, sewage sludge and the atmosphere (Bläsing and Amelung 2018; Uwamungu et al. 2022). These numerous pathways of plastics into soil have prompted the suggestion of plastic concentration in the soil to be in the range of soil organic carbon (Bläsing and Amelung 2018). Although it has been demonstrated through modelling that the concentration of MPs should not exceed the range of 2.10–14.40 g/kg soil, in order to safeguard a healthy soil status, agricultural soil serves as a bulk sink for MPs (Joos and De Tender 2022; Zeb et al. 2022; Mészáros et al. 2023). For instance, a study recorded the presence of 206 macroplastic pieces per hectare and 0.34 ± 0.36 microplastic particles per kilogram dry weight of soil on an agricultural farmland without input of microplastic-containing fertilizers and agricultural plastic applications (Piehl et al. 2018). However, study on the prevalence and fate of plastics in the soil is still very limited (Bläsing and Amelung 2018).

Despite the scarcity of evidence, Scientists have proposed the concept of "microplastic cycle", indicating the possibility of plastics' movement between different environmental media, which may also confound the initial source of plastic debris in an environment. For instance, Li and his colleagues reported on the links between microplastic communities, and that geographical proximity of sampling sites had similar pollution sources and greater ease of migration or exchange of microplastics (MPs). Therefore, MPs could actually serve as vehicles for microbial transportation across regions, which could pose a serious ecological and environmental health consequence of global impact (Li et al. 2021, 2023).

8.2.2 *Effect of Plastics on Soil Characteristics*

Studies have shown that plastic particles can potentially impact soil biophysical properties (de Souza Machado et al. 2019). For instance, depending on the type, size, shape, concentration and chemical composition, soil plastics can alter the degradation of organic matter, soil bulk density, porosity, soil permeability, water content, soil pH, and can even influence greenhouse gases emission from the soil (Lehmann et al. 2021; Krehl et al. 2022; Wang et al. 2022a, b). Lehmann et al. (2021) reported that soil biota mediates the ability of polyester microfibres to alter soil structure. Moreover, due to their high specific surface area, strong hydrophobicity, and age-related diversification of their surface physicochemical properties, MPs can accumulate hazardous contaminants such as heavy metals, antibiotics and persistent organic pollutants (POP) on their surfaces and become vectors of these contaminants in terrestrial ecosystems. These influencing factors of plastic particles on soil have an overarching effects in ecological studies, environmental health and agricultural practices (Guo et al. 2022; Wu et al. 2021; Krehl et al. 2022). Besides, plastic additives like plasticizing agents including phthalates (e.g. bis(2-ethylhexyl)) which are loosely incorporated in the polymer structure to enhance their properties might be washed out, forming leachates that could be harmful to soil dwelling organism and inhibit soil microbiological activity (Bläsing and Amelung 2018). These phthalates may exhibit carcinogenic, mutagenic and endocrine-disrupting properties (Erkekoglu and Kocer-Gumusel 2014), and are therefore regarded as harmful soil contaminants. Once released into the environment phthalates and even particulate plastics as micron- and nano-sizes can be absorbed by plants thereby endangering human and animal health via food chain (Wu et al. 2021; Huo et al. 2022).

Furthermore, accumulation of plastics residues and incidence of soil plastisphere would result in higher decomposition and degradation with resultant release of metabolites into the immediate environment, thereby offsetting the normal biogeochemical cycles, release of greenhouse gas and negatively impact soil organic matter in agricultural land (Li et al. 2023).

8.3 Plastisphere and Soil Biotic Components

Soil is a heterogeneous ecosystem comprising diverse biotic components which are essential in maintaining the integrity of the ecosystem. However, the higher plastisphere microbial biomass than the surrounding soils has the ability to impact the soil ecosystem. For instance, plastispheres has been reported to have more homogenized bacterial communities, higher proportions of *Actinomycetota* and *Bacillota*, but lower incidence of *Acidobacteriota*, *Bacteroidota*, *Gemmatimonadota* and *Pseudomonadota*, in comparison with the adjacent soils (Sun et al. 2024). Also, taxa associated with xenobiotics degradation, including Caulobacterales, Burkholderiales, *Pseudomonas*, *Rhodococcus*, and *Streptomyces*, thrived in the plastisphere of the Alpine and Arctic soils, although with a lesser impact on fungal taxa which had increased incidence of *Phlebia* and *Alternaria* genera in the plastisphere compare to the bulk soils. Suggesting that the soil microbiome of terrestrial cryoenvironments can be strongly influenced by plastic pollution (Rüthi et al. 2020).

On the other hand, several studies have documented the adverse effects of plastic leachates, nanoplastics (NPs) and MP on plant, including delayed seeds germination, obstruction in the uptake of water and nutrients, morphological changes such as decreases in aerial and subterranean length and weight due to oxidative stress, triggered by NP and MP-induced reactive oxygen species (ROS) (Li et al. 2022; Wang et al. 2022a, b). However, the effects of plastics on plants may differ based on various factors including plastic type and concentration, species of plants and environmental conditions. In another study, microplastic particles was reported to decrease the number of offspring of soil-dwelling bacterivorous nematode *Caenorhabditis elegans*, and impacted negatively on the body length under MP exposure, but without a consistent pattern (Schöpfer et al. 2020).

Plastisphere has been implicated as environmental reservoir of pathogenic microbes and antibiotic resistant bacteria (ABR). For instance, *Pseudomonas* (e.g., *Pseudomonas monteilii*, *P. mendocina*, *P. angilliseptica*, and *P. syringae*) has been observed in fresh-water and soil plastispheres (Rüthi et al. 2020; MacLean et al. 2021).

A previous study identified that the ratios of potential pathogens/bacteria in the plastisphere were higher than those in the soil (Zhu et al. 2022). Furthermore, antibiotic resistance genes (ARGs) are frequently observed with the plastisphere in diverse environments (Yang et al. 2020). For example, Yang et al. (2019) found that the abundance and diversity of ARGs in plastic microbiota were significantly greater than those in seawater microbiota in the North Pacific Gyre, which was the first report on ARGs in the plastisphere. Since this study, the roles of microplastics acting as reservoirs and refuges for ARGs are of increasing concern. As the bacteria in biofilms possess high diversity and metabolic complexity, the plastisphere is usually proposed to be a hotspot for horizontal gene transfer. Previous laboratory findings indicated that the frequency of plasmid transfer between plastic-associated bacteria was higher than that in free-living bacteria. The high level of gene changes would facilitate ARG propagation in the plastisphere. In addition, microorganisms embedded in the plastisphere are protected by extracellular polymeric substances and are more tolerant to the environmental stresses than the discrete cells. Thus, the acquisition of antibiotic tolerance by pathogens which was the first report on ARGs in the plastisphere.

There have been inconsistencies in the documented effects of plastics exposure on soil fauna and flora, probably due to variations on the plastics characteristics and concentration, and other influential factors (Huo et al. 2022). Nonetheless, assessment of the importance of these factors on the interaction between plastics and soil organisms has been limited (Huo et al. 2022). Moreover, many of these studies were conducted under controlled conditions, hence field studies are imperative in assessing the real-world impacts of plastics on plants and other soil organisms (Mészáros et al. 2023).

8.3.1 Microbial Communities Found on Plastics/Microplastics

Microplastics are described as small plastic fragments (below 5 mm), originating from plastic degradation in the natural environment (Geyer et al. 2017). Microplastics (MPs) can further be grouped into degradable and non-degradable forms which are widely distributed; therefore, it is apparent that microorganisms will possibly interact with microplastics as they are described to be ubiquitous in nature and form an important part of the natural environment (Qiu et al. 2022). These fragments present as rigid scaffold to which microorganisms, biomolecules (lipopolysaccharide, allergens, and antibiotics) can adsorb and the potential of the microplastics to absorb other substances is influenced by an elevated ratio of surface area to volume (Ateia et al. 2020). Microplastics offer an appropriate substrate growth surface area for microbial growth, supplying them with vital nutrients, including zinc, copper and iron that are metal ions recovered from the environment as the adhere unto the surfaces of MPs. The microbiome represents the combined genetic material or community of microorganisms occurring in a particular environment that is on the surface of plastics. It is termed as the "plastisphere". It is possible that microbial components discharged from dead bacteria can bind to MPs even in the absence of living organism (Yang et al. 2023). Several groups of microorganisms of diverse taxa can be recovered from the surface of plastics or MPs, belonging to bacteria, fungi, and other multicellular organisms as well as viruses etc. Accordingly, Kettner et al. (2017) and Oberbeckmann and co-authors (2014) reported the overrepresentation of bacterial phyla in the plastisphere to include *Proteobacteria*, *Bacteriodetes* and *Cyanobacteria* and fungi (e.g., *Chytridiomycota*), respectively.

The growth of bacteria is enabled by the roughness of the surface of the weathered MPs, leading to the formation of biofilms. However, the bacterial communities in the plastisphere vary between MPs (as well as their surrounding environment) being influenced by the composition and characteristics of the MPs (Zettler et al. 2013; Oberbeckmann et al. 2014). This implies the bacterial communities differ between biodegradable and non-degradable, polyethylene against polypropylene and across varying levels of hydrophobicity, thus displaying selectivity in colonisation (Frere et al. 2018).

8.3.2 Interactions Between Soil Microorganisms and Plastics/Microplastics

Microplastics found in the environment can be grouped into primary and secondary MPs based on the source points and their pathway of formation (Omidoyin and Jho 2023). Primary MPs are derived from manufacturing activities and considered for commercial purposes (e.g., cosmetic products and industrial abrasives) while secondary MPs are yielded from physical, chemical, and biological activities, including the degradation and fragmentation of larger plastics to yield smaller plastic particles (Cincinelli et al. 2019; Khalid et al. 2021). Microplastics are extensively distributed in the environment relative to the traditional plastics because of the small size and the capacity of bioretention (Rillig and Lehmann 2020). In other words, they occur in varying forms of pellets, fibers, foams, and films, serving as appropriate abiotic surfaces for colonization by microbes, owing to their small size, yet large surface area, surface texture, hydrophobic composition, and long persistence (Nath et al. 2023). Soil is viewed as one of the most valued natural resources in human society, forming the basis of all terrestrial ecosystems (Helmberger et al. 2019). Soils are known to provide ecosystem services, including removal of pathogens and pollutants from water, recycling of organic wastes, sequestering of atmospheric carbon added to delivering water and nutrients needed by crops and native plants, making food available to humans (Robinson et al. 2012). Their composition of great microbial biodiversity and animals are essential in performing the above-mentioned ecosystem services, although variation is obvious across agricultural, urban, and riparian soils (Barrios 2007). Therefore, it can be concluded that soils regulate nutrient cycling, maintains biodiversity of organisms, and provides food (Zhu et al. 2019).

The soil environment as a part of the terrestrial ecosystem, is the key carrier of MPs, with an estimated MP content higher than that of the oceans on a mass/mass or particle/mass basis (Horton et al. 2017). The reason is that the soil acts as a natural sink for MPs, originating from different sources, including landfills, sewage sludge, plastic mulching in agricultural activities, municipal wastewater effluent and fertilisers (Omidoyin and Jho 2023). The varieties of MPs in conjunction with the diverse group of microorganisms indicates the possibility of increase in the interactions between MPs and microorganisms as well as determine the complexity of the interactions (Qiu et al. 2022). The interactions between microbes and MPs are affected by certain factors, including the size, hydrophobicity, roughness, type, and shape; these factors describe the surface characteristics of MPs. Environmental factors, notably temperature, microenvironmental pH, nutrient availability, light intensity, salinity, and ionic strength equally play eminent role in the interactions, affecting microbial richness, composition, and activities (Puckowski et al. 2021). Furthermore, the duration of exposure of the microbes to MP, variation in seasons plus geographical locations have been observed to impact on the diversity of microbial communities connected to MPs. The interactions between MPs and microorganisms are said to occur at the soil -plastic interface, wherein the physicochemical characteristics (roughness, size, shape, hydrophobicity) will provoke biodiversity and activities of soil microbial communities at this site, bringing about multiplication and activities of a particular microbial taxa and cause the formation of hotspot of microorganisms in the soil.

Microbes, including bacteria can colonise the surface of plastics and microplastics. Colonisation of MPs by microbes is affected by the characteristics of the MPs, exposure time as well as environmental factors. The interaction between the plastics and microorganisms can be greatly influenced by the environment in which the plastic is found. Qiu and colleagues (2022) explained that the interaction of MPs with microorganisms is complex due to the variety and characteristics of the MPs that are found in the environment. The initial interaction of bacteria occurs via electrostatic forces, and it is reliant on the size of the MPs, chemical composition, and surface modifications (Tuson and Weibel 2013). The formation of biofilms by bacteria on the surfaces of MP encourages the interaction of MPs with numerous of the bacterial-derived molecules that form extracellular polymeric substances, consisting mainly of extracellular DNA, polysaccharides, and proteins (Dong et al. 2020).

The introduction of MPs into the soil can occur via direct or indirect channels, including agricultural, industrial, and everyday activities as well as runoff and deposition, respectively (Li et al. 2019; Evangeliou et al. 2020). Consequently, they are retained in the soil and can cause alteration of the soil properties, including porosity, bulk density, water holding capacity, pH, aggregation, and electrical conductivity. According to Tziourrou and Golia (2024) the interactions of the microbiome occurring mainly in the rhizosphere in the agricultural soils, influence the behaviour of plastics causing alterations in the availability of potassium and phosphorus, the fraction of organic matter as well as the nitrogen cycle. The release of MP into the environment is associated with direct toxic and inhibitory effects on the activities of microorganisms caused by the additives (phthalates, bisphenol A and heavy metals) incorporated into the polymers during the production of plastic (Lin 2016). However, the presence of MP creates surfaces to which other pollutants can adsorb onto, transforming them from originally clean MP to MPs hosting pollutants as shown in Fig. 8.1. Thus, increases the risk of exposure of microbes in the soil environment (Ribeiro et al. 2019). Browne et al. (2013) and Avio et al. (2015) remarked on the joint toxic effects on microorganisms, exerted by the interactions of MPs with other chemical substances (heavy metals, organic matter) in the environment. Moreover, the adsorption of pollutants to the surface of MPs causes an increase in density thereby making the MP sink through, deeper into the soil via channels bored by earthworms during burrowing.

In a rural setting, the soil is most of the time affected by using animal manure (harbouring antibiotics, microbes, and nutrients), fertilizers, and varying forms of liquid and solid wastes, originating from agriculture (Golia et al. 2023). Antibiotics are part of the other pollutants binding via hydrophobic and electrostatic forces and they include sulphonamides, tetracyclines, macrolides, chloramphenicol, fluoroquinolones. The occurrence of antibiotics is often linked with antibiotic resistance (Manyi-Loh et al. 2019), therefore the presence of MPs provokes the emergence of antibiotic resistant genes in the surrounding bacteria, enhancing the possibility of

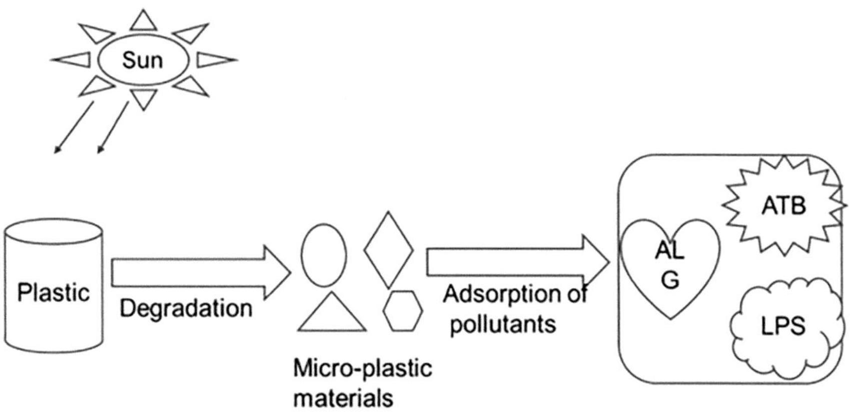

Fig. 8.1 Plastic degradation and adsorption unto microplastics of pollutants. *ATB* antibiotics, *ALG* allergens, *LPS* lipopolysaccharide

transfer of the said genes between bacterial communities and different species (Arias-Andres et al. 2018; Liu et al. 2021).

On the other hand, microorganisms will exert impact on the MPs based on degradation, transfer, adsorption added to deposition (Qiu et al. 2022). This is because microorganisms exhibit a custom of aggregation of membranes when they attach to both living and non-loving surfaces (e.g., plastics), forming biofilms.

8.4 Functional Potentials and Enzymes of Soil Plastisphere

Plastisphere possess unique functional potentials compared to its host natural environment. A variety of pathways in soil plastisphere through functional predications has been reported to include carbon, nitrogen, and sulfur metabolism, biosynthesis of secondary metabolites, ABC transporters and metabolic pathways for drug resistance and xenobiotics biodegradation and metabolism, among others (Zhu et al. 2022). Zhu et al. (2022) also documented CAZyme annotation for higher abundance of carbohydrate esterase in the plastispheres compared to the soil samples. Also in another study, Rüthi and co-workers reported the presence of genes involved in cellular processes and signalling such as vesicular transport, intracellular trafficking and secretion; organic molecule- (carbohydrate and lipid) and secondary metabolism. In addition, they recorded the presence of gene families which encode enzymes necessary in degradation of plastic, and carbohydrate-active genes involved in lignin and murein degradation; nitrogen fixation, degradation and synthesis genes were all increased on Ecovio and BI-OPL films buried on alpine soils for 5 months.

Microorganisms have the potential to reduce plastic polymer to its basic constituents through aerobic and anaerobic metabolism, in which case the polymer serves as a carbon source (Kumar et al. 2017). Microbial MPs degradation involves the steps of biosynthesis, bio-fragmentation and mineralization, and biodeterioration (Tiwari et al. 2020). Biodegradation is the microbial transformation of plastic wastes by decomposing them into basic compounds of CO_2, water, methane, inorganic compounds, and cellular biomass, depending on the organisms, polymer type and molecular weight of the plastics and environmental conditions (Singh et al. 2022).

When used as substrates by microbes or their enzymes, plastics can be regarded as biodegradable or non-biodegradable for those that can be metabolized and the non-metabolized ones respectively. Algae-derived bio-plastics (PHA and PLA) for instance, while having similar characteristics and properties as fossil-based plastics, are quite biodegradable. On the other hand, while most petroleum-based plastics (PE, PP, PS and PVC) need to be subjected to abiotic biodegradation e.g. using UV-light-initiated oxidative degradation, before being biodegradable, some such as Polycaprolactone (PCL), and poly-butylene succinate (PBS) also petroleum based, can be degraded by microorganisms (Tokiwa et al. 2009; Mohanan et al. 2020). However, the degradability of these polymers is mostly determined by the pure carbon backbones, carbon bond, presence of aromatic groups among other variables (Webb et al. 2013).

8.4.1 Plastisphere Enzymes

The possibility of enzymatic degradation of different types of plastic by microbial plastisphere has been documented. Li and colleagues reported a higher incidence of functional potentials for degradation or decomposition of organic compounds, including persistent organic pollutants (POP), plastics, hydrocarbon and aromatic hydrocarbon degradation, ligninolysis and oil bioremediation, in the plastisphere of all environments, including terrestrial ecosystems (Li et al. 2024).

The plastisphere microbial community is dominated by autotrophic microbes, biofilm-forming microbes including Gammaproteobacteria and Alphaproteobacteria, which can efficiently degrade microplastic in the environment by hydrolytic enzymes such as hydroxylase, esterase and lipase. These microbes have shown great potential for the bioconversion of microplastic to polyhydroxyalkanoates (PHA), an energy efficient and sustainable alternative to the fossil based plastics (Singh et al. 2022; Behera and Das 2003). Other enzymes enrich in soil plastisphere are several depolymerizing enzymes of α/β-hydrolase families such as polyesterase-lipase-cutinase, cutinase-like and carboxylesterase enzymes.

8.5 Biofilm Formation

The soil environment is a natural habitat harbouring microorganisms that seems to be an integral component of the environment. The microorganisms occurring in the plastisphere are diverse phenotypically, requiring a broad range of environmental conditions (e.g., aerobes, anaerobes, motile, non-motile plus extremophiles) in order to thrive or survive (Rogers et al. 2020). Micro-organisms tend to develop biofilms as a vital strategy to survive in natural or man-made niches in response to environmental stress factors such as UV radiation, desiccation, limited nutrients, antibiotics etc. (Muhammad et al. 2020). Biofilms are formed on both biotic and abiotic surfaces in the environments, including plastics as a lifestyle in which they adopt a multicellular behaviour to enable and lengthen survival in varied environmental niches (Kostakioti et al. 2013). Biofilms are unique structures developed by microorganisms, existing almost everywhere and consist of an arrayed and ordered assemblage of microorganisms, thriving within an extracellular polymeric matrix (EPS) produced by them (Rather et al. 2021). The formation of biofilms by microbes is a complex process that takes place via five stages, including reversible attachment of cells, irreversible attachment, microcolony formation, maturation, and dispersal; resulting in a heterogenous structure that constitutes largely of microbial cells (10–25%) and extracellular polymeric substance (70–90%) (Muhammad et al. 2020). Figure 8.2 shows the five phases involve in biofilm formation.

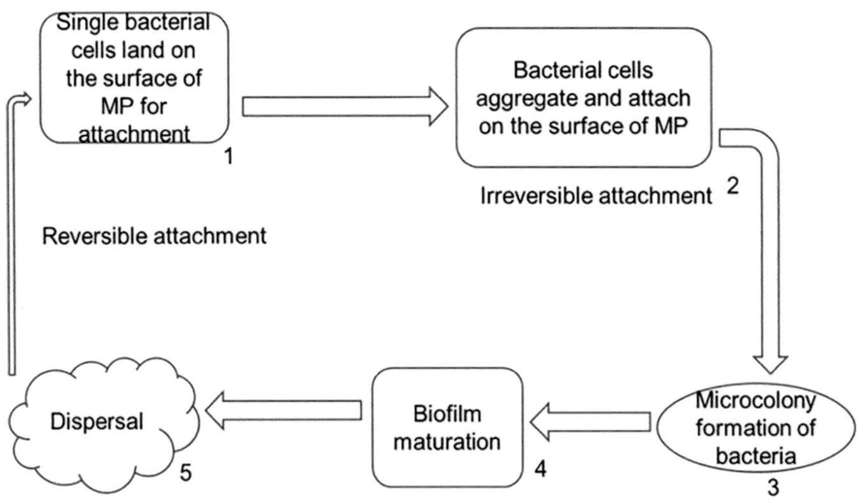

Fig. 8.2 Five phases of biofilm formation

8.5.1 Implication of MP-Associated Biofilm in Antimicrobial Resistance

The formation of biofilms by the microbes offer protection against environmental pressure, contributing to the persistence of microorganisms and toxicity (Liu et al. 2023). Biofilms tend to aggravate the problem of antimicrobial resistance, which is a global public health menace. Apparently, the biofilms occur as a distinct layer over the surface of the MPs (Valentine and Boxall 2022) and over time, they become platform for the association of other microbes and pathogens, enabling the interaction and transfer of antimicrobial resistance genes between the microbes (Nath et al. 2023). The accumulation of bacteria on the surface of MPs lead to the expression of quorum sensing genes that facilitate further cell-cell communication in adherence, maturation, and dispersal of biofilms (Liu et al. 2021). Thus, causing rising levels of bacterial resistance. Biofilm formation by bacteria on MPs is determined by the surface characteristics, the type of polymer and the size of the particles. The residing microbial communities can cause a drastic change in the physical properties of the MP, transforming it from its original state as well as modifying its properties such as stiffness and crystallinity (McGivney et al. 2020). In addition, the adsorption of antibiotics, heavy metals, and other pollutants (xenobiotics compounds) to the surface of MPs may provoke cross resistance or co-resistance in bacterial communities. Also, the extracellular polymeric substance (EPS), a viscous liquid secreted by microbes in biofilms increases adsorption of the afore mentioned pollutants to the MPs. Therefore, the long-term abuse or misuse of antibiotics could lead to its abundance in the environment, which in turn promotes the selection and persistence of antimicrobial resistance genes, thus affecting horizontal gene transfer and probably changes the resistome profile of different categories of microbes (Zhang et al. 2020).

Microorganisms, the major life forms in the soil environment demonstrate primary functions in mineralization of organic matter, cycling of nutrients and even driving climate (Finley et al. 2018). Partial degradation of MPs occurs as microbes colonise their surfaces and develop biofilms, resulting in alterations in density and mobility in the soil environment, thereby releasing leachates and toxic substances into the environment. In details, the microbes attached on the MPs release enzymes that act on the chemical bonds of the MPs, inducing hydrolysis or oxidation-reduction reactions, releasing oligomers, dimers, and monomers (small, decomposed products) that enter the microbial cells and are subsequently mineralized into inorganic products including carbon dioxide, water, and methane (Yuan et al. 2020). The above-mentioned processes could be described as biofilm-mediated degradation of MPs (termed as partial degradation) as presented in Fig. 8.3. The characteristics of the MPs, the soil properties, the type of microbial species and climate factors are amongst the factors influencing the biodegradation processes (Awet et al. 2018; Liang et al. 2019). In addition, Guo and colleagues (2020) highlighted that MPs exert selective effects on microhabitat which in turn affect the ecological balance of the soil alongside soil diversity and nutrient cycling. Similarly, Zhang et al.

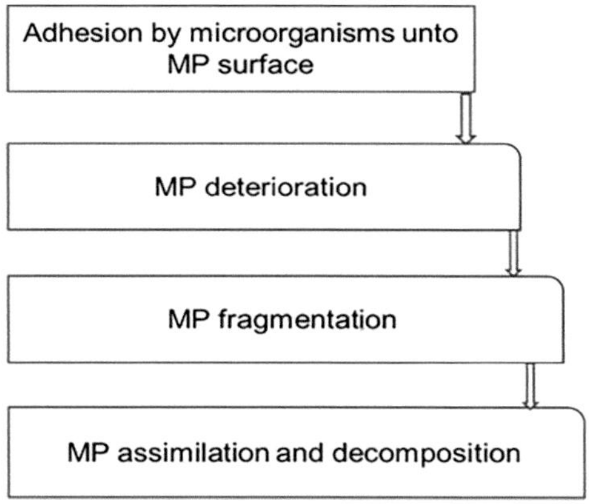

Fig. 8.3 Biofilm-mediated degradation of microplastics

(2021) purported that MPs create a great danger to the survival, reproduction, and growth of soil microbiota which in turn can affect terrestrial ecosystems in terms of services, biodiversity, and function. Li et al. (2023) demonstrated a difference in the impact of polypropylene and polyethylene MP on microbial richness, whereby the former caused a significant increase as opposed to polyethylene-derived MP. The authors further mentioned that plastics increased microbial functionality but hastened carbon loss and reduced the microbial biomass of the soil.

Also, natural weathering activities exert a series of physicochemical alterations in plastic when introduced into the soil, which in turn could affect the microbial communities therein (Pickett 2018). Photooxidation and thermal degradation of plastics take place, when plastic is exposed to sunlight, creating radicals. The radicals react with oxygen, oxidising the polymer surface and in turn lead to a rise in hydrophilic and carbonyl groups that improve the primary attachment, colonisation of microbes on the surface of MPs to facilitate microbial community formation, biofilm formation and degradation (Dong et al. 2020). Pickett (2018) mentioned that mechanical stresses in the soil result in MP fragmentation and together with biogeochemical attacks decrease the characteristics of MP (e.g., hydrophobicity, brittleness, and stiffness) thereby improving colonization by microbes and the development of plastic-associated microbial communities. Microbial degradation of MPs is carried out by diverse group of microorganisms including Actinomycetes (*Streptomyces* sp., *Saccharothrix waywayandensis*) bacterial strains (*Bacillus cereus*, *Anoxybacillus rupiensis*, *Achromobacter denitrificans* and *Bacillus* sp. YP1), fungi (*Aspergillus*, *Penicillium*, *Cladosporium*) and algae (*Oscillatoria*, *Spirogyra*, *Anabaena*) amongst others, that utilize MP as their carbon source and their metabolic activities exert changes on the morphology and chemical structure of the MPs alongside weight loss (Omidoyin and Jho 2023). MP pollution of agricultural soils can exert adverse effects across the agroecosystem, ground water resources, bioaccumulation in plants and trophic transfer to food chains through food crops and ultimately, to the health of the population (Nath et al. 2023). An overall presentation of Microorganism-Microplastic interactions is presented in Fig. 8.4.

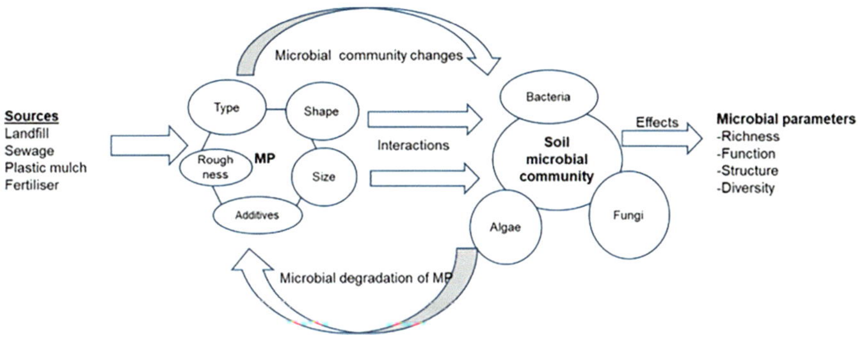

Fig. 8.4 Microorganism-microplastic interactions in the soil environment

8.6 Conclusion

Plastic wastes and the attendant characteristics is a global threat with no foreseeable reduction in the near future. Until recent time, much attention and studies have been on the marine plastisphere and a paucity of study on terrestrial plastisphere. Given the central position of soil in regulating and maintaining nutrient cycling and biodiversity respectively, and is fundamental to food security and sustainability, this resource must be guarded at all points. MP pollution of agricultural soils can exert adverse effects across the agroecosystem, ground water resources, bioaccumulation in plants and trophic transfer to food chains through food crops and ultimately, to the health of the population (Nath et al. 2023). Whatever impacts the soil has a ripple effect on the ecosystem, therefore attention is shifting to soil plastisphere, having been recognized as a major hub of microbial diversity and functions. Nevertheless, much information is still desired in 'managing' this neo-microbial niche for the betterment of the ecosystem.

References

Alassali A, Barouta D, Tirion H, Moldt Y, Kuchta K (2020) Towards a high quality recycling of plastics from waste electrical and electronic equipment through separation of contaminated fractions. J Hazard Mater 387:121741. https://doi.org/10.1016/j.jhazmat.2019.121741

Aransiola SA, Ikhumetse AA, Babaniyi BR, Abioye OP, Oyedele OJ, Falade NO (2023a) Phytoaccumulation of micro and nanoplastics: root uptake. In: Maddela NR, Reddy KV, Ranjit P (eds) Micro and nanoplastics in soil. Springer, Cham. https://doi.org/10.10007/978-3-031-21195-9_8

Aransiola SA, Victor-Ekwebelem MO, Ajiboye AE, Leh-Togi Zobeashia SS, Ijah UJJ, Oyedele OJ (2023b) Ecological impacts and toxicity of micro and nanoplastics in agroecosystem. In: Maddela NR, Reddy KV, Ranjit P (eds) Micro and nanoplastics in soil. Springer, Cham. https://doi.org/10.10007/978-3-031-21195-9_10

Arias-Andres M, Klumper U, Rojas-Jimenez K, Grossart HP (2018) Microplastic pollution increases gene exchange in aquatic ecosystems. Environ Pollut 237:253–261. https://doi.org/10.1016/j.envpol.2018.02.058

Ateia M, Zheng T, Calace S, Tharayil N, Pilla S, Karanfil T (2020) Sorption behavior of real microplastics (MPs): insights for organic micropollutants adsorption on a large set of well-characterized MPs. Sci Total Environ 720:137634

Auta SH, Abioye OP, Aransiola SA, Bala JD, Chukwuemeka VI, Hassan A, Aziz A, Fauziah SH (2022) Enhanced microbial degradation of PET and PS microplastics under natural conditions in mangrove environment. J Environ Manag 304(2022):114273. https://doi.org/10.1016/j.jenvman.2021.114273

Avio CG, Gorbi S, Milan M, Benedetti M, Fattorini D, d'Errico G et al (2015) Pollutants bioavailability and toxicological risk from microplastics to marine mussels. Environ Pollut 198:211–222

Awet TT, Kohl Y, Meier F, Straskraba S, Grun AL, Ruf T et al (2018) Effects of polystyrene nanoparticles on the microbiota and functional diversity of enzymes in soil. Environ Sci Eur 30:11–10. https://doi.org/10.1186/s12302-018-0140-6

Barrios E (2007) Soil biota, ecosystem services and land productivity. Ecol Econ 64:269–285

Behera S, Das S (2003) Environmental impacts of microplastic and role of plastisphere microbes in the biodegradation and upcycling of microplastic. Chemosphere 334(2023):138928

Bläsing M, Amelung W (2018) Plastics in soil: analytical methods and possible sources. Sci Total Environ 612:422–335. https://doi.org/10.1016/j.scitotenv.2017.08.086

Cincinelli A, Martellini T, Guerranti C, Scopetani C, Chelazzi D, Giarrizzo T (2019) A potpourri of microplastics in the sea surface and water column of the Mediterranean Sea. TrAC Trends Anal Chem 110:321–326. https://doi.org/10.1016/j.trac.2018.10.026

Danso D, Chow J, Streit WR (2019) Plastics: environmental and biotechnological perspectives on microbial degradation. Appl Environ Microbiol 85(19):14

de Souza MAA, Lau CW, Kloas W, Bergmann J, Bachelier JB, Rillig C et al (2019) Microplastics can change soil properties and affect plant performance. Environ Sci Technol 53(10):6044–6052. https://doi.org/10.1021/acs.est.9b01339

Dong M, Zhang Q, Xing X, Chen W, She Z, Luo Z (2020) Raman spectra and surface changes of microplastics weathered under natural environments. Sci Total Environ 739:1–9

Erkekoglu P, Kocer-Gumusel B (2014) Genotoxicity of phthalates. Toxicol Mech Methods 24(9):616–626. https://doi.org/10.3109/15376516.2014.960987

European Bioplastics (2018) Bioplastics market data 2018. European Bioplastics, Berlin

Finley K et al (2018) Soil mineral assemblage and substrate quality effects on microbial priming. Geoderma 322:38–47

Frere L, Maignien L, Chalopin M, Huvet A, Rinnert E, Morrison H, Reveillau J et al (2018) Microplastic bacterial communities in the Bay of Brest: influence of polymer type and size. Environ Pollut 242:614–625

Geyer R (2020) Chapter 2—production, use, and fate of synthetic polymers. In: Letcher TM (ed) Plastic waste and recycling. Academic, Oxford, pp 13–32

Geyer R, Jambeck JR, Law KL (2017) Production, use, and fate of all plastics ever made. Sci Adv 3(25–29):e1700782

Golia EE, Kantzou O-D, Chartodiplomenou M-A, Papadimou SG, Tsiropoulos NG (2023) Study of potentially toxic metal adsorption in a polluted acid and alkaline soil: influence of soil properties and levels of metal concentration. Soil Syst 7:16

Guo A, Pan C, Su X, Zhou X, Bao Y (2022) Combined effects of oxytetracycline and microplastic on wheat seedling growth and associated rhizospherebacterial communities and soil metabolite profiles. Environ Pollut 302:119046. https://doi.org/10.1016/j.envpol.2022.119046

Hartmann NB, Hüffer T, Thompson RC, Hassellöv M, Verschoor A, Daugaard AE (2019) Are we speaking the same language? Recommendations for a definition and categorization framework for plastic debris. Environ Sci Technol 53:1039–1047. https://doi.org/10.1021/acs.est.8b05297

Helmberger MS, Tiemann K, Grieshop (2019) Towards an ecology of soil microplastics. Funct Ecol 7:550–560

Horton AA, Walton A, Spurgeon DJ, Lahive E, Svendsen C (2017) Microplastics in freshwater and terrestrial environments: evaluating the current understanding to identify the knowledge gaps and future research priorities. Sci Total Environ 586:127–141. https://doi.org/10.1016/j.scitotenv.2017.01.190

Huo Y, Dijkstra FA, Possell M, Singh B (2022) Ecotoxicological effects of plastics on plants, soil fauna and microorganisms: a meta-analysis. Environ Pollut 310:119892. https://doi.org/10.1016/j.envpol.2022.119892

Joos L, De Tender C (2022) Soil under stress: the importance of soil life and how it is influenced by (micro)plastic pollution. Comput Struct Biotechnol J 20:1554–1566. https://doi.org/10.1016/j.csbj.2022.03.041

Kettner MT, Rojas-Jimenez K, Oberbeckmann S, Labrenz M, Grossart H-P (2017) Microplastics alter composition of fungal communities in aquatic ecosystems. Environ Microbiol 19:4447–4459

Khalid N, Aqeel M, Noman A et al (2021) Linking effects of microplastics to ecological impacts in marine environments. Chemosphere 264:1–13. https://doi.org/10.1016/j.chemosphere.2020.128541

Kostakioti M, Maria Hadjifrangiskou M, Hultgren SJ (2013) Bacterial biofilms: development, dispersal, and therapeutic strategies in the dawn of the postantibiotic era. Cold Spring Harb Perspect Med 3:a010306

Krehl A, Schöllkopf U, Májeková M, Tielbörger K, Tomiolo S (2022) Effects of plastic fragments on plant performance are mediated by soil properties and drought. Sci Rep 12:17771

Kumar RV, Kanna G, Elumali S (2017) Biodegradation of polyethylene by green photosynthetic microalgae. J Bioremed Biodegr 8:2

Kuzyakov Y, Blagodatskaya E (2015) Microbial hotspots and hot moments in soil: concept and review. Soil Biol Biochem 83:184–199

Lambert S, Wagner M (2016) Characterisation of nanoplastics during the degradation of polystyrene. Chemosphere 145:265–268. https://doi.org/10.1016/j.chemosphere.2015.11.078

Lear G, Kingsbury M, Franchini S, Gambarini V, Maday SDM, Wallbank JA, Weaver L, Pantos O (2021) Plastics and the microbiome: impacts and solutions. Environ Microbiome 16:2. https://doi.org/10.1186/s40793-020-00371-w

Lehmann A, Leifheit EF, Gerdawischke M, Rillig MC (2021) Microplastics have shape- and polymer-dependent effects on soil aggregation and organic matter loss—an experimental and meta-analytical approach. Microplast Nanoplast 7:2662–4966. https://doi.org/10.1186/s43591-021-00007-x

Li C, Gan Y, Zhang C, He H, Fang J, Wang L, Wang Y, Liu J (2021) "Microplastic communities" in different environments: differences, links, and role of diversity index in source analysis. Water Res 188:116574. https://doi.org/10.1016/j.watres.2020.116574

Li J, Yu S, Yu Y, Xu M (2022) Effects of microplastics on higher plants: a review. Bull Environ Contam Toxicol 109:241–265. https://doi.org/10.1007/s00128-022-03566-8

Li Y, Hou Y, Hou Q, Long M, Wang Z, Rillig MC, Liao Y, Yong T (2023) Soil microbial community parameters affected by microplastics and other plastic residues. Front Microbiol 14:1258606

Lithner D, Larsson A, Dave G (2011) Environmental and health hazard ranking and assessment of plastic polymers based on chemical composition. Sci Total Environ 409(18):3309–3324

Liu S, Shi J, Wang J, Dai Y, Li H, Li J, Liu X, Chen X, Wang Z, Zhang P (2021) Interactions between microplastics and heavy metals in aquatic environments: a review. Front Microbiol 12:652520

Liu X, Yao H, Zhao X, Ge C (2023) Biofilm formation and control of foodborne pathogenic bacteria. Molecules 28:2432. https://doi.org/10.3390/molecules28062432

Li K, Xu L, Bai X, Zhang G, Zhang M, Huang Y (2020) Differential fungal assemblages and functions between the plastisphere of biodegradable and conventional microplastics in farmland. Sci Total Environ 906:167478.https://doi.org/10.1016/j.scitotenv.2023.167478

MacLean J, Mayanna S, Benning LG, Horn F, Bartholomäus A, Wiesner Y, Wagner D, Liebner S (2021) The terrestrial plastisphere: diversity and polymer-colonizing potential of plastic-associated microbial communities in soil. Microorganisms 9(9):1876. https://doi.org/10.3390/microorganisms9091876

McGivney E, Cederholm L, Barth A, Hakkarainen M, Hamacher-Barth E, Ogonowski M, Gorokhova E (2020) Rapid physicochemical changes in microplastic induced by biofilm formation. Front Bioeng Biotechnol. https://doi.org/10.3389/fbioe.2020.00205

Mészáros E, Bodor A, Kovács E, Papp S, Kovács K, Perei K, Feigl G (2023) Impacts of plastics on plant development: recent advances and future research directions. Plants (Basel) 12(18):3282. https://doi.org/10.3390/plants12183282. PMID: 37765446; PMCID: PMC10538034

Mohanan N, Montazer Z, Sharma PK, Levin DB (2020) Microbial and enzymatic degradation of synthetic plastics. Front Microbiol 11:580709. https://doi.org/10.3389/fmicb.2020.580709

Muhammad MH, Idris AL, Fan X, Guo Y, Yu Y, Jin X, Qiu J, Guan X, Huang T (2020) Beyond risk: bacterial biofilms and their regulating approaches. Front Microbiol 11:928. https://doi.org/10.3389/fmicb.2020.00928

Manyi-Loh CE, Mamphwell SN, Meyer EL, Okoh AI (2019) Microbial anaerobic digestion: Process dynamics and implications from renewable energy, environmental and agronomy perspectives. Int J Envt Sci Tech 16:3913–3934. https://doi.org/10.1007/s13762-019-02380w

Musa IO, Auta HS, Ilyasu US, Aransiola SA, Makun HA, Adabara NU, Abioye OP, Aziz A, Jayanthi B, Maddela NR, Prasad R (2024) Micro and nanoplastics in environment—degradation, detection, ecological impact. Int J Environ Res 18:1. https://doi.org/10.1007/s41742-023-00551-9

Nath J, De J, Sur S, Banerjee P (2023) Interaction of microbes with microplastics and nanoplastics in the agroecosystems—impact on antimicrobial resistance. Pathogens 12:888. https://doi.org/10.3390/pathogens120708

Oberbeckmann S, Loeder MG, Gerdts G, Osborn AM (2014) Spatial and seasonal variation in diversity and structure of microbial biofilms on marine plastics in Northern European waters. FEMS Microbiol Ecol 90:478–492

Omidoyin KC, Jho EH (2023) Effect of microplastics on soil microbial community and microbial degradation of microplastics in soil: a review. Environ Eng Res 28(6):220716

Pickett JE (2018) Weathering of plastics. In: Kutz M (ed) Handbook of environmental degradation of materials, 3rd edn. William Andrew Publishing, Oxford, pp 163–184

Piehl S, Leibner A, Löder MGJ et al (2018) Identification and quantification of macro- and microplastics on an agricultural farmland. Sci Rep 8:17950. https://doi.org/10.1038/s41598-018-36172-y

Puckowski A, Cwiek W, Mioduszewska K, Stepnowski P, Bialk-Bielinska A (2021) Sorption of pharmaceuticals on the surface of microplastics. Chemosphere 263:127976

Qiu Y, Zhou S, Zhang C, Zhou Y, Qin W (2022) Soil microplastic characteristics and the effects on soil properties and biota: a systematic review and metaanalysis. Environ Pollut 313:120183

Rather MA, Gupta K, Mandal M (2021) Microbial biofilm: formation, architecture, antibiotic resistance, and control strategies. Braz J Microbiol 52:1701–1718

Rillig MC, Lehmann A (2020) Microplastic in terrestrial ecosystems. Science 368(6498):1430–1431

Robinson DA, Hockley N, Dominati E, Lebron I, Scow KM, Reynolds B et al (2012) Natural capital, ecosystem services, and soil change: why soil science must embrace an ecosystems approach. Vadose Zone J 11:vzj2011.0051

Rogers KL, Carreres-Calaburg JA, Gorokhova E, Posth MCY (2020) Micro-by-micro interactions: how microorganisms influence the fate of marine microplastics. Limnol Oceanogr Lett 5:18–36

Rüthi J, Bölsterli D, Pardi-Comensoli L, Brunner I, Frey B (2020) The "Plastisphere" of biodegradable plastics is characterized by specific microbial taxa of Alpine and Arctic soils. Front Environ Sci 8:562263. https://doi.org/10.3389/fenvs.2020.562263

Rillig MC, Kim SW, Zhu YG (2024) The soil plastisphere. Nat Rev Microbiol 22(2):64–74. https://doi.org/10.1038/s41579-023-00967-2

Schöpfer L, Menzel R, Schnepf U, Ruess L, Marhan S, Brümmer F, Pagel H, Kandeler E (2020) Microplastics effects on reproduction and body length of the soil-dwelling nematode Caenorhabditis elegans. Front Environ Sci 8:41. https://doi.org/10.3389/fenvs.2020.00041

Singh SP, Sharma P, Bano A, Nadda AK, Varjani S (2022) Microbial communities in plastisphere and free-living microbes for microplastic degradation: a comprehensive review. Green Anal Chem 3:100030. https://doi.org/10.1016/j.greeac.2022.100030

Sun Y, Wu M, Xie S, Zang J, Wang X, Yang Y, Li C, Wang J (2024) Homogenization of bacterial plastisphere community in soil: a continental-scale microcosm study. ISME Commun 4(1):1–12. https://doi.org/10.1093/ismeco/ycad012

Tiwari N, Santhiya D, Sharma JG (2020) Microbial remediation of micro-nano plastics: current knowledge and future trends. Environ Pollut 265:Article 115044

Tokiwa Y, Calabia BP, Ugwu CU, Aiba S (2009) Biodegradability of plastics. Int J Mol Sci 10(9):3722–3742. https://doi.org/10.3390/ijms10093722

Tuson HH, Weibel DB (2013) Bacteria-surface interactions. Soft Matter 9:4368–4380

Tziourrou P, Golia EE (2024) Plastics in agricultural and urban soils: interactions with plants, micro-organisms, inorganic and organic pollutants: an overview of polyethylene (PE) litter. Soil Syst 8(1):23

Uwamungu JY, Wang Y, Shi G, Pan S, Wang Z, Wang L, Yang S (2022) Microplastic contamination in soil agro-ecosystems: a review. Environ Adv 9:100273. https://doi.org/10.1016/j.envadv.2022.100273

Valentine KL, Boxall ABA (2022) Interactions between plastic, microbial biofilms and Gammarus pulex: an initial investigation. Bull Environ Contam Toxicol 108:609–615. https://doi.org/10.1007/s00128-021-03448-5

Van Cauwenberghe L, Devriese L, Galgani F, Robbens J, Janssen CR (2015) Microplastics in sediments: a review of techniques, occurrence and effects. Mar Environ Res 111:5–17. https://doi.org/10.1016/j.marenvres.2015.06.007

Viljakainen VR, Hug LA (2021) New approaches for the characterization of plastic-associated microbial communities and the discovery of plastic-degrading microorganisms and enzymes. Comput Struct Biotechnol J 19(2021):6191–6200

Wang C, Wang L, Ok YS, Tsang DCW, Hou D (2022a) Soil plastisphere: exploration methods, influencing factors, and ecological insights. J Hazard Mater 430:128503, ISSN 0304-3894. https://doi.org/10.1016/j.jhazmat.2022.128503

Wang W, Yuan W, Xu EG, Li L, Zhang H, Yang Y (2022b) Uptake, translocation, and biological impacts of micro(nano)plastics in terrestrial plants: progress and prospects. Environ Res 203:111867. https://doi.org/10.1016/j.envres.2021.111867

Webb HK, Arnott J, Crawford RJ, Ivanova EP (2013) Plastic degradation and its environmental implications with special reference to poly(ethylene terephthalate). Polymers 5:1–18. https://doi.org/10.3390/polym5010001

Weber CJ, Santowski A, Chifflard P (2022) Investigating the dispersal of macro-and microplastics on agricultural fields 30 years after sewage sludge application. Sci Rep 12(1):6401. https://doi.org/10.1038/s41598-022-10294-w

Wu X, Lu J, Du M, Xu X, Beiyuan J, Sarkar B, Bolan N, Xu W, Xu S, Chen X, Wu F, Wang H (2021) Particulate plastics-plant interaction in soil and its implications: a review. Sci Total Environ 792:148337. https://doi.org/10.1016/j.scitotenv.2021.148337

Yang Y, Liu G, Song W, Ye C, Lin H, Li Z, Liu W (2019) Plastics in the marine environment are reservoirs for antibiotic and metal resistance genes. Environ Int 123:79–86. https://doi.org/10.1016/j.envint.2018.11.061

Yang K, Chen Q-L, Chen M-L, Hong-Zhe Li H, Liao QP, Zhu Y-G, Cui L (2020) Temporal dynamics of antibiotic resistome in the plastisphere during microbial colonization. Environ Sci Technol 54:11322–11332. https://doi.org/10.1021/acs.est.0c04292

Yang L, Zhang Y, Kang S, Wang Z, Wu C (2021) Microplastics in soil: a review on methods, occurrence, sources, and potential risk. Sci Total Environ 780:146546. https://doi.org/10.1016/j.scitotenv.2021.146546

Yang W, Li Y, Boraschi D (2023) Association between microorganisms and microplastics: how does it change the host–pathogen interaction and subsequent immune response? Int J Mol Sci 24:4065

Yuan J, Ma J, Sun Y, Zhou T, Zhao Y, Yu F (2020) Microbial degradation and other environmental aspects of microplastics/plastics. Sci Total Environ 715:1–9. https://doi.org/10.1016/j.scitotenv.2020.136968

Zeb A, Liu W, Shi R, Lian Y, Wang Q, Tang J, Lin D (2022) Evaluating the knowledge structure of micro- and nanoplastics in terrestrial environment through Scientometric assessment. Appl Soil Ecol 177:104507. https://doi.org/10.1016/j.apsoil.2022.104507

Zettler ER, Mincer TJ, Amaral-Zettler LA (2013) Life in the "plastisphere": microbial communities on plastic marine debris. Environ Sci Technol 47:7137–7146

Zhang Y, Lu J, Wu J, Wang J, Luo Y (2020) Potential risks of microplastics combined with superbugs: enrichment of antibiotic resistant bacteria on the surface of microplastics in mariculture system. Ecotoxicol Environ Saf 187:109852

Zhang X, Li Y, Ouyang D, Lei J, Tan Q, Xie L, Li Z, Liu T, Xiao Y, Farooq TH, Wu X, Chen L, Yan W (2021) Systematical review of interactions between microplastics and microorganisms in the soil environment. J Hazard Mater 418:126288

Zhu F, Zhu C, Wang C, Gu C (2019) Occurrence and ecological impacts of microplastics in soil systems: a review. Bull Environ Contam Toxicol 102:741–749

Zhu D, Ma J, Li G, Rillig MC, Zhu Y-G (2022) Soil plastispheres as hotspots of antibiotic resistance genes and potential pathogens. ISME J 16:521–532. https://doi.org/10.1038/s41396-021-01103-9

Chapter 9
Plastisphere and Environmental Sustainability

Samuel Oluwatobi Agboola, Oluwakemi Christiana Alao, Gideon Oluranti Ojediran, Adedamola Titi Ojedokun, Oluwatosin Stephen Ajayi, Kayode Adesina Adegoke, Mohd Azmier Ahmad, and Olugbenga Solomon Bello

Abstract The widespread occurrence of plastic waste in the environment has led to the formation of a distinct microbial ecosystem known as the plastisphere. This intricate community lives on and within plastic surfaces, posing challenges as well as opportunities for environmental sustainability. This chapter explores the dynamics of the plastisphere and how it affects environmental equilibrium, looking at how

---

S. O. Agboola · O. C. Alao · A. T. Ojedokun · O. S. Ajayi · O. S. Bello (✉)
LAUTECH SDG 6 (Clean Water and Sanitation Research Group), Ogbomoso, Oyo State, Nigeria

LAUTECH SDG 11 (Sustainable Cities and Communities Research Group), Ogbomoso, Oyo State, Nigeria

Department of Pure and Applied Chemistry, Ladoke Akintola University of Technology, Ogbomoso, Oyo State, Nigeria
e-mail: osbello@lautech.edu.ng

G. O. Ojediran
LAUTECH SDG 6 (Clean Water and Sanitation Research Group), Ogbomoso, Oyo State, Nigeria

LAUTECH SDG 11 (Sustainable Cities and Communities Research Group), Ogbomoso, Oyo State, Nigeria

Department of Chemistry, Faculty of Physical Sciences, Alex Ekwueme Federal University, Ndufu-Alike, Ebonyi State, Nigeria

K. A. Adegoke
LAUTECH SDG 6 (Clean Water and Sanitation Research Group), Ogbomoso, Oyo State, Nigeria

LAUTECH SDG 11 (Sustainable Cities and Communities Research Group), Ogbomoso, Oyo State, Nigeria

Department of Industrial Chemistry, First Technical University, Ibadan, Oyo State, Nigeria

M. A. Ahmad
School of Chemical Engineering, Engineering Campus, Universiti Sains Malaysia, Penang, Malaysia

S. A. Aransiola et al. (eds.), *White Pollution: Biodiversity and hazards in marine plastisphere*, Environmental Contamination Remediation and Management,
https://doi.org/10.1007/978-3-031-95547-1_9

plastic pollution changes terrestrial and marine habitats, creating niches for microbial colonization and possibly upsetting ecosystems. Additionally, the potential for plastics to biodegrade within the plastisphere, taking into account the function of microbial enzymes in plastic degradation and the implications for waste management strategies is discussed. Furthermore, the movement of microplastics and related pollutants through the food chain, pointing out possible threats to biodiversity and human health are discussed. In order to address plastic pollution and advance environmental sustainability, it is imperative that mitigation and remediation strategies be developed with a thorough understanding of the complexities of the plastisphere ecosystem. The urgent need for interdisciplinary research and creative solutions to lessen the negative effects of plastic pollution on ecosystems and human health are highlighted in this chapter.

Keywords Environment · Plastic · Ecosystem · Plastisphere · Pollution

9.1 Introduction

The global surge in plastic production can be attributed to its widespread availability at low cost, corrosion resistance and remarkable durability (Xu et al. 2020). In 2018, plastics production was estimated to be 359 million metric tons and is projected to soar to 33 billion tons by 2050 (Roland et al. 2018). The prevalent form of plastic contaminating the environment is polyolefins, exemplified by polyethene (PE) and polypropylene (PP). This reflects their significant share in worldwide plastic manufacturing, with PE and PP making up 30% and 19.7% of the worldwide plastic production, respectively (Plastics Europe 2021; Auta et al. 2022). Additionally, products composed of PE and PP exhibit lower density than water, causing them to drift and facilitating their movement within water bodies (Erni-Cassola et al. 2019). The detrimental effect of plastics on our ecosystem stems from slow degradation, improper disposal practices and the fact that only a small portion of the synthetic organic polymers undergo recycling or incineration (Huang et al. 2021; Daizong and Yuzhu 2024; João et al. 2024; Musa et al. 2024). Around 4.8–12.7 million tons of plastics was projected to enter the Sea annually. This trend is expected to escalate massively over the next decade (Jambeck et al. 2015). Plastic aggregation in marine ecosystems raises environmental concerns as it can be ingested and subsequently passed along the food chain to higher trophic-level organisms (de Souza et al. 2018; Carbery et al. 2018; Xu et al. 2020; Aransiola et al. 2023b).

Like any substance that infiltrates an aquatic environment, plastic waste swiftly undergoes biofouling, wherein it becomes inhabited by living organisms that gradually accumulate (Thiel and Gutow 2005). Plastispheres are the biofilms produced by diverse microorganisms that inhabit the surface of plastics (Amaral-Zettler et al. 2020). The microbial community is made of various bacteria, single-celled algae, and fungi, additionally, macro-organisms like barnacles, bryozoans, hydroids, or multicellular algae (De Tender et al. 2017). Larger organisms exclusively inhabit sizable particles, while microbial biofilms occupy any surface, including tiny plastic particles measuring less than 1 mm, commonly referred to as "microplastics" (Hartmann et al. 2019). Recent research indicates variation in the microbial community among the plastisphere and surrounding water environments (Zhang et al. 2021; Li 2021). Moreover, Plastisphere microbial species could lead to increased growth of pathogens and antibiotic resistance in aqua ecosystems (Yang 2020). Considering that plastic debris can move a great distance and is located far away from its origin due to water currents, there exists a potential threat that invasive and potentially harmful species may become associated with plastic debris (João et al. 2024). Due to the following factors, it is important to study the microbial ecosystem within plastisphere and its effect on the surrounding environment (Aransiola et al. 2023a). In addition, increased microbial exposure resulting from the introduction of plastic debris into the food cycle poses a risk to both ecological integrity and human wellbeing (Li et al. 2020; Amaral-Zettler et al. 2020).

A more comprehensive understanding is needed to unravel how this emerging habitat forms distinctive microbiomes, along with the consequential functional implications on ecosystem and human health (Changchao et al. 2023). Numerous studies have demonstrated the presence of pathogenic variants of microorganisms in plastispheres. There are apprehensions regarding the potential of plastic to serve as a vector for introducing pathogens into food and drinking water chains (Kirstein et al. 2016; Silva et al. 2019; Yang et al. 2020). There is a high tendency for climate change to influence the microbial co-occurrence network within the plastisphere (Ji et al. 2022). Conversely, the microbiota within the plastisphere can exhibit adaptability to environmental changes by modifying their composition and functions (Wang et al. 2022). While information about the plastisphere is conclusive, studies primarily concentrate on the aquatic environment. Data on the plastisphere community in terrestrial biome are scarce, and the mechanisms shaping the ecosystem diversity in soil plastispheres still need to be determined (Yuanze et al. 2022). This review delves into recent research on the Plastisphere, exploring its multifaceted influence on environmental sustainability.

9.2 Overview in the Context of Environmental Sustainability

Disposed plastic waste material undergo reactions with their surroundings, leading to the release of various degradation products, including additives and polymers that constitute their composition, over their lifespan (Verma et al. 2016). The environmental and human health impact is determined by the type of plastics, their size, and the chemical composition released (Alabi et al. 2019). Diverse health conditions like thyroid dysfunction, obesity, diabetes, and reproductive impairment can be attributed to plastic pollution (Forrest et al. 2019). Incinerating plastic waste is environmentally detrimental, and the release of smoke, carbon monoxide, carbon dioxide, and nitrous oxide contributes to global warming (Prata et al. 2020). The non-biodegradable nature of plastics makes landfilling undesirable as it poses the risk of leaching into the environment (Agyeman et al. 2019). Leaked plastic waste causes havoc to the environment through blockage of waterways, creating a breeding environment for disease-carrying insects like mosquitoes, contaminating natural cycles essential for healthy ecosystems (Van Raamsdonk et al. 2020) (Fig. 9.1). The degradation of plastics on oceans' surfaces pollutes the ecosystem and contributes to global warming through the release of toxic greenhouse gases like ethylene and methane (Akdis 2021). However, due to the escalating volume of plastic waste generated daily, more than existing recycling measure are needed to reduce the adverse effects of plastic pollution (Shen et al. 2020). Hence, it is essential to explore sustainable applications for plastic waste management to address these challenges (Dey and Jamal 2021). Various recycling strategies, such as mechanical recycling, solvent separations, deconstruction or depolymerization (advanced chemical recycling), and biodegradation, are familiar choice to enhance environmentally friendly plastics (Fagnani et al. 2021; Vollmer et al. 2020; Zhao et al. 2018; Schyns and Shaver 2021; Walker et al. 2020).

Fig. 9.1 Schematic illustration of possible plastisphere havoc. (Source: Authors)

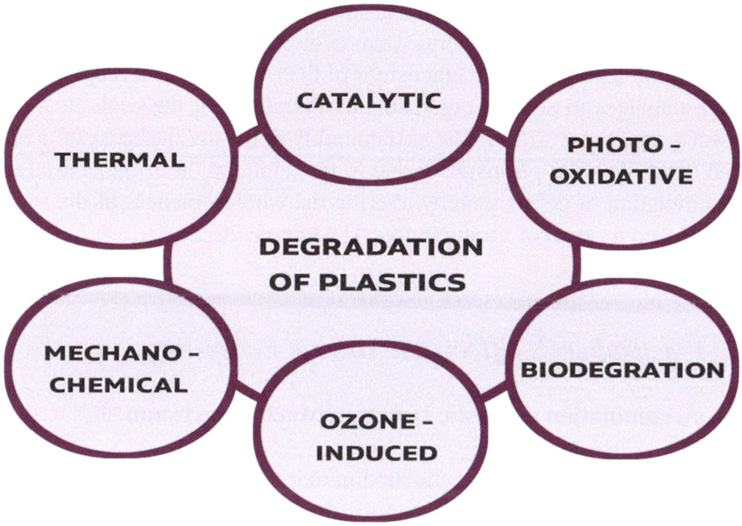

Fig. 9.2 Various form of degradation of plastics. (Source: Authors)

9.2.1 Global Views of Sustainability and Plastisphere

Environmental sustainability refers to the commitment to ensuring a balance between the ecosystem and natural resources to establish long-term health, meeting needs for both present and future demand (Noella 2024). Approximately 14 million tons of plastic enter the ocean yearly, primarily originating from Asian coastlines, with over 700 kinds of marine organisms impacted by the deadly effects of plastic pollution overtime (Schmaltz et al. 2020). In natural habitats, the plastisphere structure is influenced by temperature, seasonal variation, geographical factors, and plastic type (Oberbeckmann et al. 2014). Plastic pollution can be mitigated with the help of eco-friendly plastisphere bacteria like the bacterium Pseudomonas Putida, which breaks down polyvinyl chloride (PVC), and Rhodococcusruber degrades polystyrene (PS) (Fig. 9.2) (Rehse et al. 2016). In addition, the plastisphere displays the biosynthesis of secondary metabolites, a crucial aspect in advancing pharmaceutical and enzyme systems, providing solutions that optimize clinical and industrial processes (Mahnoor et al. 2023).

9.3 Impacts of the Plastisphere on Sustainability

The impacts of plastisphere on the ecosystem is predominant in the marine environment, these ecological niches are particularly problematic due to the persistent nature of plastic surfaces. The hydrophobic and exogenous nature of plastics wastes surfaces fosters their proliferation in marine environments which has resulted to the emergence of a unique ecological niche known as the plastisphere (Zettler et al. 2013; Koelmans et al. 2022). Generally, this ecological niche has significant impacts on sustainability across various dimensions of the aquatic habitat ranging from posing health damages to aquatic organisms to alterations of the ecological environment (McCormick et al. 2014). The sustainability of the marine ecosystem has been impaired extensively by plastisphere due to the adulteration of the coastal ecosystem, accumulation of debris thereby altering the wholesomeness of the ecosystem and has reduced its level of productivity (Al-Mamum 2023).

9.3.1 Plastisphere's Effects on Marine Ecosystems

9.3.1.1 Accumulation of Plastic Debris in Marine Environments

A large portion of the debris accumulated in marine environments through various pathways, including direct littering, runoff from land-based sources, and improper waste management includes the non-biodegradable plastics which is ubiquitous in nature (Dabrowka 2021). Pollution of marine habitats by plastics can be traced to land-based sources, marine activities and atmospheric deposition. The predominant land-based pollution pathway is from erosion run offs, wind transfer and effluents from waste water treatment plant accounting for 66%, 7% and 25% of the pollution (Plastics Europe 2016). Usually, the mobility of plastics in the marine ecosystem is predetermined by their buoyancy. Plastics which include PP and PE which are denser than water will float enhancing their transportation over wide range depending on the intensity of surface currents and wind (Woodall et al. 2014). Others which are less dense than water will sink in aquatic habitat, this includes the commonly available Polyvinylchloride (PVC). Plastics which are originally denser than water undergoes physicochemical alterations overtime resulting to increased density. Non-biodegradable plastics samples have been found to be persistent in marine ecosystem for several thousands of years due to their extensive durability (Cook et al. 2020). Plastics persist in the environment for extended periods, with estimates suggesting that a significant portion of plastic waste ends up in the oceans. Once in the marine environment, plastics undergo fragmentation due to physical and chemical processes, leading to the formation of microplastics (<5 mm) and nano plastics (<1 μm). Also, it has been reported that the Pacific Ocean presently possess an infamous great pacific garbage patch resulting from the aggregation of plastic wastes (Dey et al. 2022).

9.3.1.2 Alteration of Habitat Structure and Function

The presence of plastic debris can alter habitat structure and function in marine ecosystems. Large plastic items, such as bags and fishing gear, can entangle marine organisms, leading to injury, suffocation, or death. The leakage of plastics into marine habitat beyond creating an unpleasant and horrible water surface sight also results in the obstruction of water ways. This consequently gives rise to the formation of an undesired dam which serves as incubation spot for organisms that transmits water borne diseases (Nadal et al. 2016). This in turn results into secondary pollution of the water habitat. Plastic debris can also smother benthic habitats such as coral reefs and seagrass beds, affecting the recruitment and survival of sessile organisms and altering sediment dynamics (Van Raamsdonk et al. 2020). Furthermore, floating plastics can provide substrates for the colonization of invasive species, potentially disrupting native communities and ecosystem processes. Interactions of microplastics with the marine ecosystem results into chemical phenomenon such as biodegradation, photodegradation, uv-degradation which causes their disintegration into micro and nano plastics. The inherent characteristics of these products accounts for their persistence in the environment and their adsorption of potential contaminants hence furthering the adulteration of the ecosystem (Kumar et al. 2021).

9.3.1.3 Implications for Marine Organisms and Food Webs

Plastic pollution poses various threats to marine organisms and food webs. Many marine species mistake plastic debris for food, leading to ingestion and subsequent health effects. Ingested plastics can cause internal injuries, blockages in the digestive tract, and the leaching of toxic chemicals into tissues (Amalia 2023). These smaller particles are of particular concern as they can be ingested by a wide range of marine organisms from zooplankton to apex predators. Organisms such as sea turtles ingests plastic bags with the assumption that it is jelly fish and this results into their mortality as a result of internal blockages (Devi et al. 2022) Additionally, plastics can adsorb and concentrate persistent organic pollutants (POPs) and other contaminants, which can bioaccumulate and biomagnify through the food web, posing risks to higher trophic levels, including marine mammals and humans consuming seafood (Gola et al. 2021). The ingestion of plastics by aquatic wildlife results into the creation of a recurring problem for the wider food chain.

9.3.2 Possible Effects on Ecosystem Health and Biodiversity

9.3.2.1 Bioaccumulation and Biomagnification of Plastic-Associated Toxins

Plastics have a high surface area-to-volume ratio, making them efficient sorbents for hydrophobic pollutants such as polychlorinated biphenyls (PCBs), polycyclic aromatic hydrocarbons (PAHs), and heavy metals (Haque and Fan 2023). When ingested by marine organisms, plastics can release these adsorbed contaminants, which can then accumulate in tissues over time, leading to toxic effects. As predators consume contaminated prey, the concentration of pollutants can increase through biomagnification, posing risks to organisms at higher trophic levels (Du et al. 2022).

9.3.2.2 Disruption of Ecological Interactions and Nutrient Cycles

The presence of plastics in marine ecosystems offers a new environment for microorganisms. It causes disruption of ecological interactions and nutrient cycles. Greenhouse gases which result into climate change are consistent in the lifecycle of plastics (Kumar et al. 2021). Microorganisms involved in the nitrogen cycle are also affected in the plastisphere, especially in those associated with freshwater where there is a high abundance of denitrifiers which may increase the release of nitrite (aquatic toxicant) and nitrous oxide (greenhouse gas). Also, plastics can alter the behavior, feeding patterns, and reproductive success of marine organisms, leading to changes in population dynamics and community structure (Sun et al. 2023). Furthermore, plastics can serve as substrates for microbial colonization, potentially altering microbial communities and biogeochemical processes such as carbon and nitrogen cycling. These disruptions can have cascading effects on ecosystem functioning and services, including fisheries, nutrient recycling, and carbon sequestration.

9.3.2.3 Shifts in Species Composition and Biodiversity Loss

Plastic pollution can lead to shifts in species composition and biodiversity loss in marine ecosystems. Certain species may be more resilient to the presence of plastics, while others may suffer declines due to direct harm, habitat alteration, or competition with plastic-associated organisms (Tang 2024). Moreover, plastics can act as vectors for the dispersal of non-native species, leading to the introduction of invasive species in new habitats. These changes in species composition can have far-reaching consequences for ecosystem stability, resilience, and the provisioning of ecosystem services essential for human well-being.

9.3.3 Relationships Between Plastisphere and General Knowledge of Environmental Sustainability

9.3.3.1 Interconnections Between Plastic Pollution and Broader Sustainability Challenges

Plastic pollution is intricately linked to broader sustainability challenges, including resource depletion, pollution, and climate change. Plastics are derived from fossil fuels and require significant energy and resources for production, contributing to carbon emissions and environmental degradation (Kumar et al. 2021). Moreover, the persistence of plastics in the environment exacerbates pollution problems, impacting ecosystems, wildlife, and human health. Addressing plastic pollution requires systemic changes in production, consumption, and waste management practices, aligning with the goals of sustainable development and environmental stewardship (Thushari and Senevirathna 2020).

In line with SDG 6 which is focused on clean water and sanitation, efforts to remediate the aquatic environment is going to result into a long-term attainment of clean marine ecosystem. It has been reported by (Mason et al. 2018) that 90% of bottled water samples had plastic pollutants discovered in them. The extent of the plastics pollutant detected depends on the water treatment facility and the nature of the bottles used. Ensuring that plastics and disposed appropriately and recycled effectively will drastically reduce the presence of plastics in marine habitats (Krause et al. 2020). Hence, the evolution of plastisphere will be seemingly impossible and the safety of fresh water habitats which serve as the major source of drinking water will be guaranteed.

9.3.3.2 Role of Public Awareness and Education in Addressing Plastisphere Impacts

Public awareness and education play crucial roles in addressing plastisphere impacts and promoting sustainable behavior change. In the developing countries where plastics waste management is minimal, increasing public awareness about the sources, impacts, and solutions to plastic pollution can empower individuals to make informed choices and advocate for policy changes (Li et al. 2024). Educational initiatives targeting schools, communities, and stakeholders can foster environmental literacy, encourage waste reduction and recycling efforts, and promote alternative materials and consumption patterns that minimize plastic use. Moreover, they can be channels through which the extent of the threat that plastics possess especially through the emerging ecosystem—Plastisphere can be communicated to the society, hence, fostering a global effort to combat this salient emerging danger. Furthermore, public engagement and advocacy can mobilize collective action and support for policies and initiatives aimed at mitigating plastic pollution and promoting environmental sustainability (Leah et al. 2021).

9.3.3.3 Opportunities for Interdisciplinary Research and Collaboration

In order to globally attain a sustainable ecosystem that is void of plastics pollution, there is a need to implement workable mechanisms. Plastisphere research presents opportunities for interdisciplinary collaboration and innovation across various fields, including natural sciences, engineering, social sciences, and policy studies (Wen et al. 2021). Interdisciplinary approaches can facilitate a holistic understanding of plastic pollution dynamics, encompassing ecological, socio-economic, and governance dimensions (Yap et al. 2022). Collaborative efforts can foster the development of innovative technologies for plastic waste management, pollution prevention, and environmental monitoring (Borrelle et al. 2020). Moreover, interdisciplinary research can inform evidence-based policies and strategies that address plastic pollution at local, regional, and global scales, contributing to the advancement of sustainability goals and the protection of marine ecosystems for future generations (Yang et al. 2021). Likewise, researches can be directed to the development of eco-friendly substitutes for plastics and the sustainable recycling of the previously manufactured plastics.

9.4 Interactions of Microbes in the Plastisphere

The community of microbes within the plastisphere exhibits significant functions in the biodegradation of microplastics in aquatic environment. This microbial community in the plastisphere comprises of autotrophs, heterotrophs, predators, and pathogens which is complex, however produce a profound synergy (Borrelle et al. 2020). When plastics firstly go into the environment, they are immediately shielded by many organic materials and pollutants, known as the eco-corona (Galloway et al. 2017), expediting the progress of the colonization of microbes by diminishing the hydrophobicity of the surface (Wright et al. 2020). Microbes can fasten to the outer exposed space in the microplastics and thereby producing biofilms, that can eventually lead to accumulation of organic matter and nutrients on the surface of microplastics (Borrelle et al. 2020). Plastisphere comprises of associations of both microbes and mineral nutrients as a function of the end-result of plastic. The interactions between microbes and minerals, for instance the utilization of mineral substrate and collective formation, are of utmost importance particularly at revolving of building blocks in the course of the progression of the terrain.

These communities of microbes are originally comprised of fast-growing class of organisms that are of high beneficial surface colonization (Fig. 9.3). As the surface experiences degradation, the initial colonizers are slowly substituted by another organisms with capability to initiate disintegration at topmost layer, and subsequently dislocated by microbes incapable to decompose the topmost layer but capable to utilize intermediate minor products known as cheaters or cross-feeders (Datta et al. 2016). The process of biogeochemical processes dominates the elemental composition and microbial of the environment during transport and storage, in plastic environment forming relationship around the plastic during the movement from down to bottom (Rogers et al. 2020). The community of microbes in the plastisphere can also secrete enzymes that break down the plastic polymers thereby aiding the degradation of microplastics (Borrelle et al. 2020). Types of microbial interaction in the plastisphere and the development and progression in the plastisphere is shown in Table 9.1 and Fig. 9.3 respectively.

Fig. 9.3 The development and progression in the plastisphere. (Source: Authors)

Table 9.1 Types of microbial interaction in the plastisphere

S/N	Positive interaction	Negative interaction
1	Mutualism	Predation
2	Syntrophism	Parasitism
3	Pommensalism	Competition
3	Protocooperation	Amensalism (antagonism)

9.5 Mitigation of Plastisphere and Sustainable Practices

Plastics, with their versatile applications and durability, have become an integral part of modern life. However, the unintended consequences of plastic usage have become increasingly evident, with environmental concerns rising due to the persistence of plastic waste. The plastisphere, consisting of diverse microbial communities colonizing plastic surfaces, poses ecological challenges and potential threats to aquatic and terrestrial ecosystems. Understanding and addressing the mechanisms driving plastisphere formation is paramount in devising effective mitigation strategies.

9.5.1 Need for Early Warning Systems in Plastic Pollution Monitoring

Plastic pollution is a global environmental challenge with severe consequences for ecosystems, wildlife, and human health. In addressing this crisis, the implementation of early warning systems is imperative to enhance the effectiveness of plastic pollution monitoring and management. Some reasons why early warning systems are needed in plastic pollution monitoring are:

Timely Intervention and Mitigation: Early detection of plastic pollution allows for prompt intervention and mitigation strategies, minimizing the potential damage to ecosystems. Timely actions can prevent the escalation of the issue and enhance the success of remediation efforts.

Prevention of Cumulative Environmental Impact: Plastic pollution often accumulates over time, leading to long-term environmental damage. Early warning systems enables the identification of pollution hotspots before they become critical, allowing for preventive measures and reducing the cumulative impact on ecosystems.

Optimizing Resource Allocation: Early warning systems provide valuable data that can optimize resource allocation for cleanup operations. By identifying areas with higher plastic concentrations, authorities can allocate resources efficiently, ensuring a targeted and cost-effective response.

Minimizing Economic Impact: The economic consequences of plastic pollution, including damage to fisheries, tourism, and public health, can be substantial. Early warning systems contribute to minimizing these economic impacts by enabling swift responses that prevent the exacerbation of pollution-related problems.

Scientific Understanding and Data-driven Policies: Early warning systems generate valuable scientific data that enhances our understanding of the dynamics of plastic pollution. This information forms the basis for evidence-based policymaking, facilitating the development of effective regulations and strategies to address the issue at its roots.

Public Awareness and Participation: Early warning systems play a crucial role in raising public awareness about plastic pollution. Timely information empowers communities to actively participate in pollution prevention and cleanup efforts, fostering a sense of responsibility and environmental stewardship.

Adaptation to Changing Environmental Conditions: Environmental conditions are dynamic, and plastic pollution patterns may change over time. Early warning systems provide the adaptability needed to monitor and respond to shifts in pollution trends, ensuring that strategies remain effective in the face of evolving challenges.

9.5.2 Current State of Plastic Pollution Monitoring

The current state of plastic pollution monitoring involves a multifaceted approach, using various methods and technologies to assess the extent of plastic pollution in different environments.

Traditional Monitoring Methods: Traditional methods for assessing plastic pollution often involve field surveys and sampling. Researchers and environmental agencies conduct on-site inspections to quantify and characterize plastic waste in water bodies, beaches, and other ecosystems. This approach provides valuable qualitative data but is limited by its labor-intensive nature, time delays, and the inability to capture real-time dynamics.

Remote Sensing Technologies: Remote sensing technologies, such as satellite imagery and aerial surveys, have gained prominence in plastic pollution monitoring. These tools offer a broad-scale perspective, enabling the observation of large areas over time. Satellite data can be used to identify and track floating plastic debris in oceans and map the accumulation of plastic waste in coastal regions. However, challenges persist, including limitations in detecting smaller plastic particles and difficulties in distinguishing between types of debris.

Citizen Science Initiatives: Citizen science is a valuable resource in plastic pollution monitoring. Engaging the public in data collection through apps and community-based projects allows for a more extensive and diverse dataset. While this approach enhances spatial coverage and involves local communities, ensuring data accuracy and standardization remains a challenge.

Machine Learning and Automated Detection: Advances in machine learning and computer vision technologies have facilitated automated detection and classification of plastic debris in images. These algorithms can analyze large datasets quickly, providing a cost-effective and efficient means of monitoring. However, the reliability of these systems heavily relies on the quality and diversity of training data, and there is ongoing research to improve their accuracy.

Sensor Networks and IoT Devices: The use of sensor networks and Internet of Things (IoT) devices in water bodies enables real-time monitoring of plastic pollution. These devices can detect and quantify plastic particles, providing continuous data streams for better understanding temporal patterns. Challenges include the maintenance and deployment of such networks in diverse and often harsh environments.

9.5.3 Integration of Data Sources

Plastic pollution is a complex and dynamic environmental issue that requires a multifaceted approach to monitoring. The integration of various data sources plays a key role in enhancing the effectiveness of early warning systems, providing a more holistic understanding of the distribution, movement, and impact of plastic waste. The significance of integrating data from diverse sources are highlighted below:

Satellite Imagery: Satellite imagery offers a valuable tool for large-scale monitoring of plastic pollution. High-resolution satellite images provide insights into the extent of plastic debris in oceans, rivers, and coastal areas. Integration of satellite data allows for the identification of hotspots, tracking the movement of floating plastic, and assessing the impact on marine ecosystems.

Ground-based Sensors: Ground-based sensors, deployed in strategic locations, contribute real-time data on plastic concentrations in terrestrial environments. These sensors can detect microplastics in soil, sediment, and air, providing critical information for understanding the sources and pathways of plastic pollution. Integration of sensor data enhances the spatial resolution of monitoring systems (Jambeck 2019).

Citizen Science Initiatives: Engaging citizens in data collection expands the geographical coverage and fosters community involvement. Integrating data from citizen science initiatives not only supplements traditional monitoring methods but also promotes public awareness and participation in plastic pollution mitigation efforts.

Machine Learning and Data Analytics: Leveraging machine learning algorithms and data analytics facilitates the processing and analysis of vast datasets. These technologies enable the identification and classification of plastic debris in images, improving the accuracy and efficiency of monitoring systems. Integration of these tools enhances the interpretation of complex datasets (Cózar et al. 2014).

9.5.4 *Sustainable Plastic Alternatives*

Sustainable plastic alternatives are becoming popular due to the escalating environmental challenges associated with traditional plastics. This growing awareness has prompted the exploration of various innovative materials and technologies aimed at mitigating plastic pollution and reducing the ecological footprint. Some of the numerous sustainable plastic alternatives include:

Bioplastics: Bioplastics, derived from renewable sources such as corn starch, sugarcane, or algae, represent a sustainable alternative. Polylactic acid (PLA) is a prominent bioplastic derived from fermented plant sugars (Sharif et al. 2019), while polyhydroxyalkanoates (PHA) are produced by bacteria consuming plant sugars (Chen 2009).

Mushroom-based materials: Mycelium, the root structure of mushrooms, is harnessed to create biodegradable packaging. Researchers are exploring the potential of mycelium-based packaging to replace traditional materials (Digafe et al. 2022).

Algae-based plastics: Plastics extracted from algae offer a biodegradable alternative. Algae, being abundant and renewable, can be grown using minimal resources. Research highlights the potential of algae-based plastics for various applications (Beauty et al. 2023).

Paper and cardboard packaging: Natural, renewable, and biodegradable materials like paper and cardboard can be recycled or composted. Studies such as "Life Cycle Assessment of Beverage Packaging" underscore the environmental benefits of paper packaging (Vignali 2015).

Edible packaging: Made from edible materials like seaweed, starch, or proteins, edible packaging provides a sustainable alternative to single-use plastic. Seaweed-based packaging, for instance, has gained attention for its potential to reduce plastic waste (Waseem et al. 2023).

Biodegradable polymers: Biodegradable polymers, like polybutylene adipate-co-terephthalate (PBAT), naturally break down into harmless compounds over time, offering an alternative to persistent traditional plastics (Mohamed et al. 2020).

Glass and metal: Reusable and recyclable materials like glass and metal, though not universally applicable, provide sustainable options for certain products. The recyclability and environmental benefits of glass packaging are highlighted in studies such as "Life Cycle Assessment of Food Packaging Systems" (Vignali 2016). While these alternatives show promise, it is important to consider factors such as production scalability, cost-effectiveness, and end-of-life management. Continued research and technological advancements are needed to improve the performance and viability of sustainable plastic alternatives and drive the transition toward a more environmentally friendly future.

9.5.5 *Implications for Policy*

9.5.5.1 Role of Policy Frameworks in Early Warning Systems

The role of policy frameworks in supporting the development and implementation of early warning systems for plastic pollution is crucial in addressing and mitigating the environmental challenges posed by plastic waste. Policies play a fundamental role in shaping the strategies, guidelines, and mechanisms necessary for the establishment and effective operation of early warning systems.

Regulatory Frameworks: Policies provide the legal foundation for the establishment and operation of early warning systems. Governments can enact laws that mandate the creation of systems to monitor and provide timely warnings about plastic pollution.

Funding and Resource Allocation: Policy frameworks determine the allocation of financial resources and manpower for the development, maintenance, and improvement of early warning systems. Adequate funding is essential for the proper functioning and sustainability of such systems (United Nations Environment Programme 2018).

Interagency Collaboration: Policies can facilitate collaboration between various government agencies, research institutions, and non-governmental organizations involved in monitoring and combating plastic pollution. An integrated approach supported by policies ensures a comprehensive and efficient early warning system.

Data Collection and Standardization: Policy frameworks can establish standards for data collection, storage, and sharing. Standardization is crucial for ensuring that data from various sources can be integrated into a coherent early warning system, providing a more accurate picture of plastic pollution trends.

Public Awareness and Education: Policies can include provisions for public awareness and education campaigns regarding the impacts of plastic pollution. An informed public is more likely to participate in and support early warning initiatives.

International Cooperation: Given the transboundary nature of plastic pollution, international cooperation is essential. Policy frameworks can set the stage for collaboration between countries in sharing information, best practices, and technology to enhance global early warning efforts (European Environment Agency 2019).

Monitoring and Evaluation: Policies can establish frameworks for monitoring and evaluating the effectiveness of early warning systems. Regular assessments ensure that the system is adapting to emerging challenges and that resources are used efficiently.

9.5.5.2 Role of Research Findings on Policy Decisions

Research findings are important in shaping policy decisions related to plastic pollution at various levels—local, national, and international. These findings provide policymakers with valuable insights into the extent of the problem, its impacts on ecosystems and human health, and the effectiveness of potential interventions.

Local Level

Identifying Local Sources and Impacts: Local research can help pinpoint specific sources of plastic pollution, such as industries, urban areas, or specific types of waste. It can also assess the direct impact on local ecosystems, wildlife, and communities.

Tailoring Interventions: Research findings can inform the development of targeted policies that address local challenges. For example, if a particular type of plastic is prevalent, regulations can be designed to restrict its use or improve recycling infrastructure.

Community Engagement: Research can provide data on community perceptions and behaviors related to plastic use. Policymakers can use this information to design public awareness campaigns and community engagement programs.

National Level

Policy Formulation and Revision: Research findings contribute to the formulation and revision of national policies on plastic pollution. Policymakers can use data on the most significant sources, types, and pathways of plastic pollution to design effective regulations.

Economic Impact Assessment: Research can help policymakers understand the economic impact of plastic pollution, including costs associated with cleanup, healthcare, and damage to industries like tourism and fisheries. This information can be crucial for making informed decisions on regulatory measures.

Monitoring and Enforcement: Research can guide the development of monitoring and enforcement strategies to ensure that national policies are effectively implemented. It can help identify gaps in existing regulations and inform adjustments to improve their efficacy.

International Level

Global Cooperation: Research findings on the transboundary nature of plastic pollution can inform international cooperation efforts. Collaborative initiatives can be developed to address issues such as marine litter and plastic waste trade.

Setting Global Standards: Scientific evidence can guide the establishment of international standards for plastic production, use, and disposal. These standards can be adopted by countries worldwide, ensuring a more consistent and effective approach to tackling plastic pollution.

Resource Allocation: Research helps identify regions or countries that may require additional support or resources in addressing plastic pollution. International funding and assistance programs can be targeted towards areas with the greatest need.

9.6 Challenges, Future Directions and Research Needs

The intricate ecosystem that lives on and inside plastic waste, known as the plastisphere, has drawn attention as a serious environmental issue with far-reaching effects. Although some insights into this distinct microbial community have been gained through research, there are still a number of unanswered questions that require further investigation in order to promote environmental sustainability. The composition, diversity, and functionality of the plastisphere are still largely unknown. To fully understand the important microbial actors in this ecosystem and their interactions, more research is required. It's still unclear how the plastisphere might affect the environment and human health. There are still unanswered concerns about the plastisphere's function in the biodegradation of plastics, the possibility of antibiotic resistance gene transfer, and the potential for plastic debris to harbor invasive species and spread them to new areas. Numerous organisms have been known to consume the ubiquitous microplastics, which could endanger their health and have an effect on the food chain. It is essential to comprehend the relationship between the plastisphere and swallowed microplastics. It is essential to create efficient and standardized procedures for plastisphere sampling and analysis in order to accurately assess the plastisphere's dynamics and composition in a variety of environmental settings. By applying cutting-edge sequencing technologies, more information about the genetic composition and potential functions of the plastisphere microbiome can be obtained. Undertaking direct research in natural settings where plastic pollution transpires, facilitates a comprehensive comprehension of the plastisphere's function in ecosystems. Certain facets of the plastisphere, such as interactions with ingested microplastics or biodegradation processes, can be isolated and studied in controlled laboratory settings.

In order to inform mitigation strategies and regulations, it is imperative to evaluate the potential ecological and health risks associated with the plastisphere. Changing societal perceptions of the plastisphere and its link to plastic pollution can influence policy choices as well as consumer behavior. Establishing consistent procedures for data sharing, analysis, and sampling is crucial to producing data that is comparable across various study teams and geographical areas. To ascertain whether the plastisphere has the capacity to promote plastic degradation and to pinpoint methods for enhancing this procedure, more investigation is required. A multifaceted approach is needed to develop effective mitigation strategies to minimize plastic pollution and the ecological and health risks that it entails. Some of these strategies include reducing plastic use, improving waste management, and potentially utilizing the plastisphere's biodegradation potential. It is possible to learn more about the plastisphere and its significance for environmental sustainability by tackling these issues and exploring new research avenues. This information can help develop policies that reduce plastic pollution, encourage biodegradation, and safeguard the wellbeing of living things and ecosystems.

References

Agyeman S, Obeng-Ahenkora NK, Assiamah S, Twumasi G (2019) Exploiting recycled plastic waste as an alternative binder for paving blocks production. Case Stud Constr Mater 11:e00246

Akdis CA (2021) Does the epithelial barrier hypothesis explain the increase in allergy, autoimmunity and other chronic conditions? Nat Rev Immunol 21:739–751

Alabi OA, Ologbonjaye KI, Awosolu O, Alalade OE (2019) Public and environmental health effects of plastic wastes disposal: a review. J Toxicol Risk Assess 5:1–13

Al-Mamum M (2023) Plastic pollution threatens biodiversity and human health. Acta Sci Agric 7(11):3–4

Amalia E (2023) Uncovering the plastisphere: a closer look to plastic pollution. Knowledge: Seven Clean Seas

Amaral-Zettler LA, Zettler ER, Mincer TJ (2020) Ecology of the plastisphere. Nat Rev Microbiol 18(3):139–151. https://doi.org/10.1038/s41579-019-0308-0

Aransiola SA, Ikhumetse AA, Babaniyi BR, Abioye OP, Oyedele OJ, Falade NO (2023a) Phytoaccumulation of micro and nanoplastics: root uptake. In: Maddela NR, Reddy KV, Ranjit P (eds) Micro and nanoplastics in soil. Springer, Cham. https://doi.org/10.10007/978-3-031-21195-9_8

Aransiola SA, Victor-Ekwebelem MO, Ajiboye AE, Leh-Togi Zobeashia SS, Ijah UJJ, Oyedele OJ (2023b) Ecological impacts and toxicity of micro and nanoplastics in agroecosystem. In: Maddela NR, Reddy KV, Ranjit P (eds) Micro and nanoplastics in soil. Springer, Cham. https://doi.org/10.10007/978-3-031-21195-9_10

Auta SH, Abioye OP, Aransiola SA, Bala JD, Chukwuemeka VI, Hassan A, Aziz A, Fauziah SH (2022) Enhanced microbial degradation of PET and PS microplastics under natural conditions in mangrove environment. J Environ Manag 304(2022):114273. https://doi.org/10.1016/j.jenvman.2021.114273

Beauty A, Mashura S, Zobaidul K (2023) Algae for plastic biodegradation: emerging approach in mitigating marine pollution. In: Microbiology for cleaner production and environmental sustainability, pp 347–362. https://doi.org/10.1201/9781003394600-18

Borrelle SB, Ringma J, Law KL, Monnahan CC, Lebreton L, McGivern A, Murphy E, Jambeck J, Leonard GH, Hilleary MA (2020) Predicted growth in plastic waste exceeds efforts to mitigate plastic pollution. Science 369:1515–1518

Carbery M, Wayne O-C, Thavamani P (2018) Trophic transfer of microplastics and mixed contaminants in the marine food web and implications for human health. Environ Int 115:400–409. https://doi.org/10.1016/j.envint.2018.03.007

Changchao L, Michael R, Gillings CZ, Qinglin C, Dong Z, Jie W, Kankan Z, Qicheng X, Polly HL, Xiangdong L, Jian L, Ling J (2023) Ecology and risks of the global plastisphere as a newly expanding microbial habitat. Innovation 5(1):100543. https://doi.org/10.1016/j.xinn.2023.100543

Chen G-Q (2009) A microbial polyhydroxyalkanoates (PHA) based bio- and materials industry. Chem Soc Rev 38(8):2434. https://doi.org/10.1039/b812677c

Cook S, Chan H-L, Abolfathi S, Bending GD, Schäfer H, Pearson JM (2020) Longitudinal dispersion of microplastics in aquatic flows using fluorometric techniques. Water Resour 170:115337. https://doi.org/10.1016/j.watres.2019.115337

Cózar A, Fidel E, Ignacio JG, Xabier I, Bárbara Ú, Santiago H, Álvaro TP, Sandra N, Juan G, Andrea R, María LF, Carlos MD (2014) Plastic debris in the open ocean. Proc Natl Acad Sci 111(28):10239–10244

Dabrowska A (2021) A roadmap for a plastisphere. Mar Pollut Bull 167:112322

Daizong M, Yuzhu L (2024) Assessing the settling velocity of biofilm-encrusted microplastics: accounting for biofilms as an equivalent to surface roughness. Environ Sci Technol 58(2):1329–1337. https://doi.org/10.1021/acs.est.3c07147

Datta MS, Sliwerska E, Gore J, Polz MF, Cordero OX (2016) Microbial interactions lead to rapid micro-scale successions on model marine particles. Nat Commun 7:1–7

de Souza M, Anderson A, Chung WL, Jennifer T, Werner K, Anika L, Roland B, Matthias C (2018) Impacts of microplastics on the soil biophysical environment. Environ Sci Technol 52:9656–9665. https://doi.org/10.1021/acs.est.8b02212

De Tender C, Devriese LI, Haegeman A, Maes S, Vangeyte J, Cattrijsse A, Dawyndt P, Ruttink T (2017) Temporal dynamics of bacterial and fungal colonization on plastic debris in the North Sea. Environ Sci Technol 51(13):7350–7360. https://doi.org/10.1021/acs.est.7b00697

Devi KN, Raju P, Santhanam P, Perumal P (2022) Impacts of microplastics on marine organisms: present perspectives and the way forward. Egypt J Aquat Res 48:205–209

Dey TK, Jamal M (2021) Separation of microplastics from water—what next? J Water Process Eng 44:102332

Dey S, Rout AK, Behera BK, Ghosh K (2022) Plastisphere community assemblage of aquatic environment: plastic-microbe interaction, role in degradation and characterization technologies. Environ Microbiome 17:32

Digafe A, Mesfin T, Ajoy KM (2022) Mycelium-based composite: the future sustainable biomaterial. Int J Biomater 2022:8401528. https://doi.org/10.1155/2022/8401528

Du Y, Liu X, Dong X, Yin Z (2022) A review on marine plastisphere: biodiversity, formation, and role in degradation. Comput Struct Biotechnol J 20:975–988

Erni-Cassola G, Zadjelovic V, Gibson MI, Christie-Oleza JA (2019) Distribution of plastic polymer types in the marine environment; a meta-analysis. J Hazard Mater 369:691–698. https://doi.org/10.1016/j.jhazmat.2019.02.067

European Environment Agency (2019) Preventing plastic waste in Europe: policy measures to address environmental and economic challenges. European Environment Agency

Fagnani DE, Tami JL, Copley G, Clemons MN, Getzler YD, McNeil AJ (2021) 100th anniversary of macromolecular science viewpoint: redefining sustainable polymers. ACS Macro Lett 10(1):41–53. https://doi.org/10.1021/acsmacrolett.0c00789

Forrest A, Giacovazzi L, Dunlop S, Reisser J, Tickler D, Jamieson A, Meeuwig JJ (2019) Eliminating plastic pollution: how a voluntary contribution from industry will drive the circular plastics economy. Front Mar Sci 6:627. https://doi.org/10.3389/fmars.2019.00627

Galloway TS, Cole M, Lewis C (2017) Interactions of microplastic debris throughout the marine ecosystem. Nat Ecol Evol 1:1–8

Gola D, Tyagi PK, Arya A, Chauchan N, Agarwal M, Singh K, Gola S (2021) The impact of microplastics on marine environment: a review. Environ Nanotechnol Monit Manag 16:100552

Haque F, Fan C (2023) Fate and impacts of microplastics in the environment: hydrosphere, pedosphere, and atmosphere. Environment 10:70

Hartmann N, Hüffer T, Thompson RC, Hassellöv M, Verschoor A, Daugaard AE, Rist S, Karlsson TM, Brennholt N, Cole M, Herrling MP, Heß M, Ivleva NP, Lusher AL, Wagner M (2019) Are we speaking the same language? Recommendations for a definition and categorization framework for plastic debris. Environ Sci Technol 53(3):1039–1047. https://doi.org/10.1021/acs.est.8b05297

Huang D, Xu Y, Lei F, Yu X, Ouyang Z, Chen Y (2021) Degradation of polyethylene plastic in soil and effects on microbial community composition. J Hazard Mater 416:126173. https://doi.org/10.1016/j.jhazmat.2021.126173

Jambeck JR (2019) Plastic waste inputs from land into the ocean. Science 347(6223):768–771

Jambeck J, Geyer R, Wilcox C, Siegler T, Perryman M, Andrady A, Narayan R, Law K (2015) Plastic waste inputs from land into the ocean. Science (Washington, DC, U S) 347(6223):768–771. https://doi.org/10.1126/science.1260352

Ji L, Tanunchai B, Wahdan SFM, Schädler M, Purahong W (2022) Future climate change enhances the complexity of plastisphere microbial co-occurrence networks but does not significantly affect the community assembly. Sci Total Environ 844:157016. https://doi.org/10.1016/j.scitotenv.2022.157016

João F, Müllertonia C, Capuano T (2024) Policy brief plastic pollution and the plastisphere: findings and recommendations. Science 369:1515–1518. https://doi.org/10.5281/zenodo.10560911

Kirstein IV, Kirmizi S, Wichels A, Garin-Fernandez A, Erler R, Löder M (2016) Dangerous hitchhikers? Evidence for potentially pathogenic Vibrio spp. on microplastic particles. Mar Environ Res 120:1–8

Koelmans AA, Paula ER, Nur HM, Vera N, Svenja MM, Merel K (2022) Risk assessment of microplastic particles. Nat Rev Mater 7:138–152. https://doi.org/10.1038/s41578-021-00411-y

Krause S, Baranov V, Nel HA, Drummond J, Kukkola A, Hoellein T, Smith GS, Lewandowski J, Bonnet B, Packman AI (2020) Gathering at the top? Environmental controls of microplastic uptake and biomagnification in freshwater food webs. Environ Pollut 268:115750

Kumar R, Verma A, Shome A, Sinha R, Sinha S, Jha PK, Kumar R, Kumar P, Das S, Sharma P, Vara PV (2021) Impacts of plastic pollution on ecosystem services, sustainable development goals, and need to focus on circular economy and policy interventions. Sustainability 13:9963

Leah W, Hunt J, Kovaleva M (2021) Garbage patches and their environmental implications in a plastisphere. J Mar Sci Eng 9:1289

Li C (2021) The ecology of the plastisphere: microbial composition, function, assembly, and network in the freshwater and seawater ecosystems. Water Resour 202:117428

Li L, Luo Y, Li R (2020) Effective uptake of submicrometre plastics by crop plants via a crack-entry mode. Nat Sustain 3:929–937

Li C, Gillings MR, Zhang C, Chen Q, Zhu D, Wang J, Zhao K, Xu Q, Leung PH, Li X, Liu J, Jin L (2024) Ecology and risks of the global plastisphere as a newly expanding microbial habitat. Innovation 5(1):100543

Mahnoor S, Akhtar A, Safina AR, Saira Y (2023) Metagenomic insights into the plastisphere microbiome: unveiling potential for clinical and industrial enzymes production along with plastic degradation. Arch Microbiol. https://doi.org/10.21203/rs.3.rs-3333696/v1

Mason SA, Welch VG, Neratko J (2018) Synthetic polymer contamination in bottled water. Front Chem 6:407

McCormick A, Timothy JH, Sherri AM, Joseph S, John JK (2014) Microplastic is an abundant and distinct microbial habitat in an urban river. Environ Sci Technol 48:11863–11871. https://doi.org/10.1021/es503610

Mohamed D, Yesmine F, Quim T, Marc D, Peré M, Sami B (2020) Blends of PBAT with plasticized starch for packaging applications: mechanical properties, rheological behaviour and biodegradability. Ind Crop Prod 144:112061. https://doi.org/10.1016/j.indcrop.2019.112061

Musa IO, Auta HS, Ilyasu US, Aransiola SA, Makun HA, Adabara NU, Abioye OP, Aziz A, Jayanthi B, Maddela NR, Prasad R (2024) Micro and nanoplastics in environment—degradation, detection, ecological impact. Int J Environ Res 18:1. https://doi.org/10.1007/s41742-023-00551-9

Nadal M, Alomar C, Deudero S (2016) High levels of microplastic ingestion by the semipelagic fish bogue *Boopsboops* (L.) around the Balearic Islands. Environ Pollut 214:517–523

Noella J (2024) Urbanization and its impact on environmental sustainability. J Appl Geogr Stud 3(1):54–66. https://doi.org/10.47941/jags.1624

Oberbeckmann S, Loeder MGJ, Gerdts G, Osborn AM (2014) Spatial and seasonal variation in diversity and structure of microbial biofilms on marine plastics in Northern European waters FEMS. Microb Ecol 90:478–492. https://doi.org/10.1111/1574-6941.12409

Plastics Europe (2016) The plastics facts 2016 (An analysis of European plastics production, demand and waste data). Plastics Europe

Plastics—The Facts (2021) An analysis of European plastics production, demand and waste data. Plastics Europe

Prata JC, da Costa JP, Lopes I, Duarte AC, Rocha-Santos T (2020) Environmental exposure to microplastics: an overview on possible human health effects. Sci Total Environ 702:134455

Rehse S, Kloas W, Zarfl C (2016) Short-term exposure with high concentrations of pristinemicroplastic particles leads to immobilisation of daphnia magna. Chemosphere 153:91–99. https://doi.org/10.1016/j.chemosphere.2016.02.133

Rogers KL, Carreres-Calabuig JA, Gorokhova E, Posth NR (2020) Special issue-current evidence micro-by-micro interactions: how microorganisms influence the fate of marine microplastics. Limnol Oceanogr Lett 5:18–36. https://doi.org/10.1002/lol2.10136

Roland G, Jenna RJ, Kara LL (2018) Production, use, and fate of all plastics ever made. Sci Adv 3:7. https://www.science.org/doi/10.1126/sciadv.1700782

Schmaltz EC, Melvin Z, Diana EF, Gunady D, Rittschof JA, Somarelli MM (2020) Dunphy-Daly plastic pollution solutions: emerging technologies to prevent and collect marine plastic pollution. Environ Int 144:106067

Schyns ZOG, Shaver MP (2021) Mechanical recycling of packaging plastics: a review. Macromol Rapid Commun 42(3):2000415. https://doi.org/10.1002/marc.202000415

Sharif A, Mondal S, Md EH (2019) Polylactic acid (PLA)-based nanocomposites: processing and properties. In: Bio-based polymers and nanocomposites. Springer, pp 233–254

Shen M, Song B, Zeng G, Zhang Y, Huang W, Wen X, Tang W (2020) Are biodegradable plastics a promising solution to solve the global plastic pollution? Environ Pollut 263:114469

Silva MM, Maldonado GC, Castro RO, de Sa Felizardo J, Cardoso RP, Anjos RMD (2019) Dispersal of potentially pathogenic bacteria by plastic debris in Guanabara Bay, RJ, Brazil. Mar Pollut Bull 141:561–568

Sun Y, Wu M, Zang J, Du L, Huang M, Cheng C, Wang J (2023) Plastisphere microbiome: methodology, diversity, and functionality. iMeta Wiley 2770596x:1–16

Tang KHD (2024) Terrestrial and aquatic plastisphere: formation, characteristics, and influencing factors. Sustainability 16:2163

Thiel M, Gutow L (2005) The ecology of rafting in the marine environment. II. The rafting organisms and community. Oceanogr Mar Biol 43(1):279–418. https://doi.org/10.1201/9781420037449.ch7

Thushari GGN, Senevirathna JDM (2020) Plastic pollution in the marine environment. Heliyon 6:e04709

United Nations Environment Programme (UNEP) (2018) Single-use plastics: a roadmap for sustainability. United Nations Environment Programme (UNEP)

Van Raamsdonk LW, van der Zande M, Koelmans AA, Hoogenboom RL, Peters RJ, Groot MJ, Peijnenburg AA, Weesepoel YJ (2020) Current insights into monitoring, bioaccumulation, and potential health effects of microplastics present in the food chain. Food Secur 9:72

Verma R, Vinoda K, Papireddy M, Gowda A (2016) Toxic pollutants from plastic waste—a review. Procedia Environ Sci 35:701–708. https://doi.org/10.1016/j.proenv.2016.07.069

Vignali G (2015) Life-cycle assessment of food-packaging systems. In: Environmental footprints and eco-design of products and processes, pp 1–22. https://doi.org/10.1007/978-981-287-913-4_1

Vignali G (2016) Life-cycle assessment of food-packaging systems. In: Muthu S (ed) Environmental footprints of packaging. Environmental footprints and eco-design of products and processes. Springer, Singapore. https://doi.org/10.1007/978-981-287-913-4_1

Vollmer I, Jenks MJF, Roelands MCP, White RJ, van Harmelen T, de Wild P, van der Laan GP, Meirer F, Keurentjes JTF, Weckhuysen BM (2020) Beyond mechanical recycling: giving new life to plastic waste. J German Chem Soc 59(36):15402–15423. https://doi.org/10.1002/anie.201915651

Walker TW, Frelka N, Shen Z, Chew AK, Banick J, Grey S, Kim MS, Dumesic JA, Van Lehn RC, Huber GW (2020) Recycling of multilayer plastic packaging materials by solvent-targeted recovery and precipitation. Sci Adv 6(47):eaba7599. https://doi.org/10.1126/sciadv.aba7599

Wang C, Wang L, Ok YS, Tsang CW, Hou D (2022) Soil plastisphere: exploration methods, influencing factors, and ecological insights. J Hazard Mater 430:128503. https://doi.org/10.1016/j.jhazmat.2022.128503

Waseem M, Khan MU, Majeed Y, Ntsefong GN, Kirichenko I, Klopova A, Trushov P, Lodygin A (2023) Seaweed-based films for sustainable food packaging: properties, incorporation of essential oils, applications, and future directions. Potr S J Food Sci 17:899–917. https://doi.org/10.5219/1908

Wen Z, Xie Y, Chen M, Dinga CD (2021) China's plastic import ban increases prospects of environmental impact mitigation of plastic waste trade flow worldwide. Nat Commun 12:1–9

Woodall LC, Sanchez-Vidal A, Canals M, Paterson GLJ, Coppock R, Sleight V (2014) The deep sea is a major sink for microplastic debris. R Soc Open Sci 1(4):140317. https://doi.org/10.1098/rsos.140317

Wright RJ, Erni-Cassola G, Zadjelovic V, Latva M, Christie-Oleza JA (2020) Marine plastic debris—a new surface for microbial colonization. Environ Sci Technol 54:11657–11672

Xu S, Ma J, Ji R, Pan K, Miao A-J (2020) Microplastics in aquatic environments: occurrence, accumulation, and biological effects. Sci Total Environ 703:134699. https://doi.org/10.1016/j.scitotenv.2019.134699

Yang K (2020) Temporal dynamics of antibiotic resistome in the plastisphere during microbial colonization. Environ Sci Technol 54:11322–11332

Yang K, Chen Q-L, Chen M-L, Li H-Z, Liao H, Pu Q (2020) Temporal dynamics of antibiotic resistome in the plastisphere during microbial colonization. Environ Sci Technol 54(18):11322–11332

Yang H, Chen G, Wang J (2021) Microplastics in the marine environment: sources, fates, impacts and microbial degradation. Toxics 9:41

Yap HS, Shukor MY, Yasid NA (2022) The microplastics occurrence and toxic effects in marine environment. J Environ Microbiol Toxicol 10(2):1–6

Yuanze S, Jia S, Xiang W, Changfeng D, Jie W (2022) Deciphering the mechanisms shaping the plastisphere microbiota in soil. mSystems 7:e00352–e00322. https://doi.org/10.1128/msystems.00352-22

Zettler ER, Tracy JM, Linda AA (2013) Life in the "plastisphere": microbial communities on plastic marine debris. Environ Sci Technol 47:7137–7146. https://doi.org/10.1021/es401288x

Zhang S-J, Zeng Y-H, Zhu J-M, Cai Z-H, Zhou J (2021) The structure and assembly mechanisms of plastisphere microbial community in natural marine environment. J Hazard Mater 421:126780–126780

Zhao Y-B, Lv X-D, Ni H-G (2018) Solvent-based separation and recycling of waste plastics: a review. Chemosphere 209:707–720. https://doi.org/10.1016/j.chemosphere.2018.06.095

Chapter 10
White Pollution Biodiversity and Methods of Detection

Emmanuel Kuyet Adejokun, Sylvester Chukwudi Ugoh, Sesan Abiodun Aransiola, Sarah Chioma Nwakanma, Nasser Belboukhari, Alissar AlJerf, and Naga Raju Maddela

Abstract The proliferation and disposal of plastics pose a significant challenge to the planet's sustainability. Over the recent years, around 70–80% of plastic waste has been deposited in landfills or the environment. This Plastic waste that enters the environment undergoes fragmentation, weathering and other processes, breaking down into microplastics (plastic particles smaller than 5 mm in size). While much research and attention has been given to microplastic pollution in marine/aquatic ecosystems, it is important to also pay attention to the terrestrial ecosystems which are also vulnerable. Microplastics have been found in various terrestrial environments, originating from sources such as the textile industries which produces microfibres, sewage sludge, pharmaceutical industries, plastic mulch and many others. Plastic disrupts environmental balance by acting as physical barriers, degrading soil quality and being ingested by organisms, leading to their demise. Therefore, effective management of plastic pollution is crucial due to its detrimental impact on the environment. This work discusses microplastics in the ecosystems as well as methods of detection and identification such as the use of visual inspections, Fourier transform infrared, SEM-EDS amongst others.

E. K. Adejokun (✉) · S. C. Ugoh · S. C. Nwakanma
Department of Microbiology, Faculty of Science, University of Abuja, Abuja, Nigeria

S. A. Aransiola
Department of Microbiology, Faculty of Science, University of Abuja, Abuja, Nigeria

Bioresources Development Centre, National Biotechnology Research Development Agency, Ogbomoso, Nigeria

N. Belboukhari
Bioactive Molecules and Chiral Separation Laboratory, University of Bechar, Bechar, Algeria

A. AlJerf
Haitham Abdel-Salam School, Al-Jama'at District, Jaramanah, Ministry of Education, Damascus, Syrian Arab Republic

N. R. Maddela
Departamento de Ciencias Biológicas, Facultad la Ciencias de la Salud, Universidad Técnica de Manabí, Portoviejo, Manabí, Ecuador

S. A. Aransiola et al. (eds.), *White Pollution: Biodiversity and hazards in marine plastisphere*, Environmental Contamination Remediation and Management,
https://doi.org/10.1007/978-3-031-95547-1_10

Keywords Microplastics · Biodiversity · Ecosystems · Biodegradation · Plastics

10.1 Introduction

In today's world, there is a rapid increase in population size and with this comes urbanization leading to the generation of immense quantity of waste. The waste generated in large quantities can be treated by incineration, disposal in landfills and others ends up being dumped in water bodies (rivers, oceans). Some of the waste produced (plastic waste) are non-biodegradable because they do not breakdown or decompose naturally overtime. The accumulation of plastic waste leads to white pollution which causes an environmental issue (Auta et al. 2022; Musa et al. 2024).

The term "plastic" originates from the Greek word "plastico", meaning something that can be shaped or molded. Plastics are ubiquitous human-made products, prevalent globally. Annually, the average person generates at least 52 kg of plastic waste (Jambeck et al. 2015). Plastics are polymeric materials like natural polymers such as rubber and silk. It is the synthetic plastics that contributes to environmental pollution due to their longevity in nature. These synthetic materials are designed to resist natural decomposition because they are largely non-biodegradable, leading to their persistence in the environment and presenting a significant issue. Plastics are used in various industries because of their durability, affordability, strength and other properties, making them one of the most commonly used materials worldwide. The packaging industry is the largest consumer of plastic, accounting for over a third of usage (Schmidt et al. 2017). A significant portion of plastic waste ends up in the oceans, leading to the degradation of the marine ecosystem (Andrady and Neal 2009).

Polymers, the broad category encompassing all types of plastics, are primarily derived from the petroleum industry. These materials accumulate in different settings and break down into microplastics under environmental stresses. Common polymers includes Polypropylene, Nylon, Polystyrene, Polyurethane, among others. Microbial actions can fragment plastics into microplastics. The term "polymer" comes from Greek, with "poly" meaning many, indicating that polymers, which have a molecular weight usually above 1000 g/mol, are made of many smaller units called monomers, with each less than 100 g/mol. Polymers are high-molecular weight compounds formed by linking many low-molecular weight monomers through covalent bonds (Sharma 2014). These can be homopolymers, with identical repeating units, or copolymers, with varying sequences of different sub-units. Plastics, a subgroup of polymers that soften upon heating and can be molded, have become extensively used due to their unique properties (Robert et al. 2021). From 1950 to 2017, an estimated 9 billion metric tons of plastic were produced, with annual production reaching 420 million metric tons (William and Rangel-Buitrago 2022). If current trends continue, it is estimated that annual plastic production could exceed 1500 million metric tons by 2060. The surge in plastic demand since the early twentieth century has led to significant environmental issues, primarily due to their slow decomposition rate. Most plastics are not recycled and end up as environmental pollution, affecting both the aquatic and terrestrial ecosystems worldwide. The constant introduction of plastic waste to the environment poses a growing challenge. Therefore, there is need to investigate the impact of microplastics and how they affect our ecosystems. This chapter reviewed the origin and sources of microplastics, effects of microplastics in the ecosystems, methods of detection and possible solutions to tackle the impact of microplastics.

10.1.1 Origin of Microplastics

The origin and increasing awareness of microplastics as a significant issue are deeply rooted in scientific research and discoveries over several decades. Microplastics, defined as small plastic pieces which are smaller than 5 mm in size have become a focal point of environmental research due to their widespread presence in marine and terrestrial ecosystems and potential harm to wildlife and human health. Microplastics origin, impact and distribution involves contributions from numerous scientists across various disciplines around the globe.

In 1972, Edward Carpenter and K.L. Smith Jr. identified microplastics and published a study in the journal "science" that detailed their observation of small plastic particles which they referred to as "plastic particles" in the Sargassso sea. This study marked one of the first documented instances of microplastics in the marine environment highlighting their synthetic origin and raising concerns about their potential harm to marine life. This observation drew attention to the issue of plastic pollution in the oceans, specifically the presence of tiny plastic fragments that had broken down from larger ones. This discovery shifted the focus from larger pieces of the marine debris and litter towards understanding the sources, persistence and ecological impacts of these minuscule particles. Studies began to investigate the abundance, distribution and effects of microplastics in various environments (aquatic and terrestrial ecosystems) with scientists across the globe contributing to the growing body of knowledge. Richard Thompson and his team at the university of Plymouth coined the term "Microplastic" in 2004 in a study that analyzed plastic debris accumulated on beaches. Their research demonstrated that microplastics were not only prevalent in the marine environments but could originate from the breakdown of larger plastic items over time. No ecosystem in the world seems to have avoided microplastic contamination (Taylor et al. 2016).

In recent times, there has been a global surge in demand for plastics, leading to an increase in their production to around 359 million tons annually. The COVID-19 pandemic in 2020 significantly contributed to this trend with a spike in plastic waste production due to the widespread use of personal protective equipment (PPE) such as face shields, surgical gloves and face mask. China leads in plastic production accounting for 30% of the global total, followed by members of the North American Free Trade Agreement (NAFTA)—the United States, Canada and Mexico which collectively produce 18%. The African continent contributes 7%, Latin America 4%

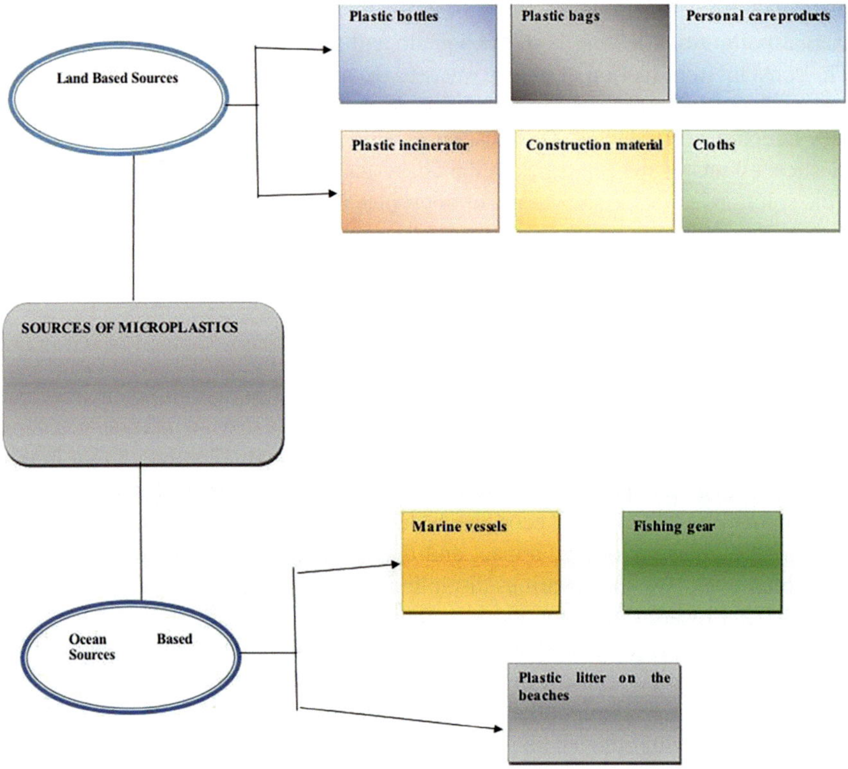

Fig. 10.1 Sources of microplastics

and the 9 countries of the Commonwealth of Independent States (CIS) contributes 3% (Fig. 10.1). Unfortunately, due to inadequate recycling processes, a significant proportion of these plastics ends up in rivers, oceans and soil. Estimates suggest that the amount of plastic waste entering water bodies annually ranges between 1.15 and 241 million tons with the majority originating from Asian countries.

10.1.2 Classification of Microplastics

Microplastics can be classified based on various criteria including origin, size, composition and environmental compartment.

10.1.2.1 Classification Based on Size

The size of microplastics determines their availability and potential toxicity to different organisms in the environment. Mesoplastics and nanoplastics are of particular concern because of their ability to penetrate biological membranes and potentially cause harm at cellular level. The size also affects the transport dynamics of microplastics in water and air, influencing their distribution and accumulation in different environments.

Macroplastics (Large Microplastic)

Macroplastics are larger than 1 mm up to 5 mm. They are closer to the upper size limit of microplastics and can be fragments from larger plastic items that have degraded. They are easily ingested by a wider range of animals including fishes, birds and some mammals which potentially causes physical harm or blockages in the digestive tract. Examples of some of these macroplastics includes;

Plastic bottles (used for beverages and personal care items)
Plastic bags (includes shopping bags, garbage bags and produce bags)
Plastic Utensils and Straw (used in fast food and takeout dining)
Toys (durable goods made partially or solely of plastic)
Packaging materials (includes plastic wraps, Styrofoam containers and large industrial wrapping)
Fishing gear (includes nets, ropes and fishing lines)
Vehicle parts (includes bumpers and dashboard components)
Construction materials (includes pipes and insulation materials made of plastics)

Mesoplastics (Small Microplastics)

Mesoplastics are less than 0.1 mm (100 micrometers) down to 1 micrometer (μm) in size. These includes the smallest particles resulting from the further breakdown of plastics or from products containing extremely fine plastic particles. Their tiny size allows them to be ingested by an even wider range of organisms including the planktons and filter feeders. They can penetrate the tissues, potentially leading to more direct physiological effects. Examples of some of these mesoplastics includes; Microbeads in personal care products, Fragmented plastic packaging, Shredded plastic bags, Abraded plastic fishing gear, Disintegrated plastic toys.

Nanoplastics

Nanoplastics are less than 1 micometer (μm) down to the nanometer scale. They are at the extreme end of the microplastic size spectrum and represent the finest particles which results from further degradation or are specifically manufactured at this scale for certain applications. Due to their minuscule size, nanoplastics can cross cellular membranes and enter organisms' circulatory systems, posing unknown risk to wildlife and humans. Their impacts and environmental behaviors are an area of ongoing research. Nanoplastics are now widely detected in environmental samples and food chain (Lai et al. 2022); bringing about the exploration of their potential toxic effects. Examples of some of these nanoplastics includes; Nano-sized fibers, Nanoplastic beads, Polymer additives (includes nanoscale fillers and nanoscale stabilizers).

10.1.2.2 Classification Based on Origin

Microplastics are categorized into two main groups based on their origin: primary microplastics and secondary microplastics. Primary microplastics are those intentionally produced to that small size, while secondary microplastics result from the fragmentation of larger plastic items.

Primary Microplastics

Primary microplastics are particles that are intentionally produced to be of microscopic size. Primary microplastics are intentionally added to consumer and commercial products (Duis and Coors 2016). They can enter the environment directly from their various applications. These includes:

Personal care products: Microbeads used in cosmetics such as exfoliating face scrubs and toothpaste. These tiny particles are designed to perform specific functions but can end up in waterways through domestic wastewater due to their microscopic size which makes it difficult for them to be effectively captured by many wastewater treatment processes.

Plastic pellets or nurdles: These are the raw material for nearly all plastic manufacturing processes. Nurdles are small pellets that can be spilled during transport or handling and end up in the environment, particularly the aquatic ecosystems

Synthetic fibers: Microfibers released from the washing of synthetic textiles like polyester and nylon contribute significantly to primary microplastics. These fibers are detached during laundry and are carried into water systems through domestic wastewater.

Industrial scrubbers: They are used in air-blasting techniques. These microplastics are utilized for cleaning industrial equipment and surfaces. Similar to microbeads, they are designed to be small for their abrasive properties.

Secondary Microplastics

Secondary microplastics results from the breakdown of larger plastic items or materials into smaller fragments once they enter the environment. This degradation occurs due to many factors such as UV radiation, physical abrasion and biological degradation. Secondary microplastics includes:

Films: Thin plastics such as plastic bags and wrap can degrade into smaller pieces resulting to secondary microplastics. These films often become brittle and break down into smaller pieces over time.

Foam: Polyestyrene foam products commonly used in packaging and as disposable food containers can fragment into microplastics. These lightweight particles can easily be carried by wind and water, spreading across various environments.

Fragments: These are broken pieces from larger plastic waste such as bottles, bags and wrappers that degrade over time due to environmental exposure. They usually vary greatly in shape and size, often depending on the original shape of the source material and the degradation process.

10.1.2.3 Classification Based on Composition

Microplastics, owing to their diverse applications and origins are made from a variety of polymers with each of them having their distinct physical and chemical properties. The composition of microplastics determines their physical properties, environmental fate and potential toxicity. Polymers that can float can travel long distances on water currents while denser polymers tend to sink and accumulate in sediments. Some polymers, due to their chemical structure can adsorb and concentrate environmental pollutants, potentially becoming vectors for contaminant transport. The degradation rates of microplastics vary significantly with their composition, influencing their persistence in the environment.

Polyethylene (PE)

Polyethylene is the most commonly produced plastic widely used in packaging such as plastic bags, bottles and films. It has a lightweight and it is resistant to chemicals and moisture. Due to its extensive use, Polyethylene is one of the most prevalent types of microplastics found in the environment. It tends to float in water making it a significant component of marine and freshwater microplastic pollution.

Polypropylene (PP)

Polypropylene is used in packaging, textiles and various containers. It is known for its versality, chemical resistance and high melting point. Polypropylene microplastics are commonly found in aquatic environments. Polypropylene also tends to float in water, contributing to the surface accumulation of microplastics in water bodies.

Polyethylene Terephthalate (PET)

Polyethylene Terephthalate is widely used for manufacturing beverage bottles, food packaging and synthetic fibers. It is favored for its strength, thermos-stability and clarity. Polyethylene Terephthalate can degrade into microplastics through physical and chemical processes. Polyethylene Terephthalate fibers from textiles and fragments from bottles commonly detected in both terrestrial and aquatic environments.

Polystyrene (PS)

Polystyrene is used in disposable food containers, packaging materials and insulation. It has a lightweight and can either be rigid or foamed (expanded polystyrene). Polystyrene microplastics are prevalent in the environment especially from fragmented expanded polystyrene containers. They are resistant to degradation and can persist in environments for extended periods.

Polyvinyl Chloride (PVC)

Polyvinyl Chloride is used in construction materials, pipes and medical devices. It is known for its durability and resistance to environmental degradation. Polyvinyl Chloride degrades slowly, releasing microplastics into the environment. It can contain harmful additives such as plasticizers which may leach out and pose additional environmental risks.

Nylon (Polyamides)

Nylon is a synthetic polymer used extensively in textiles, fishing lines and brushes. It is durable, resistant to abrasion and has high tensile strength. Nylon microfibers are a significant concern in the aquatic environments, particularly from the washing of synthetic textiles. These fibers can adsorb and concentrate environmental pollutants.

Acrylic and Polyester

These synthetic polymers are primarily used in the production of textiles. They offer durability, resistance to wrinkles and ease of care. Microfibers from Acrylic and Polyester garments are released into waterways through laundry effluents. These fibers are also persistent and can transport adsorbed pollutants, contributing to microplastic pollution.

10.1.2.4 Classification Based on Environmental Compartment

Microplastics due to their small size and persistence have been discovered virtually every corner of the globe from mountaintops to the deep sea. They have been distributed across different environmental compartments such as the aquatic (marine and freshwater), terrestrial and atmospheric ecosystems.

Aquatic Environment

Marine ecosystems: Marine ecosystems includes open oceans, coastal waters, seabeds and even the deep sea. Microplastics in these environments originate from various sources including land runoff, shipping, fishing activities and direct dumping. Microplastics in marine environments can be ingested by a wide range of organisms from plankton to large mammals, leading to physical harm and potential chemical exposure.

Freshwater ecosystems: This includes rivers, lakes, streams and wetlands. Sources of microplastics in freshwater ecosystems are similar to marine environments with additional contributions from wastewater treatment plants and stormwater runoff. Freshwater organisms including fishes, invertebrates and birds can ingest microplastics leading to potential health impacts. Microplastics here can also affect the quality of drinking water sources.

Terrestrial Environment

Terrestrial environments encompass urban and rural landscapes, agricultural lands, forests and deserts. Sources of microplastics in these areas includes landfills, plastic mulch used in agriculture, sewage sludge applied as fertilizer and the breakdown of larger plastic debris. Microplastics can affect soil health, water retention and the activity of soil organisms. They can also be ingested by terrestrial animals including insects, birds and mammals leading to potential ecological and health impacts.

Atmospheric Environment

Microplastics have been detected in the atmosphere where they can be transported over long distances by winds. They originate from urban areas, industrial sites and agricultural activities and can settles out of the atmosphere through dry or wet decomposition. Atmospheric microplastics can contribute to air pollution and respiratory health issues when inhaled by humans and animals. They represent a pathway for the deposition of microplastics in remote areas including mountains and polar regions.

Cryospheric Environment

The cryosphere includes all frozen water parts of the earth such as glaciers, ice caps and sea ice. Microplastics can become trapped in ice and snow, transported through atmospheric pathways or through the melting of this frozen compartments. The presence of microplastics in the cryosphere can affect albedo (reflectivity) and contribute to the melting of ice due to increased absorption of sunlight and can also be a potential release source of microplastics as global temperature rise.

10.1.3 Sources of Microplastics

There are many sources of microplastics, but they are mainly classified into land based sources and ocean based sources (Fig. 10.2).

10.1.3.1 Land-Based Sources of Microplastics

Land-based origins account for about 70–80% of microplastics found in aquatic environments. These sources encompass a wide array of items such as bottles, personal care items, clothing, plastic bags and building materials. Plastic incineration also contributes to this issue by generating bottom ash containing microplastics. Major sources of larger plastic items on land includes food and beverages packaging waste, construction materials, packaging materials, household goods, and waste from ship construction (Aransiola et al. 2023a). Sewage sludge and industrial processes, particularly those involving granules and small resin pellets, are additional potential sources of microplastics released into water bodies. Certain cosmetics and personal care products containing microplastics, used as drug carriers or ingredients, also contributes to plastic pollution. Examples includes toothpaste, makeup products, face washes, hand soaps and laundry detergents. Synthetic fibers such as acrylics shed from clothing, polyester and nylon enter into water bodies through wastewater discharge. Additionally, tire abrasion from vehicles releases microplastics into the environment. Thus, effective control and reduction of various sources

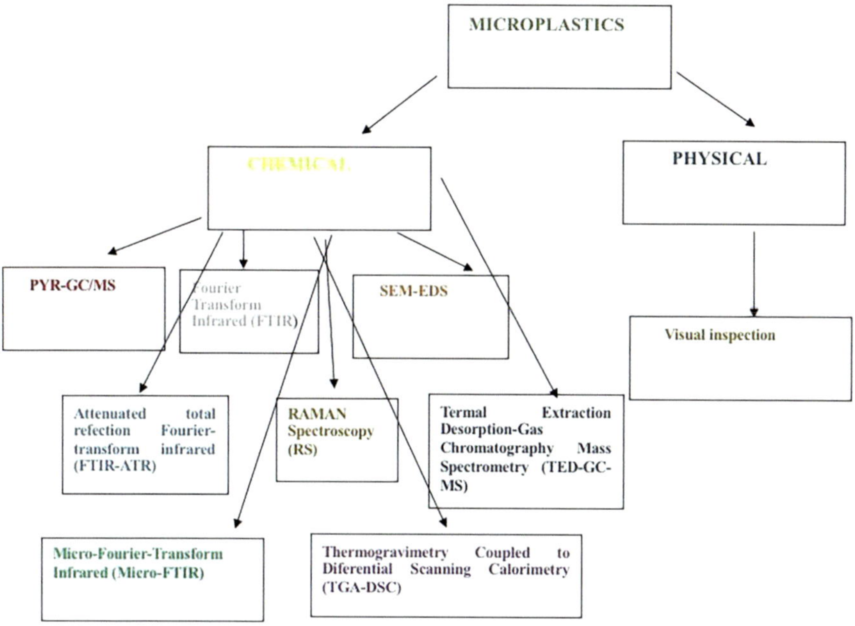

Fig. 10.2 The detection of microplastics, methods and technique involved

of microplastics are very crucial. The single-use polymeric plastic products such as straws, cutlery, bottles, coffee cups and bags are identified as significant contributors to environmental plastic pollution (Aransiola et al. 2023b).

10.1.3.2 Ocean-Based Sources of Microplastics

Around 15–30% of microplastics introduced into aquatic environments originate from sources based in the ocean, which include coastal tourism, commercial fishing, offshore industries and marine vessels. Lost or abandoned fishing equipment such as nylon nets and plastic monofilaments, poses a notable threat as they release microplastics that can remain at various ocean depths. An estimated 500,000 tons of fishing gear are discarded into the ocean annually, exacerbating the issue. Microplastic waste from shipping activities, commonly emitted by commercial and naval vessels, further contributes to marine pollution. Significant quantities of plastic waste from offshore industries, particularly those involved in petrochemicals, are discharged into marine ecosystems. While ocean-based sources of microplastic pollution may not be as high as land-based sources, they still represent a considerable concern that warrants effective management and control (Fig. 10.1).

10.2 Impact of White Pollution (Plastics) on Biodiversity

Plastic pollution poses significant threats to biodiversity, impacting ecosystems and species at various levels of organization, from individual organisms to entire ecosystems. The global impact of plastic pollution on the environment is widely acknowledged (Meeker et al. 2009; Oehlmann et al. 2009). Although plastics play a crucial role in various aspects of human life, their disposal poses significant threats to both living organisms and their habitats. The widespread availability and low cost of plastics have led to a surge in its usage, particularly in the food industry for packaging various products, consequently escalating plastic waste production (ECOSS 2014). Plastic bags in particular have become a major pollutant due to their toxic composition, contributing to water, soil and air pollution (ECOSS 2014). Chemicals used in plastic production, including carcinogenic substances like vinyl chloride and benzene (Swan et al. 2005; Swan 2008), as well as gases released during combustion such as oxide, ethylene and xylenes, further degrades the environment (Rudel et al. 2001, 2003; Oehlmann et al. 2009). Plastic's inherent stability and durability prevents natural degradation (ECOSS 2014). This leads to its accumulation in landfills, where toxic pollutants persist indefinitely in surrounding water and soil (Kaewlue 2012). Plastic degradation is extremely slow, taking years or even centuries under favorable conditions like high humidity and oxygen levels (Allen et al. 1988). Plastic bottle for instance, can remain intact for about 30–180 years in wet soil before gradually breaking down into microplastics (Allen et al. 1994; Edge et al. 1991). In marine environments, degradation begins within weeks (Kedzierski et al. 2018) but complete breakdown can take decades or even centuries due to factors like oxygen levels, sunlight exposure and temperature (Barnes et al. 2009; Ioakeimidis et al. 2016).

Animals, especially marine species like seabirds, turtles and seals are at the risk of entanglement in plastic debris, resulting in injuries, drowning or suffocation. Fishing gears such as nets and lines are extensively problematic in causing entanglement. Marine and terrestrial species or organisms often mistake plastic debris for food, leading to ingestion. This can result in internal injuries, blockage of digestive tracts, malnutrition and finally starvation. Plastics can release variety of harmful chemicals including additives used in manufacturing (such as phthalates, bisphenol A), persistent organic pollutants (PCBs, DDT) and heavy metals. These chemicals get leached into the environment and cause contamination in soil, water and air. Plastic particles can adsorb and concentrate pollutants from the surrounding environment. As these particles are ingested by organisms, the pollutants accumulate in their tissues and organs leading to toxic effects and potential disruptions of endocrine systems. The accumulation of plastic debris on land and in water can alter habitats and ecosystems, smothering plants, corals and other benthic organisms. Plastic pollution can modify substrate composition and block sunlight penetration, affecting photosynthesis. Microplastics ingested by filter feeding organisms such as planktons and bivalves can impair feeding efficiency and growth rates, affecting higher trophic levels. Thus, it is very important not to disturb the balance of aquatic ecosystem and terrestrial ecosystems so that the various species can be conserved.

10.2.1 Microplastics in the Marine/Aquatic Environment

Microplastics have been discovered in oceans worldwide, encompassing even remote areas like the deep sea and Antarctica (Derraik 2002). The distribution of microplastics in the ocean is uneven, with elevated concentrations notably observed near industrial hubs and urban centers. The entry of microplastics into aquatic ecosystems primarily stems from domestic runoff, containing microbeads and fragmented plastic particles, as well as the breakdown of larger plastic debris. Plastic industries discharge pellets and resin powders through air blasting, further contaminating aquatic habitats. Human activities such as fishing practices and tourism also contribute to this pollution. Marine industries are the primary contributors to plastics in marine ecosystems. Over time, plastics from these industries degrade into microplastics due to various processes like photo-oxidative degradation, thermal degradation, and biodegradation (Singh and Sharma 2008). Human activities also introduce microplastics into aquatic environments, such as washing synthetic fiber clothing, which releases microplastics into rivers through domestic sewage. Agricultural plastic products, including films and pipelines, weather over time, generating microplastics that are transported into rivers via wind and rain. These pathways contribute to a high abundance of microplastics in rivers. In aquatic environments, granules, fragments, fibers, films and styrofoam particles are the most prevalent types of particles, with polystyrene (PS), polypropylene (PP), and polyethylene (PE) being the most common polymers. It's estimated that 90% of the 5.2 trillion particles in the global marine environment are microplastics (Eriksen et al. 2014). Between 1.1 and 2.4 million metric tons of plastic waste enter the marine environment annually from rivers (Lebreton et al. 2017). The influx of plastic waste into marine and aquatic habitats has led to a surge in microplastic production in these environments, and inadequate waste management has resulted in the accumulation of plastic waste in the oceans, leading to adverse effects on aquatic ecosystems and posing a subsequent threat to aquatic and human life.

10.2.1.1 Effects of Microplastics on Aquatic Environment

Microplastics are non-biodegradable and they contain additives used in their production and as a result pose a direct or indirect threat to aquatic organisms and ecosystems. The small surface area of microplastics allows these additives to leach into waterways, contaminating them and harming aquatic lives. Direct effect occurs when marine or freshwater organisms ingest microplastics, leading to harmful biological consequences. Sea turtle for example, mistake plastic bags for jellyfish and ingest them, resulting in plastic accumulation in their bodies. Similarly, fishes, marine mammals and seabirds often consume plastic debris accidentally, mistaking it for food (Thompson et al. 2009). For example, birds often mistake plastic pellets for fish eggs. This extends the impact beyond animals to plants and non-mobile organisms such as coral reefs damaged by discarded fishing gear (Mato et al. 2001). Plants can choke plastic bags and fishing nets. The movement of derelict ships or other marine debris can damage and alter seabed ecosystems. Microplastics can release various harmful chemicals into waterways including phthalates, heat stabilizers, antioxidants platicizers, flame retardants and many others. These impurities are toxic and are leached into the aquatic environment when larger plastics are broken down due to wave action and sunlight (photodegradation) into microplastics which are absorbed by marine organisms due to the small surface area of microplastics. This contamination ultimately affects higher organisms including humans who consume seafood contaminated with microplastics and associated toxins from the aquatic environment.

10.2.2 Microplastics in the Terrestrial Environment (Soil)

Microplastics in the terrestrial environment are a growing problem because they have been found in several terrestrial ecosystems. Inefficient waste management and unauthorized dumping of plastic waste are critical causes of soil pollution. Plastic waste in soil breaks down into microplastics through biological processes of soil fauna, including eating habits and digestive and excretory processes. Its sources are proportionally related to anthropogenic uses such as cosmetics, mulching, washing and care, textiles (microfibres), automotive (tyre friction/tyre wear) and plastic products, covering all agricultural, industrial and manufacturing properties. Therefore, the inevitable presence of microplastics in the biome and in everyday products makes human exposure to microplastics certain. It has been reported that the main sources of microplastics in soil ecosystems are mulching with plastic film, sewage sludge, improper disposal of plastic waste, farming, etc., which pose a serious environmental threat to various ecosystems. Microplastics act as a vector of soil contamination because they are unavoidable and ubiquitous in the environment. The release of microplastics into the environment can affect their interactions with several animal, plant and microbial species and change the behavior of microplastics in the soil. Microplastics are incorporated into soil aggregates through soil fragmentation, plant root extension, agronomic practices and burrowing animals, allowing vertical transport of microplastics in the soil. The introduction of microplastics into soils destined for deposition, migration, erosion, pollution and groundwater filtration subsequently endangers microorganisms and ultimately affects all living organisms through indirect usc. Recent studies have also shown that tire wear particles (tire beads) and microfibers (textiles) are the largest microplastic pollutant on earth, and these microplastics are gradually leaching from land into the oceans.

10.2.2.1 Effects of Microplastics on the Soil

The quantity, shape, composition and size of microplastics exhibit significant variability within soil. Their presence in soil can disrupt soil structure and compromise its physical and chemical properties. Plastics alter soil's water retention capacity, diminish rainwater and irrigation infiltration, solute movement, impede soil water and may induce hypoxia (Wang et al. 2015). Plastics disrupt soil aggregate formation, reducing soil aeration and permeability, leading to soil degradation marked by increased compaction and diminished porosity, ultimately impacting plant and microorganism growth (Table 10.1) (Jiang et al. 2017; Zeng et al. 2013; Zhang et al. 2018). Due to their extensive surface area, microplastics have the potential to adsorb various metals such as lead, zinc, copper aluminum, silver as well as hazardous pollutants like pesticides, dichlorodiphenyltrichloroethane (DDT), polycyclic aromatic hydrocarbons (PAHs) and polychlorinated biphenyls (PCBs). Microplastics hydrophobic nature facilitates the absorption of chemical contaminants, posing risks to ecosystem health. Acting as carriers, microplastics can transport pesticides across trophic levels within ecosystems, amplifying their impact. Furthermore, microplastics can serve as catalysts for chemical pollutants, either as additives during polymer production or sourced directly from the environment. Previous studies have explored the effects and potential ramifications of various microplastics on organisms and vegetation within soil ecosystems (Table 10.2).

10.2.2.2 Effects of Microplastics on Soil Microorganisms

Soil microorganisms are one of the most active agents in the soil and thus play an important role in nutrient cycling, soil health and ecosystem functioning. Microplastics can physically interact with soil microorganisms, changing their habitat and microbial communities (Table 10.2). Microplastics can create microhabitats that influence microbial colonization and activity. Microplastics can adsorb and accumulate nutrients and organic substances in soil, changing the availability of nutrients to soil microorganisms, which can affect microbial growth, metabolism, and community composition. Microorganisms can adsorb on the surface of microplastics for a long time, forming a biofilm and a unique bacterial community (Harrison et al. 2014; Zettler et al. 2013). Microplastics can act as substrates for microbial biofilm formation. Biofilms are complex microbial communities that adhere to surfaces, including microplastics, and can alter interactions between soil microorganisms and their environment. Some microplastics contain dirt and contaminants that can leach into the soil and pose toxicological risks to the soil. The diversity, composition and structure of the bacterial community are also affected by the amounts and types of microplastics (Li et al. 2022; Seeley et al. 2020) (Table 10.1). Unique bacterial communities were found in microplastics that could be potential hosts for antibiotic resistance genes (ARGs). Although microplastics are resistant to degradation, certain soil microorganisms have been found to have the ability to degrade and metabolize plastic polymers under certain conditions.

Table 10.1 Effects of microplastics on various organisms

Microplastic type	Size	Specie	Effects	Country of study/ research	References
PE	<50–100 μm, >100 μm	Lumbricus terrestris	Increased concentrations of microplastics may impact the growth rate of L. terrestris.	Netherlands	Lwanga et al. (2016)
PE	2400–710 μm	Lumbricus terrestris	The presence of earthworms significantly enhances the accumulation of microplastic particles at the soil's bottom layers.	Berlin	Rillig et al. (2017)
PE	<400 μm	Lumbricus terrestris	Breeding inhibition and decreased bacterial diversity were observed in the springtail gut.	China	Ju et al. (2019)
LDPE	0.20 μm, 1–4 mm	Lumbricus terrestris	Microplastics do not act as carriers of organic pollutants to earthworms.	Spain	Rodríguez et al. (2016)
LDPE	1 mm–40 μm	Wheat (Triticum aestivum)	Plastic remnants affected both the upper and lower parts of wheat plants.	Netherlands	Qi et al. (2018)
PVC	200–90 μm	Hypoaspis aculeifer, Folsomia candida	Trophic predator-prey interactions facilitate the transfer of microplastics by 40%.	China	Zhu et al. (2018)
PVC	80–200 μm	Folsomia candida	Changes and suppression of microbiota occur in collembolan gut.	China	Zhu et al. (2018)
Polystyrene	0.1–0.05 μm	Enchytraeus crypticus	Animals fed 10% PS experienced reduced biomass, while those fed 0.025% showed increased breeding rates.	China	Zhu et al. (2018)

(continued)

Table 10.1 (continued)

Microplastic type	Size	Specie	Effects	Country of study/ research	References
Urea-formaldehyde	1500–400 μm	Folsomia candida, Proisotoma minuta	Microarthropods contribute to the movement and distribution of microplastics.	Berlin	Maaß et al. (2017)
PA, PET, PEHD, PES, PS, PP	20–12 μm	Spring onions (Allium istulosum)	Microplastics can impact leaf characteristics, root systems and overall plant biomass.	Berlin	Machado et al. (2019)
PE, PS	<200 μm–<250 μm	Eisenia fetida	The concentration of hydrophobic organic compounds in E. fetida decreased by over 1% in the presence of microplastics.	China	Wang et al. (2019)

Microplastic can indirectly affect the life activity of soil microbes by changing the physical, chemical and aerobic properties of the soil (Veresoglou 2015; Zhang et al. 2021).

10.3 Method of Detection of Microplastics

The identification and detection of microplastics in the ecosystem remains a challenge due to the internal and physicochemical characteristics of microplastics, which makes accurate detection difficult. Identifying microplastics has now become a priority due to its environmental damage. There are different methods to detect microplastics (Fig. 10.2). One is the physical method, which involves inspection under a microscope. This method may not be completely accurate due to the small size and highly variable shape of microplastics. This alone can cause confusion and lead to incorrect quantification due to the presence of other materials similar in shape and size to some microplastics. However, this method can be used to analyze large plastic particles (Xiang et al. 2021). The second method, known as the chemical method, offers more precise outcomes compared to the physical method. Utilizing vibrational techniques such as Raman spectroscopy and Fourier transform infrared spectroscopy, along with microscopic variations (μRaman or μFTIR), enhances microscopic identification by referencing existing data. Additionally, the pyrolysis gas chromatography/pyrolysis mass spectrometry (Pyr-GC-MS)

Table 10.2 Effects of microplastics on soil microorganisms

Polymer Type	Concentration	Size	Effects	References
PE	0.1% (w/w)	0.02 mm	The presence of pathogenic microorganisms increased, leading to a decrease in microbial diversity.	Li et al. (2021)
PE	3% (w/w)	100–250 μm	There was an increase in acidobacteria abundance, resulting in an increase in the Shannon index	Rong et al. (2021)
PE	6% (w/w)	150–200 μm	An increase in acidobacteria abundance was observed, accompanied by a decrease in the Shannon index.	Rong et al. (2021)
PLA, PS	1% (w/w)	150–200 μm	There was an increase in alpha diversity.	Sun et al. (2022)
PU	1% (w/w)	4.5 mm	The abundance of fibrobacteres, firmicutes, bacteriodes and verrucomicrobia increased.	Lian et al. (2021)
PS	2% (w/w)	30.5 nm	There was a decrease in soil microbial biomass.	Awet et al. (2018)
PE, PS, PA, PLA, PBS, PHB	0.1%, 2% (w/w)	35–80 μm	Abundance of ktedonobacterales decreased while rhizobiales abundance increased, leading to a decrease in bacterial alpha diversity.	Feng et al. (2022)
PE, PS, PP	1% (w/w)	160–200 μm	Both Shannon and chao indices showed a decrease.	Yu et al. (2021)
PVC	6.55, 35 (gm^{-2})	15 mm	Simpson indices and Shannon-weaver also exhibited a decrease.	Wang et al. (2016)
PE, PVC, PET	1% (w/w)	<2 mm	There was no significant effect observed on soil microbial function	Judy et al. (2019)
PS	0.5% (w/w)	330–650 μm	No significant effect was observed on the Simpson indices and Shannon.	Xu et al. (2021)

technique integrates these methods with pyrolysis for comprehensive analysis (Mccormick et al. 2016). Another tool employed is the scanning electron microscope (SEM), providing detailed morphological insights through high-resolution imaging. This research is often complimented by energy dispersive X-ray spectroscopy (EDS), which assesses particle chemical composition by utilizing a high-energy electron beam to generate X-rays specific to each element. Thermogravimetry and differential scanning calorimetry (TGA-DSC) is another method used for analyzing polymer properties and thermal reactions, along with gas chromatography, thermal, desorption and mass spectrometry (TED-GC-MS) (Mansa and Zou 2021). Combining multiple methods can synergize their capabilities and address the challenges associated with microplastic detection (Table 10.3) (Zhang and Chen 2020).

Table 10.3 Sources of microplastics and corresponding identification methods

Plastic bottled water	FTIR	Sieve and wash with ethanol, freeze and oxidize with H_2O	6.5–100 μm	Polypropylene	Singh et al. (2021)
Plastic water bottles	SEM-EDS	Deposit on an electric conductive surface, using a thin layer of conductive material such as carbon or gold	≥3 μm	PET, PE	Zhang and Chen (2020)
Bottled mineral water	RAMAN spectroscopy	Clean with 30% H_2O_2 solution and filter using 47 mm Whatsman glass fiber filter then rinse with deionized water.	1–10 μm	PET, PE, PP	Muhib et al. (2023)
Wastewater	TGA-DSC	Add $ZnCl_2$ solution then oxidize biomass residue using 30% H_2O_2	<12 μm	PE, PP	Majewsky et al. (2016)
Rivers	TED-GC-MS	Pretreat with 20% H_2O_2 for 48 h and 8 days	–	PE, PP, PS, PET, PA	D¨umichen et al. (2017)
Marine mangrove	FTIR-ATR	Oxidize using density separation (30% H_2O_2)	<20 μm	PE, PP, PVC	Park and Park (2021)
Wastewater in Germany	PYR-GC-MS	Filtration	10 μm, 100 μm, 50 μm	PE, PS	Park and Park (2021)

10.3.1 Physical Method

1. Visual inspection method: Microplastics can be detected either by visual observations or by microscopy to determine their presence in the analyzed sample. This method is strictly based on the different physical characteristics of microplastics that make them distinguishable from other particles. This usually requires recognizing their color and shape. This method has several advantages, such as the fact that it does not require expensive equipment, extensive training or toxic materials. However, it may lack precision, especially when the particles to be analyzed are smaller than 400 μm (Zhou et al. 2022). Therefore, due to the high error rate, this method is recommended only for training or only for initial procedures.

10.3.2 Chemical method

1. Pyr-GC/MS: Polymers undergo thermal degradation into smaller particles via pyrolysis at elevated temperatures, a process that can range from 400 to 800 °C (Hendrickson et al. 2018). Following pyrolysis, the resulting materials can be distinguished using a gas chromatography column, which separates them based on their retention time that differs according to their unique chemical and physical characteristics. Mass spectrometry is then employed to match the samples against a spectral database for the identification of microplastic particles (Peters et al. 2018).
2. Fourier transform infrared (FTIR): FTIR spectroscopy operates in three unique modes: transmission, reflection and attenuated total reflection, each tailored to analyze specific properties of a test sample. Transmission mode identifies functional groups and chemical constituents by comparing infrared spectra. Reflection mode is applied to samples too opaque for transmission analysis and attenuated total reflection facilitates the capture of clear, interpretable signals. FTIR excels in identifying microplastics by examining the vibrational frequencies of polymer bonds and provides insights into material aging through the study of specific functional groups, rendering it crucial for the characterization of plastic materials (Campanale et al. 2023).
3. SEM-EDS: combining scanning electron microscopy (SEM) with energy dispersive X-ray spectroscopy (EDS), this technique offers a robust method for microplastic identification and characterization. SEM reveals high-resolution images showing microplastic's surface textures and sizes while EDS analyzes chemical compositions to determine microplastic types based on the X-rays emitted by sample elements. Together, SEM-EDS serves as a vital tool for understanding microplastic degradation and for strategies aimed at mitigating their impacts (Girão 2020).
4. Attenuated Total Reflection Fourier Infrared (FTIR-ATR): FTIR-ATR is a rapid, non-destructive analytical approach primarily for microplastics detection and characterization through molecular vibrations. Optimal detection accuracy requires sample-specific pre-treatment (Fojt et al. 2022). Despite its technological advancements, challenges remain in detecting small particles within complex matrices or obscured by biological coatings. Studies have highlighted it efficacy in identifying specific polymers, such as polyester and polyethylene vinyl acetate, with high precision (M¨uhlschlegel et al. 2017; D¨umichen et al. 2017).
5. Raman spectroscopy (RS): As a non-invasive technique, Raman spectroscopy provides insights into molecular structures without altering sample integrity, employing a high-energy laser to analyze materials. Each polymer's unique Raman spectrum aids in its identification and characterization, making this technique indispensable for polymer analysis (Chakraborty et al. 2022).
6. Thermal Extraction-Desorption Gas Chromatography Mass Spectrometry (TED-GC-MS): TEDGC-MS stands out for its ability to handle samples and accurately identify microplastic types and masses without the need for time-consuming,

expensive pre-treatments. This method's efficiency and cost-effectiveness makes it a valuable option for microplastics detection and analysis across various samples (Boyle and Ormeci 2020; Pathan et al. 2020).

7. Micro-Fourier Transform Infrared (micro-FTIR): Utilizing infrared microscopy, this technique analyzes molecular vibrations to detect and characterize microplastics in samples, with each bond type showcasing distinct IR spectral features. Micro-FTIR offers a meticulous approach for identifying microscopic particles and their chemical compositions within diverse ecosystems, proving essential for microplastic research. This technique is precise and can be used to detect microplastics below 100 mm (Zhong et al. 2022).
8. Thermogravimetry combined with differential scanning calorimetry (TGA-DSC): This method employs thermal analysis to identify microplastics by monitoring weight changes in TGA or phase transitions in DSC (Silva et al. 2022). Although it is effective for analyzing small samples and detecting thermal degradation, distinguishing between polymers with similar thermal properties poses challenges, potentially complicating the differentiating process (Ivleva et al. 2017). This technique can accurately measure PP and PE particles, but distinguishing microplastics from other municipalities has proven difficult.

10.4 Future Direction in the Detection of Microplastic

The future direction of microplastic detection may see a significant shift to the use of artificial intelligence (AI) and machine learning (ML) technologies. These technologies offer promising advances in the efficiency and accuracy of microplastic detection and address some of the current challenges of traditional methods. There are various methods AI can be used in the detection of microplastics in both terrestrial and marine ecosystems. Some of them are:

Integration With Imaging Techniques: AI can be integrated with advanced imaging techniques such as microscopy, hyperspectral imaging and remote sensing to identify or for the detection and quantification of microplastics in different environments. Artificial intelligence algorithms can quickly process large images and distinguish microplastics from natural materials based on shape, size, color and texture. This approach can significantly reduce the time required for manual detection and can be adapted to detect microplastics in water and soil.

Improved Spectroscopic Analysis: AI can also improve the performance of spectroscopic analysis methods such as FTIR and Raman spectroscopy. By training machine learning models with spectral data from known plastics, artificial intelligence can help quickly identify types of microplastics in environmental samples. This can improve the accuracy of microplastic characterization and facilitate monitoring of pollution sources and distribution patterns.

Real-Time Monitoring and Detection: The deployment of AI-based sensors and devices in aquatic and terrestrial environments can enable real-time monitoring.

of microplastic pollution. These devices can continuously collect data on microplastic concentrations, types and movements, feeding that data into artificial intelligence models for ongoing analysis. Such systems can provide immediate warnings of pollution hotspots or trends, enabling faster responses and mitigation measures.

Data Analysis and Forecasting: AI can analyze massive data sets from various sources to identify patterns and trends in microplastic pollution. Forecast models can be used to predict future changes in distribution and concentrations under various environmental and human activity scenarios. This information can be invaluable to policymakers, environmentalists and researchers as they develop targeted strategies to reduce microplastic pollution.

Citizen Science and Commons Acquisition: AI can support citizen science initiatives by providing tools and applications that allow the public to participate in microplastics data collection. Image recognition algorithms can help non-experts identify and report microplastics in local environments, enrich global datasets and promote community engagement in reducing pollution.

While AI offers many opportunities to advance microplastic detection, there are still several challenges. These include the need for large data sets to train AI models, the potential for biased data collection and analysis methods, and ensuring the accuracy and reliability of AI predictions. Integrating artificial intelligence into microplastic research and monitoring will require interdisciplinary collaboration and significant investment in technology development and deployment.

10.5 Conclusion

Microplastics pose a threat not just to water ecosystems, but also to larger species that consume microplastic-laden marine flora and fauna. This issue extends to humans, particularly through the consumption of fish sourced from seawater, highlighting a significant concern. Many of the world's leading contributors to plastic pollution, including countries in Africa, have yet to fully grasp the magnitude of microplastic contamination in marine and freshwater environments. Addressing the challenges posed by plastic waste requires collective efforts from individuals, civil society, and local governments. The escalating problem of plastic waste generation demands immediate action to prevent further environmental degradation. There is a need to fully understand the impact of microplastics on ecosystems, as well as the methods and techniques used to detect them. This information will further help to remove them later.

References

Allen NS, Edge M, Appleyardt JR, Jewitt TS, Rorie CV (1988) Degradation of historic cellulose triacetate cinematographic film: influence of various film parameters and prediction of archival life. J Photogr Sci 36(6):194–198. https://doi.org/10.1080/00223638.1988.11736999

Allen NS, Edge M, Mohammadian M, Jones K (1994) Physicochemical aspects of the environmental degradation of polyethylene terephthalate. Polym Degrad Stab 43(2):229–237. https://doi.org/10.1016/0141-3910(94)90074-4

Andrady AL, Neal MA (2009) Applications and societal benefits of plastics. Philos Trans R Soc B Biol Sci 364:1977–1984

Aransiola SA, Ikhumetse AA, Babaniyi BR, Abioye OP, Oyedele OJ, Falade NO (2023a) Phytoaccumulation of micro and nanoplastics: root uptake. In: Maddela NR, Reddy KV, Ranjit P (eds) Micro and nanoplastics in soil. Springer, Cham. https://doi.org/10.10007/978-3-031-21195-9_8

Aransiola SA, Victor-Ekwebelem MO, Ajiboye AE, Leh-Togi Zobeashia SS, Ijah UJJ, Oyedele OJ (2023b) Ecological impacts and toxicity of micro and nanoplastics in agroecosystem. In: Maddela NR, Reddy KV, Ranjit P (eds) Micro and nanoplastics in soil. Springer, Cham. https://doi.org/10.10007/978-3-031-21195-9_10

Auta SH, Abioye OP, Aransiola SA, Bala JD, Chukwuemeka VI, Hassan A, Aziz A, Fauziah SH (2022) Enhanced microbial degradation of PET and PS microplastics under natural conditions in mangrove environment. J Environ Manag 304(2022):114273. https://doi.org/10.1016/j.jenvman.2021.114273

Awet TT, Kohl Y, Meier F, Straskraba S, Grun AL, Ruf T, Jost C, Drexel R, Tunc E, Emmerling C (2018) Effects of polystyrene nanoparticles on the microbiota and functional diversity of enzymes in soil. Environ Sci Pollut Res Int 25(33):33666–33676. https://doi.org/10.1007/s11356-018-3134-x

Barnes DKA, Galgani F, Thompson RC, Barlaz M (2009) Accumulation and fragmentation of plastic debris in global environments. Philos Trans R Soc B Biol Sci 364(1526):1985–1998. https://doi.org/10.1098/rstb.2008.0205

Boyle K, Ormeci B (2020) Microplastics and nanoplastics in the freshwater and terrestrial environment: a review. Water 12:2633–2639

Campanale C, Savino I, Massarelli C, Uricchio VF (2023) Fourier transform infrared spectroscopy to assess the degree of alteration of artificially aged and environmentally weathered microplastics. Polymers 15(4):911

Chakraborty I, Banik S, Biswas R, Yamamoto T, Noothalapati H, Mazumder N (2022) Raman spectroscopy for microplastic detection in water sources: a systematic review. Int J Environ Sci Technol 20(9):10435–10448

Derraik JG (2002) The pollution of the marine environment by plastic debris: a review. Mar Pollut Bull 44:842–852

Duis K, Coors A (2016) Microplastics in the aquatic and terrestrial environment: sources (with a specific focus on personal care products), fate and effects. Environ Sci Pollut Res 23(3):2224–2237. https://doi.org/10.1007/s11356-015-5887-7

Dümichen E, Eisentraut P, Bannick CG, Barthel AK, Senz R, Braun U (2017) Fast identification of microplastics in complex environmental samples by a thermal degradation method. Chemosphere 174:572–584

Edge M, Hayes M, Mohammadian M, Allen NS, Jewitt TS, Brems K, Jones K (1991) Aspects of polyethylene terephthalate degradation for archival life and environmental degradation. Polym Degrad Stab 32(2):131–153. https://doi.org/10.1016/0141-3910(91)90047-U

Eriksen M, Lebreton LCM, Carson HS, Thiel M, Moore CJ, Borerro JC, Galgani F, Ryan PG (2014) Plastic pollution in the world's oceans: more than 5 trillion plastic pieces weighing over 250,000 tons afloat at sea. PLoS One 9:e111913

Feng X, Wang Q, Sun Y, Zhang S, Wang F (2022) Microplastics change soil properties, heavy metal availability and bacterial community in a Pb-Zn-contaminated soil. J Hazard Mater 424(Pt A):127364. https://doi.org/10.1016/j.jhazmat.2021.127364

Fojt J, Romaneková I, Procházka P, David J, Brtnický M, Kucerık J (2022) A simple method for quantification of polyhydroxybutyrate and polylactic acid microbioplastics in soils by evolved gas analysis. Molecules 27(6):1898

Girão AV (2020) SEM/EDS and optical microscopy analysis of microplastics. In: Handbook of microplastics in the environment. Springer, Berlin, pp 1–22

Harrison JP, Schratzberger M, Sapp M, Osborn AM (2014) Rapid bacterial colonization of low-density polyethylene microplastics in coastal sediment microcosms. BMC Microbiol 14:232. https://doi.org/10.1186/s12866-014-0232-4

Hendrickson E, Minor EC, Schreiner K (2018) Microplastic abundance and composition in western Lake Superior as determined via microscopy, pyr-GC/MS, and FTIR. Environ Sci Technol 52(4):1787–1796

Ioakeimidis C, Fotopoulou KN, Karapanagioti HK, Geraga M, Zeri C, Papathanassiou E et al (2016) The degradation potential of PET bottles in the marine environment: an ATR-FTIR based approach. Sci Rep 6(1):1–8. https://doi.org/10.1038/srep23501

Ivleva NP, Wiesheu AC, Niessner R (2017) Microplastic in aquatic ecosystems. Angew Chem Int Ed 56(7):1720–1739

Jambeck JR, Geyer R, Wilcox C, Siegler TR, Perryman M, Andrady A et al (2015) Plastic waste inputs from land into the ocean. Science 347(6223):768–771. https://doi.org/10.1126/science.1260352

Jiang XJ, Liu WJ, Wang EH, Zhou TZ, Xin P (2017) Residual plastic mulch fragments effects on soil physical properties and water flow behavior in the Minqin Oasis, Northwestern China. Soil Tillage Res 166:100–107. https://doi.org/10.1016/j.still.2016.10.011

Ju H, Zhu D, Qiao M (2019) Effects of polyethylene microplastics on the gut microbial community, reproduction and avoidance behaviors of the soil springtail, Folsomia candida. Environ Pollut 247:890–897. https://doi.org/10.1016/j.envpol.2019.01.097

Judy JD, Williams M, Gregg A, Oliver D, Kumar A, Kookana R, Kirby JK (2019) Microplastics in municipal mixed-waste organic outputs induce minimal short to long-term toxicity in key terrestrial biota. Environ Pollut 252(Pt A):522–531. https://doi.org/10.1016/j.envpol.2019.05.027

Kaewlue W (2012) Solid waste management of Tha Khon Yang Municipality, Kantharawichai District, Maha Sarakham Province. Master's thesis. Mahasarakham University, Talat, Thailand

Kedzierski M, D'Almeida M, Magueresse A, Le Grand A, Duval H, César G et al (2018) Threat of plastic ageing in marine environment. Adsorption/desorption of micro pollutants. Mar Pollut Bull 127:684–694. https://doi.org/10.1016/j.marpolbul.2017.12.059. PMID:29475712

Lai H, Liu X, Qu M (2022) Nanoplastics and human health: hazard identification and biointerface. Nanomaterials (Basel) 12(8):1298. https://doi.org/10.3390/nano12081298

Lebreton LCM, van der Zwet J, Damsteeg J-W, Slat B, Andrady A, Reisser J (2017) River plastic emissions to the world's oceans. Nat Commun 8:15611

Li HZ, Zhu D, Lindhardt JH, Lin SM, Ke X, Cui L (2021) Long-term fertilization history alters effects of microplastics on soil properties, microbial communities, and functions in diverse farmland ecosystem. Environ Sci Technol 55(8):4658–4668. https://doi.org/10.1021/acs.est.0c04849

Li C, Cui Q, Li Y, Zhang K, Lu X, Zhang Y (2022) Effect of LDPE and biodegradable PBAT primary microplastics on bacterial community after four months of soil incubation. J Hazard Mater 429:128353. https://doi.org/10.1016/j.jhazmat.2022a.128353

Lian J, Liu W, Meng L, Wu J, Zeb A, Cheng L, Lian Y, Sun H (2021) Effects of microplastics derived from polymer-coated fertilizer on maize growth, rhizosphere, and soil properties. J Clean Prod. https://doi.org/10.1016/j.jclepro.2021.128571

Lwanga EH, Gertsen H, Gooren H, Peters P, Salanki T, van der Ploeg M (2016) Microplastics in the terrestrial ecosystem: implications for Lumbricus terrestris (Oligochaeta, Lumbricidae). Environ Sci Technol 50:2685–2691

Maaß S, Daphi D, Lehmann A, Rillig MC (2017) Transport of microplastics by two collembolan species. Environ Pollut 225:456–459. https://doi.org/10.1016/j.envpol.2017.03.009
Machado A, Lau CW, Kloas W, Bermann J, Bachelier JB, Faltin E (2019) Microplastics can change soil properties and affect plant performance. Environ Sci Technol 53:6044–6052. https://doi.org/10.1021/acs.est.9b01339
Majewsky M, Bitter H, Eiche E, Horn H (2016) Determination of microplastic polyethylene (PE) and polypropylene (PP) in environmental samples using thermal analysis (TGA-DSC). Sci Total Environ 568:507–511
Mansa R, Zou S (2021) Thermogravimetric analysis of microplastics: a mini review. Environ Adv 5:100117
Mato Y, Isobe T, Takada H, Kanehiro H, Ohtake C, Kaminuma T (2001) Plastic resin pellets as a transport medium for toxic chemicals in the marine environment. Environ Sci Technol 35(2):318–324. https://doi.org/10.1021/es0010498
McCormick AR, Hoellein TJ, London MG, Hittie J, Scott JW, Kelly JJ (2016) Microplastic in surface waters of urban rivers: concentration, sources, and associated bacterial assemblages. Ecosphere 7(11):e01556
Meeker JD, Sathyanarayana S, Swan SH (2009) Phthalates and other additives in plastics: human exposure and associated health outcomes. Philos Trans R Soc Lond Ser B Biol Sci 364(1526):2097–2113. https://doi.org/10.1098/rstb.2008.0268. PMID:19528058
Muhib MI, Uddin MK, Rahman MM, Malafaia G (2023) Occurrence of microplastics in tap and bottled water, and food packaging: a narrative review on current knowledge. Sci Total Environ 865:161274–165297
Mühlschlegel P, Hauk A, Walter U, Sieber R (2017) Lack of evidence for microplastic contamination in honey. Food Addit Contam: Part A 34(11):1982–1989
Musa IO, Auta HS, Ilyasu US, Aransiola SA, Makun HA, Adabara NU, Abioye OP, Aziz A, Jayanthi B, Maddela NR, Prasad R (2024) Micro and nanoplastics in environment—degradation, detection, ecological impact. Int J Environ Res 18:1. https://doi.org/10.1007/s41742-023-00551-9
Oehlmann J, Schulte Oehlmann U, Kloas W, Jagnytsch O, Lutz I, Kusk KO et al (2009) A critical analysis of the biological impacts of plasticizers on wildlife. Philos Trans R Soc Lond Ser B Biol Sci 364(1526):2047–2062. https://doi.org/10.1098/rstb.2008.0242. PMID:19528055
Park H, Park BS (2021) Review of microplastic distribution, toxicity, analysis methods, and removal technologies. Water 13(19):2736–2764
Pathan SI, Arfaioli P, Bardelli T, Ceccherini MT, Nannipieri P, Pietramellara G (2020) Soil pollution from micro-and nanoplastic debris: a hidden and unknown biohazard. Sustain For 12(18):7255–7331
Peters CA, Hendrickson E, Minor EC, Schreiner K, Halbur J, Bratton SP (2018) Pyr-GC/MS analysis of microplastics extracted from the stomach content of benthivore fish from the Texas Gulf Coast. Mar Pollut Bull 137:91–95
Qi Y, Yang X, Pelaez AM, Lwanga EH, Beriot N, Gertsen H (2018) Macro- and micro-plastics in soil-plant system: effects of plastic mulch film residues on wheat (Triticum aestivum) growth. Sci Total Environ 645:1048–1056
Rillig MC, Ziersch L, Hempel S (2017) Microplastic transport in soil by earthworm. Sci Rep 7:1362. https://doi.org/10.1038/s41598-017-01594-7
Robert J, Strife SC, Jason MP, Safa M (2021) Normalized mass maps in three-dimensional space combining Kendrick-like values with chromatographic separation for enhanced data deconvolution. J Am Soc Mass Spectrom 32(4):989–995
Rodríguez-Seijo A, Lourenco J, Rocha-Santos TAP, Duarte AC, Vala H (2016) Histopathological and molecular effects of microplastics in Eisenia andrei Bouche. Environ Pollut 220(Pt A):495–503. https://doi.org/10.1016/j.envpol.2016.09.092
Rong L, Zhao L, Zhao L, Cheng Z, Yao Y, Yuan C, Wang L, Sun H (2021) LDPE microplastics affect soil microbial communities and nitrogen cycling. Sci Total Environ 773:145640. https://doi.org/10.1016/j.scitotenv.2021.145640

Rudel RA, Brody JG, Spengler JC, Vallarino J, Geno PW, Sun G, Yau A (2001) Identification of selected hormonally active agents and animal mammary carcinogens in commercial and residential air and dust samples. J Air Waste Manage Assoc 51(4):499–513. https://doi.org/10.1080/10473289.2001.10464292. PMID:11321907

Rudel RA, Camann DE, Spengler JD, Korn LR, Brody JG (2003) Phthalates, alkylphenols, pesticides, polybrominated diphenyl ethers, and other endocrine-disrupting compounds in indoor air and dust. Environ Sci Technol 37(20):4543–4553. https://doi.org/10.1021/es0264596. PMID:14594359

Schmidt C, Krauth T, Wagner S (2017) Export of plastic debris by rivers into the sea. Environ Sci Technol 51(21):12246–12253. https://doi.org/10.1021/acs.est.7b02368. PMID:29

Seeley ME, Song B, Passie R, Hale RC (2020) Microplastics affect sedimentary microbial communities and nitrogen cycling. Nat Commun 11(1):2372. https://doi.org/10.1038/s41467-020-16235-

Sharma BK (2014) Industrial chemistry: including chemical engineering. Goel Publishing, New Delhi, p TP 145.S53

Silva ALP, Silva SAM, Duarte A, Barcelo D, Rocha-Santos T (2022) Analytical methodologies used for screening micro(nano)plastics in (eco)toxicity tests. Green Anal Chem 3:Article ID 100037

Singh B, Sharma N (2008) Mechanistic implications of plastic degradation. Polym Degrad Stab 93(3):561–584

Singh S, Kalyanasundaram M, Diwan V (2021) Removal of microplastics from wastewater: available techniques and way forward. Water Sci Technol 84(12):3689–3704

Sun Y, Li X, Cao N, Duan C, Ding C, Huang Y, Wang J (2022) Biodegradable microplastics enhance soil microbial network complexity and ecological stochasticity. J Hazard Mater 439:129610. https://doi.org/10.1016/j.jhazmat.2022.129610

Swan SH (2008) Environmental phthalate exposure in relation to reproductive outcomes and other health endpoints in humans. Environ Res 108(2):177–184. https://doi.org/10.1016/j.envres.2008.08.007. PMID:18949837

Swan SH, Main KM, Liu F, Stewart SL, Kruse RL, Calafat AM et al (2005) Decrease in anogenital distance among male infants with prenatal phthalate exposure. Environ Health Perspect 113(8):1056–1061. https://doi.org/10.1289/ehp.8100. PMID:16079079

Taylor ML, Gwinnett C, Robinson LF, Woodall LC (2016) Plastic microfiber ingestion by deep-sea organisms. Sci Rep 6:33997. https://doi.org/10.1038/srep33997

The Ecotourism and Conservation Society of Sikkim (ECOSS) (2014) Plastics and the environment: assessing the impact of the complete ban on plastic carry bag. Retrieved from http://www.toxicslink.org

Thompson RC, Moore CJ, vom Saal SF, Swan SH (2009) Plastics, the environment, and human health: current consensus and future trends. Philos Trans R Soc Lond Ser B Biol Sci 364(1526):2153–2166. https://doi.org/10.1098/rstb.2009.0053. PMID:19528062

Veresoglou SD, Halley JM, Rillig MC (2015) Extinction risk of soil biota. Nat Commun 6:8862. https://doi.org/10.1038/ncomms9862

Wang ZC, Li XY, Shi HB, Sun M, Ding T (2015) Effects of residual plastic film on soil hydrodynamic parameters and soil structure (in Chinese). Trans Chin Soc Agric Mach 46(5):101–106

Wang J, Lv S, Zhang M, Chen G, Zhu T, Zhang S, Teng Y, Christie P, Luo Y (2016) Effects of plastic film residues on occurrence of phthalates and microbial activity in soils. Chemosphere 151:171–177. https://doi.org/10.1016/j.chemosphere.2016.02.076

Wang HT, Ding J, Xiong C, Zhu D, Li G, Jia XY, Zhu YG (2019) Exposure to microplastics lowers arsenic accumulation and alters gut bacterial communities of earthworm Metaphire californica. Environ Pollut 251:110–116. https://doi.org/10.1016/j.envpol.2019.04.054

Williams AT, Rangel-Buitrago N (2022) The past, present, and future of plastic pollution. Mar Pollut Bull 176:113429. https://doi.org/10.1016/j.marpolbul.2022.113429

Xiang S, Xie Y, Sun X, Du H, Wang J (2021) Identification and quantification of microplastics in aquaculture environment. Front Mar Sci 8:1–10

Xu M, Du W, Ai F, Xu F, Zhu J, Yin Y, Ji R, Guo H (2021) Polystyrene microplastics alleviate the effects of sulfamethazine on soil microbial communities at different CO2 concentrations. J Hazard Mater 413:125286

Yu H, Qi W, Cao X, Hu J, Li Y, Peng J, Hu C, Qu J (2021) Microplastic residues in wetland ecosystems: do they truly threaten the plant-microbe-soil system? Environ Int 156:106708

Zeng LS, Zhou ZF, Shi YX (2013) Environmental problems and control ways of plastic film in agricultural production. Appl Mech Mater 295–298:2187–2190

Zettler ER, Mincer TJ, Amaral-Zettler LA (2013) Life in the "Plastisphere": microbial communities on plastic marine debris. Environ Sci Technol 47(13):7137–7146

Zhang Z, Chen Y (2020) Effects of microplastics on wastewater and sewage sludge treatment and their removal: a review. Chem Eng J 382:122955–122970

Zhang MM, Dong BD, Qiao YZ, Yang H, Wang YK, Liu MY (2018) Effects of sub-soil plastic film mulch on soil water and salt content and water utilization by winter wheat under different soil salinities. Field Crop Res 225:130–140

Zhang X, Li Y, Ouyang D, Lei J, Tan Q, Xie L et al (2021) Systematical review of interactions between microplastics and microorganisms in the soil environment. J Hazard Mater 418:126288

Zhong Y, Bao Q, Yuan L, Liu J, Cai Y, Chen X (2022) Analysis of microplastics in aquatic shellfish by pyrolysis–gas chromatography/mass spectrometry after alkali digestion and solvent extraction. Polymers 14:3888–3918

Zhou F, Wang X, Wang G, Zuo Y (2022) A rapid method for detecting microplastics based on fluorescence lifetime imaging technology (FLIM). Toxics 10(3):118

Zhu BK, Fang YM, Zhu D, Christie P, Ke X (2018) Exposure to nanoplastics disturbs the gut microbiome in the soil oligochaete Enchytraeus crypticus. Environ Pollut 239:408–415

Part III
Sources, Prevalence and Effects of the Atmospheric Plastisphere

Chapter 11
White Pollution and Air Quality

Gbolahan Olamide Isaac, Babafemi Raphael Babaniyi, Taiwo Oluwasola Abe, and Ademola Bisi-Omotosho

Abstract This study explores the diverse impacts of white pollution on air quality, public health, and the ecosystem. White pollution, characterized by the accumulation of plastic waste and its associated pollutants, significantly impacts air quality through various mechanisms such as the emission of volatile organic compounds (VOCs), release of hazardous chemicals, and production of particulate matter (PM) during incineration. The deterioration of air quality from white pollution presents significant health hazards, including respiratory illnesses, cardiovascular conditions, cancer, neurological impacts, and immune system disorders. Moreover, white pollution has broad environmental repercussions, encompassing marine and terrestrial pollution, habitat deterioration, chemical contamination, and contributions to global climate change. Various strategies and actions are being implemented globally to address these impacts, including limitations on plastic usage, enhancements in waste management systems, public awareness initiatives, innovations in packaging and materials, air quality surveillance and governance, international accords, and corporate sustainability efforts. The long-term outlook for tackling white pollution and air quality concerns involves transitioning towards circular economy models, advancements in recycling technologies, policy and regulatory interventions, promotion of renewable energy sources, eco-friendly innovations, sustainable methodologies, and educational and public involvement. Resolving white pollution and enhancing air quality necessitate collaborative endeavors and partnerships across different sectors to safeguard public health and environmental integrity for current and future generations.

G. O. Isaac (✉) · T. O. Abe
Department of Chemistry, Federal University of Technology, Akure, Nigeria

B. R. Babaniyi
Bioresources Development Centre, National Biotechnology Development Agency, Abuja, Nigeria

A. Bisi-Omotosho
Department of Sustainability Management and Innovation, University of Westminster, London, UK

S. A. Aransiola et al. (eds.), *White Pollution: Biodiversity and hazards in marine plastisphere*, Environmental Contamination Remediation and Management,
https://doi.org/10.1007/978-3-031-95547-1_11

Keywords Climate change · Pollution · Plastics · Environment

11.1 Introduction

In recent decades, the ubiquitous presence of plastic debris across landscapes and ecosystems has emerged as a major environmental concern, aptly termed "white pollution" (Li et al. 2018; Musa et al. 2024). This pervasive issue is primarily driven by the exponential growth of single-use plastics, coupled with inadequate waste management practices. Jambeck et al. (2015) paint a stark picture, estimating that a staggering 300 million tons of plastic waste are generated annually, with only a small fraction being properly recycled or disposed of. The vast majority of this plastic waste escapes proper management, ending up in landfills, littering our environment, or accumulating in waterways.

While the initial focus of "white pollution" centered on the visual blight of plastic debris, its ramifications extend far beyond aesthetics. Mounting evidence highlights the significant ecological and health impacts associated with plastic pollution (Wright et al. 2015; Aransiola et al. 2023a). One particularly concerning aspect, often overlooked, is the detrimental effect white pollution can have on air quality. The lifecycle of plastics, encompassing production, use, and disposal, contributes to air quality degradation in several ways. This paper delves into the intricate relationship between white pollution and air quality, shedding light on the mechanisms through which plastic pollution contaminates the atmosphere. By examining the causes, impacts, and implications of white pollution on air quality, this study aims to underscore the urgency of tackling plastic pollution as a multifaceted environmental challenge.

This article will emphasize empirical evidence and scholarly literature to substantiate claims and provide a robust understanding of the subject matter. By elucidating the interconnectedness of white pollution and air quality degradation, this study aspires to inform policy interventions, technological advancements, and public awareness campaigns aimed at mitigating plastic pollution and safeguarding both environmental and human well-being.

11.2 The Connection Between White Pollution and Air Quality

The association between "white pollution" and air quality signifies a crucial overlap in environmental deterioration, where the widespread presence of plastic waste leads to air contamination, posing significant risks to both human health and ecosystems. White pollution, characterized by the accumulation of non-biodegradable plastic debris in diverse environments, goes beyond surface-level concerns to encompass substantial ecological and health implications.

Plastic pollution, a primary component of white pollution, has emerged as a global environmental issue due to the surge in plastic production, consumption, and inadequate waste management practices. While plastic pollution is commonly associated with marine and terrestrial environments, its impact on air quality is gaining recognition as a significant issue. The life cycle of plastics, from production to disposal, involves the release of various pollutants into the air, exacerbating air quality degradation and posing risks to human health and the environment.

Several studies have elucidated mechanisms through which plastic pollution contributes to air pollution. According to, the degradation of plastics can emit volatile organic compounds (VOCs) and other hazardous chemicals, particularly under conditions of sunlight exposure or heat. Additionally, the incineration of plastic waste emits toxic gases and particulate matter, further compromising air quality (Andrady 2011). Microplastics, ubiquitous components of plastic pollution, also play a role in atmospheric contamination. These minute particles can become airborne through processes like wind erosion, combustion, or volatilization, dispersing pollutants over extensive areas (Wright et al. 2013). Once airborne, microplastics can act as carriers for various pollutants, including heavy metals and organic contaminants, exacerbating air quality concerns.

11.3 Causes of White Pollution

White pollution, characterized by the pervasive presence of non-biodegradable plastic waste in the environment, is driven by a multitude of interconnected factors spanning the entire lifecycle of plastics, from production to disposal. Understanding these causes is crucial for devising effective strategies to mitigate plastic pollution and its detrimental impacts on ecosystems and human health.

I. **Extensive Generation and Usage of One-Time-Use Plastics**

The rampant production and consumption of single-use plastics have become a significant driver of white pollution, leading to severe environmental consequences. According to a study by, the widespread adoption of single-use plastics, including bags, bottles, straws, and packaging materials, has significantly contributed to the escalation of plastic pollution globally. These items are often made from inexpensive materials like polyethylene and polypropylene, which are durable and lightweight but contribute to environmental degradation due to their non-biodegradable nature. The mass production and consumption of single-use plastics have led to a surge in plastic waste generation, exacerbating the problem of white pollution. For example, in the United States alone, over 35 million tons of plastic waste were generated in 2017, with packaging accounting for the largest share (U.S. EPA 2020). Similarly, a report by the Ellen MacArthur Foundation (2016) estimated that the global production of plastics reached 311 million metric tons in 2014, with around 40% of that production dedicated to packaging, much of which is single-use.

The convenience and affordability of single-use plastics have fueled their widespread use in various industries and consumer products. Plastic bags are ubiquitous in retail stores, plastic bottles are commonly used for beverages, and plastic packaging is prevalent in food and consumer goods. Moreover, the proliferation of online shopping and home delivery services has led to an increase in the use of plastic packaging materials to ensure the safe transportation of goods. The rapid growth of the global population and urbanization has further intensified the demand for single-use plastics, as they are perceived as convenient and hygienic solutions for packaging and storage. However, the linear nature of the plastics economy, where products are designed for disposal after a single use rather than reuse or recycling, has resulted in a significant accumulation of plastic waste in landfills, oceans, and natural environments (Auta et al. 2022; Aransiola et al. 2023b).

II. **Inadequate Methods of Waste Management**

Inadequate waste management practices play a crucial role in exacerbating the problem of white pollution, leading to the accumulation of plastic waste in the environment. Insufficient infrastructure and poor waste management practices contribute to the improper disposal of plastic waste, including littering, inadequate recycling facilities, and inefficient waste collection systems. Schmidt et al. (2017) highlighted in their study that inadequate waste management practices significantly contribute to plastic pollution in the environment. When plastic waste is not properly managed, it can easily find its way into waterways, oceans, and natural habitats, where it persists for decades or even centuries, posing threats to marine life and ecosystems.

Littering is a common problem worldwide, with plastic items often ending up in streets, parks, beaches, and other public spaces. The improper disposal of plastic waste through littering not only degrades the aesthetic quality of the environment but also contributes to the fragmentation of plastics into microplastics, which can be ingested by wildlife and enter the food chain (Eriksen et al. 2014). Furthermore, inadequate recycling facilities and inefficient waste collection systems hinder efforts to properly manage plastic waste. Many regions lack sufficient infrastructure for recycling and may not have access to appropriate recycling technologies or facilities to handle various types of plastics. As a result, a significant portion of plastic waste ends up in landfills or incinerators, contributing to environmental pollution and greenhouse gas emissions (Andrady 2015). Moreover, informal waste management practices, such as open dumping and burning of waste, are prevalent in many parts of the world, particularly in low-income countries with limited resources and infrastructure. These practices not only release harmful pollutants into the air and soil but also fail to address the root causes of plastic pollution (Hoornweg and Bhada-Tata 2012).

III. **Lack of Effective Recycling and Recovery Systems**

The lack of effective recycling and recovery systems presents a significant challenge in addressing white pollution caused by plastic waste. Although recycling holds promise as a solution to reduce plastic pollution, the effectiveness of recycling systems varies widely across regions and is hindered by various challenges. Baldé et al. (2017) emphasized in their research the challenges associated with recycling and recovery systems for plastic waste. One major challenge is the contamination of recyclables, where mixed materials or non-recyclable items are inadvertently included in recycling streams, reducing the quality and value of the recycled material. Contamination not only complicates the sorting and processing of recyclables but also undermines the economic viability of recycling operations. Furthermore, limited market demand for recycled plastics poses a barrier to the establishment and expansion of recycling systems. Despite efforts to increase the use of recycled materials in manufacturing processes, the demand for recycled plastics remains insufficient to absorb the growing supply of plastic waste. This mismatch between supply and demand discourages investment in recycling infrastructure and hampers efforts to close the loop on plastic production and consumption (Jambeck et al. 2015).

Inadequate recycling infrastructure is another significant obstacle to effective plastic waste management. Many regions lack the necessary facilities and technologies for collecting, sorting, and processing plastic waste efficiently. In developing countries, in particular, the lack of investment in recycling infrastructure exacerbates the problem of plastic pollution, as much of the waste ends up in open dumps or informal recycling operations with limited capacity and environmental safeguards (Santos et al. 2019). Moreover, the complexity and diversity of plastic materials present technical challenges in the recycling process. Different types of plastics require separate recycling streams, and the presence of additives and contaminants further complicates recycling efforts. As a result, the recycling of plastics is often less cost-effective and energy-intensive compared to other materials, reducing its attractiveness as a waste management option (Plastics Europe 2020a, b).

Addressing the lack of effective recycling and recovery systems requires coordinated efforts from government, industry, and civil society. Investments in recycling infrastructure, technology, and education are essential to improve collection rates, reduce contamination, and increase the quality of recycled materials. Additionally, policies that promote extended producer responsibility, incentivize the use of recycled materials, and discourage single-use plastics can help create a more circular economy for plastics (Nixon et al. 2020).

IV. **Planned Obsolescence and Overconsumption**

Planned obsolescence and overconsumption are significant factors contributing to the proliferation of plastic waste and exacerbating the problem of white pollution. Planned obsolescence, in particular, perpetuates a cycle of continuous plastic production and disposal, while consumer culture emphasizing disposability and convenience further drives excessive consumption of plastic products. Cooper and Mok (2021) highlighted in their research the role of planned obsolescence in contributing to plastic pollution. Planned obsolescence refers to the deliberate design of products to have a limited lifespan or to become obsolete after a short period, thereby encouraging consumers to replace them frequently. In the case of plastic products, planned obsolescence often involves designing items with built-in flaws or using materials that degrade quickly, leading to premature failure and disposal. Moreover, the culture of disposability and convenience promoted by consumerism encourages the overconsumption of plastic products. Single-use plastics, in particular, epitomize this culture of convenience, offering quick and easy solutions for packaging, storage, and consumption. However, the short-term convenience of single-use plastics comes at a significant long-term cost to the environment, as these products contribute to the accumulation of plastic waste in landfills, oceans, and natural habitats (Parker 2018). Furthermore, the marketing strategies employed by companies often emphasize novelty, trends, and perceived obsolescence to stimulate consumer demand for new products. Advertising campaigns create a sense of urgency and desire for the latest gadgets, fashion items, and consumer goods, driving consumers to discard perfectly functional items in favor of newer models. This relentless pursuit of novelty and consumption fuels the demand for plastic products and contributes to the cycle of plastic production, consumption, and disposal (Hawken et al. 2013).

V. **Globalization and Industrialization of Plastic Production**

The globalization of supply chains and the industrialization of plastic production have profoundly influenced the proliferation of plastic products worldwide, contributing significantly to white pollution. This phenomenon has been extensively studied, and researchers like Lebreton et al. (2017) have highlighted the environmental consequences of the globalization and industrialization of plastic production. The globalization of supply chains has enabled the seamless movement of raw materials, intermediate products, and finished goods across borders, facilitating the widespread distribution of plastic products. Manufacturers can source materials and components from different regions, assemble products in one location, and distribute them globally to meet consumer demand. This interconnectedness has fueled the exponential growth of plastic production and consumption, particularly in emerging markets with expanding middle-class populations. Moreover, the industrialization of plastic production has led to the mass manufacturing of plastic products on a scale never seen before. Advances in technology, automation, and economies of scale have made plastic production more efficient and cost-effective, driving down prices and increasing accessibility. Plastic products have become ubiquitous in everyday life, from consumer goods and packaging to construction materials and medical devices.

However, the rapid expansion of plastic production and consumption has outpaced the development of waste management infrastructure, particularly in developing countries with rapidly growing economies. These countries often lack adequate regulations, investment, and public awareness to manage plastic waste effectively. As a result, plastic pollution has become a pressing environmental issue in many regions, with plastic waste clogging waterways, polluting oceans, and contaminating terrestrial ecosystems. Furthermore, the globalization of plastic production has shifted the environmental burden of plastic pollution from consumer countries to producer countries. Developing countries often serve as dumping grounds for plastic waste exported from developed countries, exacerbating local pollution problems and straining limited resources. This unequal distribution of environmental costs highlights the need for global cooperation and regulatory frameworks to address the transboundary impacts of plastic pollution (Wong et al. 2019).

11.4 Impact on Air Quality

The presence of white pollution, characterized by plastic waste accumulation, has significant implications for air quality, contributing to atmospheric contamination through various mechanisms. Understanding these impacts is crucial for assessing the full extent of plastic pollution's environmental consequences and devising strategies to mitigate its effects on air quality and public health.

(a) **Emission of Volatile Organic Compounds (VOCs)**

The emission of volatile organic compounds (VOCs) from plastic materials is a significant concern regarding its impact on air quality. When plastics undergo degradation processes, particularly when exposed to sunlight or heat, they release VOCs into the atmosphere. This phenomenon has been studied extensively, with researchers such as shedding light on the environmental consequences of plastic degradation.

VOCs emitted from plastics include compounds such as styrene, benzene, toluene, and ethylbenzene, among others. These compounds are known to be hazardous air pollutants with adverse effects on human health and the environment. Styrene, for example, is classified as a possible human carcinogen by the International Agency for Research on Cancer (IARC), while benzene is a known human carcinogen (ATSDR 2021).

One of the significant concerns regarding VOC emissions from plastics is their contribution to the formation of smog and ground-level ozone. VOCs react with nitrogen oxides (NOx) in the presence of sunlight to form ground-level ozone, a key component of smog. Ground-level ozone can exacerbate respiratory problems such as asthma and bronchitis, and long-term exposure to high levels of ozone is associated with adverse health effects (EPA 2021). Moreover, VOCs emitted from plastics can have a range of other environmental impacts. They can contribute to the formation of secondary organic aerosols (SOAs), which affect air quality and climate by influencing atmospheric chemistry and cloud formation (Donahue et al. 2012). Additionally, some VOCs are persistent organic pollutants (POPs) that can bioaccumulate in the environment and pose risks to ecosystems and wildlife (Scheringer et al. 2014).

(b) **Release of Toxic Chemicals**

The release of toxic chemicals from the degradation of plastics is a significant concern due to its implications for both human health and ecosystems. As plastics degrade over time, exposure to environmental stressors such as UV radiation can lead to the leaching of various toxic substances into the air. Researchers like Rochman et al. (2013) have highlighted the risks associated with the release of toxic chemicals from plastic degradation.

Phthalates, bisphenol A (BPA), and flame retardants are among the toxic chemicals commonly found in plastics. These chemicals are added to plastics during manufacturing to impart certain properties such as flexibility, durability, and flame resistance. However, they can leach out of plastics over time, especially when exposed to heat, light, or mechanical stress, posing risks to human health and the environment (Halden 2016). Phthalates, for example, are endocrine-disrupting chemicals that have been linked to adverse effects on reproductive health, hormone regulation, and development in humans and wildlife (Sathyanarayana et al. 2008). BPA, another endocrine disruptor, has been associated with various health issues, including reproductive disorders, metabolic disorders, and cancer (Rochester 2013). Flame retardants, such as polybrominated diphenyl ethers (PBDEs), can accumulate in the environment and bioaccumulate in organisms, posing risks to ecosystems and human health (de Wit 2002).

The release of these toxic chemicals from plastics into the air can occur through processes such as off-gassing, where volatile compounds evaporate from the surface of plastics, and leaching, where chemicals migrate out of plastics into the surrounding environment. Once airborne, these chemicals can be transported over long distances by wind currents, potentially contaminating the atmosphere in urban and remote areas alike (Rochman et al. 2013).

The presence of toxic chemicals in the air poses risks to human health, particularly for vulnerable populations such as children, pregnant women, and individuals with pre-existing health conditions. Inhalation of airborne pollutants can lead to respiratory issues, cardiovascular problems, and neurological disorders (Grandjean and Landrigan 2014). Additionally, these chemicals can have adverse effects on ecosystems, disrupting ecological processes and harming wildlife populations (Sunderland et al. 2019).

(c) **Particulate Matter (PM) Emissions from Incineration**

The incineration of plastic waste is a significant source of particulate matter (PM) emissions, which poses risks to both human health and the environment. Improper disposal methods, including open burning and uncontrolled incineration of plastic waste, release fine particles and toxic gases into the air. Particulate matter (PM) refers to tiny particles suspended in the air, categorized based on their size into PM10 (particles with a diameter of 10 micrometers or less) and PM2.5 (particles with a diameter of 2.5 micrometers or less). PM2.5 particles are of particular concern due to their ability to penetrate deep into the respiratory system and enter the bloodstream, posing significant health risks (Brook et al. 2010). During the incineration of plastic waste, various pollutants are released into the atmosphere, including PM2.5, volatile organic compounds (VOCs), heavy metals, and dioxins. These pollutants can have adverse effects on human health, including respiratory and cardiovascular diseases, as well as neurological and developmental disorders (Thompson et al. 2009).

Exposure to PM2.5 emissions from plastic incineration has been linked to a range of health issues, including exacerbation of asthma, chronic obstructive pulmonary disease (COPD), and lung cancer. Long-term exposure to high levels of PM2.5 has also been associated with increased mortality rates and reduced life expectancy (Lelieveld et al. 2015). Moreover, PM2.5 emissions from plastic incineration can have environmental impacts beyond human health. Fine particles can contribute to haze and smog formation, reducing visibility and degrading air quality. PM2.5 particles can also deposit onto soil and water bodies, affecting ecosystems and wildlife (Masiol et al. 2015).

(d) **Alteration of Atmospheric Chemistry**

The alteration of atmospheric chemistry due to plastic-derived chemicals released into the atmosphere is a complex and relatively understudied aspect of plastic pollution. Plastic-derived chemicals, including additives, breakdown products, and VOCs, can interact with other pollutants in the atmosphere, leading to changes in atmospheric chemistry and potentially affecting air quality and environmental health. This phenomenon has been highlighted by researchers such as Andrady (2011), shedding light on the interconnectedness of plastic pollution and atmospheric processes. Plastic-derived chemicals released into the atmosphere can undergo various chemical reactions and transformations when exposed to sunlight, reactive species, and other atmospheric constituents. For example, VOCs emitted from plastics can react with nitrogen oxides (NOx) and sunlight to form ground-level ozone, a key component of smog. Ozone formation is a complex process involving the photochemical oxidation of VOCs and NOx in the presence of sunlight, and the presence of additional VOCs from plastic sources can enhance ozone production and exacerbate air quality problems (Sillman 1999). Moreover, plastic-derived chemicals can contribute to the formation of secondary organic aerosols (SOAs), which are formed through the oxidation and condensation of VOCs and other organic compounds in the atmosphere. SOAs have implications for air quality, climate, and human health, as they can influence atmospheric chemistry, visibility, and respiratory health (Donahue et al. 2012).

The alteration of atmospheric chemistry due to plastic pollution may have far-reaching implications for environmental health and ecosystems. Changes in ozone levels, for example, can affect plant productivity, crop yields, and ecosystem dynamics. Additionally, the formation of SOAs can influence cloud formation and precipitation patterns, with potential consequences for regional and global climate (Kroll and Seinfeld 2008).

(e) **Contribution to Climate Change**

The contribution of plastic pollution to climate change is multifaceted, involving both direct and indirect impacts on air quality and environmental health. Plastic pollution indirectly affects climate change through various processes, including the production, use, and disposal of plastics, which release greenhouse gases (GHGs) into the atmosphere. Production of plastics involves the extraction, refining, and processing of fossil fuels, primarily oil and natural gas, releasing significant amounts of greenhouse gases, particularly carbon dioxide (CO_2), into the atmosphere. Additionally, the manufacturing of plastics consumes energy and resources, further contributing to carbon emissions and climate change. Improper disposal methods for plastic waste, such as incineration and open burning, release greenhouse gases such as carbon dioxide (CO_2) and methane (CH_4) into the atmosphere, along with other air pollutants like particulate matter (PM) and volatile organic compounds (VOCs), exacerbating air pollution and climate change (Thompson et al. 2009; USEPA 2021). Moreover, plastic waste that ends up in landfills undergoes decomposition, generating methane (CH_4) emissions, a potent greenhouse gas with a warming potential greater than that of carbon dioxide over a 20-year period, further contributing to climate change and exacerbating air quality issues (Raga et al. 2014). These interactions between plastic pollution and climate change have implications for atmospheric chemistry, weather patterns, and the frequency and intensity of extreme weather events, affecting air quality and environmental health (Bell et al. 2011). Addressing the contribution of plastic pollution to climate change necessitates comprehensive strategies to reduce plastic waste generation, improve waste management practices, and promote alternatives to plastic products, aiming to mitigate its impacts on air quality and climate change. In conclusion, plastic pollution indirectly contributes to climate change through various processes such as production, incineration, and decomposition, releasing greenhouse gases into the atmosphere and exacerbating air quality issues, emphasizing the urgent need for action to mitigate its environmental impacts and protect air quality and environmental health for present and future generations.

11.5 Health Risks Associated with Poor Air Quality Caused by White Pollution

The degradation of air quality resulting from white pollution, characterized by the release of pollutants from plastic waste, poses significant health risks to human populations. Exposure to these pollutants can lead to a range of adverse health effects, exacerbating respiratory problems, cardiovascular diseases, and other health conditions. Understanding these health risks is essential for developing effective strategies to mitigate the impacts of white pollution on public health.

(i) **Respiratory Diseases**

The impact of white pollution on respiratory diseases is profound, primarily due to the emission of fine particulate matter (PM2.5) during the incineration of plastic waste. PM2.5 particles are small enough to penetrate deep into the respiratory system, causing inflammation, irritation, and exacerbating existing respiratory conditions. Studies by Kelly and Fussell (2015) have emphasized the association between PM2.5 emissions from plastic incineration and respiratory ailments such as asthma, bronchitis, and chronic obstructive pulmonary disease (COPD).

The inhalation of PM2.5 particles poses significant risks to respiratory health. These particles can deposit in the lungs and airways, triggering inflammatory responses and oxidative stress. Chronic exposure to PM2.5 has been linked to the development and worsening of respiratory conditions, including asthma, a chronic inflammatory disease characterized by airway constriction and inflammation (Brook et al. 2010). Individuals with pre-existing respiratory conditions, such as bronchitis and COPD, are particularly susceptible to the adverse effects of PM2.5 exposure, experiencing exacerbations of symptoms and reduced lung function (Kelly and Fussell 2015). Moreover, PM2.5 particles can act as carriers for toxic chemicals and allergens, further exacerbating respiratory issues. Plastic incineration releases a variety of pollutants into the air, including volatile organic compounds (VOCs), polycyclic aromatic hydrocarbons (PAHs), and heavy metals, which can adhere to PM2.5 particles and be inhaled deep into the lungs (Thompson et al. 2009). Prolonged exposure to these pollutants can lead to chronic inflammation, tissue damage, and increased susceptibility to respiratory infections.

(ii) **Cardiovascular Diseases**

Exposure to air pollution stemming from white pollution has been strongly linked to heightened risks of cardiovascular diseases, encompassing heart attacks, strokes, and hypertension. The toxic chemicals emitted from plastic waste, including polycyclic aromatic hydrocarbons (PAHs) and dioxins, pose a significant threat to cardiovascular function (Shah et al. 2015). These pollutants have the capacity to infiltrate the bloodstream, initiating a cascade of vascular inflammation, endothelial dysfunction, and cardiovascular damage. Research conducted by Jiang et al. (2016) has elucidated the mechanisms through which exposure to PAHs and dioxins can induce oxidative stress, impair vascular function, and promote atherosclerosis, ultimately increasing the risk of cardiovascular event. Particulate matter (PM) pollution, a prevalent component of air pollution associated with white pollution, also plays a pivotal role in cardiovascular diseases. Fine particles, such as PM2.5, are capable of penetrating deep into the lungs and entering the systemic circulation, where they instigate inflammatory responses, oxidative stress, and vascular injury. The adverse effects of PM2.5 on cardiovascular health have been extensively documented in studies like those by Brook et al. (2010), which highlight the correlation between PM2.5 exposure and elevated risks of myocardial infarction, stroke, and arrhythmias. Furthermore, the coalescence of multiple pollutants emitted during plastic incineration and degradation synergistically exacerbates cardiovascular risks. Chemical interactions between different pollutants can potentiate oxidative stress, inflammation, and endothelial dysfunction, amplifying the burden on the cardiovascular system.

(iii) **Cancer Risk**

The risk of cancer associated with white pollution arises from the release of plastic-derived chemicals into the air during the degradation of plastic waste. Certain chemicals present in plastics have been identified as carcinogens or have been linked to cancer in animal studies, as highlighted by research conducted by. These carcinogenic compounds include substances such as polycyclic aromatic hydrocarbons (PAHs), phthalates, bisphenol A (BPA), and dioxins, which can leach from plastic materials into the environment during decomposition processes.

Long-term exposure to these carcinogens via inhalation of polluted air may significantly elevate the risk of developing various types of cancer. Epidemiological studies and toxicological assessments have indicated potential associations between exposure to plastic-derived chemicals and an increased incidence of cancers, including lung cancer, bladder cancer, and leukemia. For instance, research by Rusyn et al. (2014) has elucidated the carcinogenic properties of certain plastic additives and breakdown products, underscoring the need for vigilant monitoring and regulation of plastic pollution to mitigate cancer risks.

The mechanisms through which plastic-derived chemicals contribute to cancer development are multifaceted and may involve genotoxicity, mutagenicity, and disruption of endocrine function. Carcinogens present in plastic waste can induce DNA damage, impair cellular repair mechanisms, and promote tumor initiation and progression. Additionally, some plastic additives exhibit estrogenic or anti-androgenic properties, interfering with hormonal signaling pathways and potentially contributing to hormone-related cancers (La Merrill et al. 2013). Moreover, the synergistic effects of multiple carcinogenic compounds released during plastic degradation and incineration may amplify cancer risks. Chemical interactions between different pollutants can enhance their toxicological potency, exacerbating cellular damage and increasing the likelihood of cancer development.

(iv) **Neurological Effects**

The neurological effects of white pollution stem from exposure to airborne pollutants, including heavy metals and neurotoxic chemicals, which are associated with plastic degradation and incineration. These pollutants have the potential to impact neurological development and function, posing significant risks to cognitive function, behavior, and neurodevelopmental outcomes, particularly in vulnerable populations such as children.

Research by Costa and Giordano (2007) has elucidated the neurotoxic properties of certain pollutants emitted during plastic degradation and combustion. Heavy metals like lead, mercury, and cadmium, as well as organic chemicals such as polychlorinated biphenyls (PCBs) and brominated flame retardants (BFRs), have been identified as neurotoxicants capable of crossing the blood-brain barrier and interfering with central nervous system function. These pollutants can disrupt neuronal signaling, induce oxidative stress, and promote neuroinflammation, contributing to neurological dysfunction and impairments in cognitive and behavioral domains. Prenatal and early-life exposure to neurotoxic pollutants associated with white pollution is of particular concern due to the heightened vulnerability of the developing brain. Studies by Block and Calderón-Garcidueñas (2009) have highlighted the adverse effects of prenatal exposure to environmental toxins on neurodevelopmental outcomes in children, including deficits in cognitive function, attention, and executive control. Exposures during critical periods of brain development can disrupt neuronal migration, synaptogenesis, and myelination, leading to long-lasting alterations in brain structure and function.

Moreover, the neurotoxic effects of plastic-derived pollutants may extend beyond cognitive and behavioral domains to encompass neurodegenerative diseases and psychiatric disorders. Epidemiological studies have suggested associations between environmental exposures to certain pollutants and an increased risk of neurodevelopmental disorders such as autism spectrum disorder (ASD), attention deficit hyperactivity disorder (ADHD), and schizophrenia (Grandjean and Landrigan 2014).

(v) **Immune System Dysfunction**

The impact of white pollution on immune system function is a critical concern, as airborne pollutants emitted from plastic waste can disrupt immune responses and increase susceptibility to infections and autoimmune diseases. Chronic exposure to environmental pollutants associated with white pollution has been linked to dysregulation of immune system function, leading to inflammation, oxidative stress, and immune system dysfunction. These pollutants, including volatile organic compounds (VOCs), particulate matter (PM), and toxic chemicals such as phthalates and bisphenol A (BPA), can modulate immune cell function, impairing immune surveillance and defense mechanisms (Dietert and Dietert 2008). Exposure to these pollutants may compromise the body's ability to mount effective immune responses against pathogens, increasing the risk of infections and reducing resistance to microbial invaders. Furthermore, chronic exposure to environmental pollutants has been associated with dysregulation of inflammatory pathways and oxidative stress, both of which can contribute to immune system dysfunction. Peden (2019) discusses how pollutants such as PM, ozone, and nitrogen dioxide can trigger inflammatory responses in the respiratory system, leading to systemic inflammation and immune system activation. Prolonged activation of immune cells and inflammatory mediators can disrupt immune homeostasis, impairing immune surveillance and tolerance mechanisms and increasing the risk of autoimmune diseases. Moreover, emerging evidence suggests that environmental pollutants may modulate the balance of immune cell subsets, leading to shifts in immune cell populations and functions. For instance, exposure to certain pollutants has been associated with alterations in T-cell subsets, cytokine profiles, and regulatory T-cell function, which play crucial roles in immune regulation and tolerance (Peden 2019). Dysregulation of these immune pathways can contribute to the development and exacerbation of autoimmune diseases such as rheumatoid arthritis, lupus, and multiple sclerosis.

11.6 Environmental Impacts of White Pollution

The environmental impacts of white pollution are multifaceted and far-reaching, encompassing various ecosystems and habitats. White pollution, primarily driven by the proliferation of plastic waste, poses significant threats to terrestrial and aquatic environments, wildlife, and overall ecosystem health. Let's look into the some environmental impacts of white pollution:

(i) **Marine Pollution**: Plastic waste, encompassing items like bottles, bags, and packaging materials, frequently finds its way into marine ecosystems due to improper disposal methods, littering, and runoff from land. Once in the ocean, plastics break down into smaller particles known as microplastics, presenting grave dangers to marine organisms. Seabirds, turtles, and marine mammals are at risk of ingesting or becoming entangled in plastic debris, resulting in injury, suffocation, and mortality. Furthermore, microplastics have the potential to accumulate within the food chain, exposing marine life to harmful chemicals and disrupting the balance of ecosystems (Jambeck et al. 2015).

(ii) **Terrestrial Pollution**: Plastic pollution isn't limited to marine environments; it also affects terrestrial ecosystems, including soils, rivers, and urban areas. Inappropriately discarding plastic waste on land can degrade soil health, impede water absorption, and hinder plant development (Thompson et al. 2009). Plastic waste in rivers and water bodies can block water flow, worsen flooding, and damage aquatic habitats. Additionally, littered plastics in urban settings create visual pollution, diminishing the beauty of natural landscapes and urban environments.

(iii) **Habitat Degradation**: White pollution contributes to habitat degradation and loss, particularly in sensitive ecosystems such as wetlands, mangroves, and coral reefs. Plastic debris can smother benthic habitats, alter sediment composition, and disrupt the ecological balance of marine and terrestrial ecosystems. Coastal habitats, in particular, are vulnerable to plastic pollution, as marine litter can accumulate along shorelines, impacting nesting sites for seabirds and sea turtles, and degrading important breeding grounds for marine species (Derraik 2002).

(iv) **Chemical Pollution**: Plastics contain a variety of chemical additives and contaminants, including plasticizers, flame retardants, and persistent organic pollutants (POPs), which can leach into the environment over time (Rochman et al. 2013). These chemicals can pose risks to ecosystem health and biodiversity, affecting the physiology, behavior, and reproductive success of wildlife. Furthermore, plastics act as vectors for transporting pollutants in the environment, facilitating the spread of contaminants across ecosystems and bioaccumulating in the tissues of organisms throughout the food chain.

(v) **Climate Impacts**: The creation, transportation, and elimination of plastics all play a role in greenhouse gas emissions and thus contribute to climate change. Extracting and refining raw materials for plastic manufacturing, along with burning plastic waste, emit carbon dioxide (CO_2) and other greenhouse gases into the air (Wilcox et al. 2015). Furthermore, plastic waste in oceans and on land can affect surface reflectivity, absorb sunlight, and affect heat exchange mechanisms, worsening climate impacts at local and regional levels.

11.7 White Pollution Mitigations and Air Quality Improvement

Several initiatives and strategies are being implemented worldwide to address white pollution and improve air quality. These efforts involve a combination of policy interventions, technological innovations, community engagement, and corporate responsibility programs (UNEP 2016). Here are some existing initiatives aimed at tackling white pollution and improving air quality:

(a) **Restrictions and Policy Enforcement Regarding Plastic Materials**: Many countries and municipalities have implemented bans or restrictions on single-use plastics, such as plastic bags, straws, and Styrofoam containers. These measures aim to reduce plastic consumption and promote the use of alternative materials. Additionally, regulations on plastic production, labeling, and disposal help minimize plastic pollution and encourage sustainable practices throughout the product lifecycle.

(b) **Waste Management Infrastructure**: Allocating resources to enhance waste management infrastructure, which encompasses recycling facilities, waste collection systems, and effective landfill management, is paramount for tackling plastic waste efficiently (European Commission 2021). Enhanced waste sorting and recycling initiatives play a pivotal role in diverting plastics from landfills, thus mitigating the environmental consequences of plastic pollution. Moreover, investing in waste-to-energy technologies, such as incineration with energy recovery and anaerobic digestion, presents alternative disposal avenues for plastic waste while simultaneously producing renewable energy.

(c) **Public Awareness Campaigns**: Public awareness campaigns play a vital role in educating individuals and communities about the environmental impacts of plastic pollution and the importance of waste reduction and recycling (WWF 2019). These campaigns raise awareness about the benefits of sustainable consumption practices, encourage behavior change, and promote responsible waste disposal habits. Engaging schools, businesses, and community organizations in environmental education initiatives helps foster a culture of environmental stewardship and collective action.

(d) **Innovations in Packaging and Materials**: Advancements in packaging design and materials science are propelling the creation of eco-conscious substitutes for standard plastics. Biodegradable and compostable packaging materials, sourced from renewable origins, provide sustainable options that diminish reliance on fossil fuels and curtail environmental contamination. Furthermore, progressions in bioplastics, bio-based polymers, and circular economy frameworks advocate for resource optimization and waste minimization across various sectors (Ellen MacArthur Foundation 2019).

(e) **Air Quality Monitoring and Regulation**: Governments and environmental agencies implement air quality monitoring programs to assess ambient air pollution levels, identify sources of pollution, and inform policy decisions (WHO 2021). Regulatory measures, such as emission standards for vehicles and industrial facilities, aim to reduce air pollutant emissions and improve air quality. In addition, incentives for adopting clean technologies, such as electric vehicles and renewable energy sources, help transition to low-emission and sustainable energy systems.

(f) **International Agreements and Collaborations**: International cooperation and agreements play a crucial role in addressing global environmental challenges, including plastic pollution and air quality degradation (UNEP 2016). Initiatives such as the United Nations Environment Programme (UNEP) Clean Seas campaign and the Paris Agreement on climate change facilitate collaboration among countries, businesses, and civil society organizations to tackle pollution and promote sustainable development goals.

(g) **Corporate Sustainability Initiatives**: Businesses and industries are progressively embracing sustainability measures and corporate social responsibility (CSR) strategies to shrink their ecological impact and advocate for circular economy ideals (UN Global Compact 2020). Corporate endeavors to cut down on plastic packaging, introduce recycling schemes, and endorse waste reduction endeavors aid in alleviating plastic pollution and enhancing air purity. Additionally, collaborations among governments, non-governmental organizations (NGOs), and private enterprises stimulate innovation and expedite the shift towards a more sustainable and environmentally conscious economy.

11.8 Long-Term Prospects for Addressing White Pollution and Air Quality Issues

Addressing white pollution and air quality issues requires long-term strategies that encompass systemic changes in consumption patterns, waste management practices, and energy production. While significant challenges lie ahead, there are promising prospects for mitigating these environmental concerns and creating a more sustainable future. Here are some long-term prospects for addressing white pollution and air quality issues:

(i) **Circular Economy Transition**: Embracing a circular economy framework presents a hopeful, enduring solution for combatting white pollution. Within this model, resources are maximized through recycling, reusing, and remanufacturing, thereby curtailing waste production and lessening environmental harm (Ellen MacArthur Foundation 2020). Through product redesign aimed at durability, recyclability, and resource optimization, as well as the promotion of closed-loop material cycles, the circular economy holds the potential to diminish the need for new materials and alleviate plastic pollution.

(ii) **Advancements in Recycling Technologies**: Continuous progress in recycling technologies shows potential for enhancing the efficiency and efficacy of plastic waste recycling. Developments like chemical recycling, mechanical recycling, and biodegradable polymers present alternatives to conventional waste management approaches, facilitating the retrieval and reuse of valuable materials from plastic waste streams (Plastics Europe 2020a, b). By investing in the research and development of innovative recycling technologies, we can move towards closing the loop on plastic production and consumption, thereby lessening dependence on finite resources and reducing environmental pollution.

(iii) **Policy and Regulatory Measures**: Enhancing policy and regulatory frameworks both domestically and internationally is crucial in tackling white pollution and air quality concerns. Measures such as implementing extended producer responsibility (EPR) schemes, imposing taxes or levies on plastic products, and establishing rigorous emission standards for air pollutants can incentivize the adoption of sustainable practices and catalyze widespread transformation within industries (European Commission 2021). Furthermore, international agreements and partnerships, such as the Basel Convention on the Control of Transboundary Movements of Hazardous Wastes, foster collaboration among nations to address overarching environmental issues on a global scale.

(iv) **Renewable Energy Transition**: Transitioning to renewable energy sources, such as solar, wind, and hydroelectric power, offers long-term benefits for improving air quality and reducing greenhouse gas emissions. By phasing out fossil fuels and investing in clean energy infrastructure, societies can decrease reliance on polluting energy sources and mitigate the adverse impacts of air pollution on human health and the environment (IEA 2021). Additionally, integrating energy-efficient technologies and promoting energy conservation measures can further enhance air quality and reduce environmental pollution.

(v) **Green Innovation and Sustainable Practices**: Encouraging green innovation and sustainable practices across industries and sectors can drive long-term solutions for addressing environmental challenges. Investing in research and development of eco-friendly materials, renewable technologies, and low-carbon solutions fosters innovation and entrepreneurship, leading to the development of sustainable products and services (UNEP 2021). By promoting eco-design principles, eco-labeling schemes, and eco-certification programs, businesses can reduce their environmental footprint and contribute to a more sustainable future.

(vi) **Education and Public Engagement**: Education and public engagement play a crucial role in fostering environmental awareness, promoting behavior change, and building support for sustainable policies and initiatives. Integrating environmental education into school curricula, raising awareness through media campaigns, and engaging communities in environmental conservation projects empower individuals to take action and advocate for positive change

(UNESCO 2020). By fostering a culture of environmental responsibility and collective action, societies can mobilize efforts to address white pollution and air quality issues effectively.

11.9 Conclusion

The challenge of white pollution, stemming from widespread single-use plastic production and consumption, poses significant environmental concerns. Insufficient waste management practices exacerbate this issue, leading to the accumulation of plastic waste in ecosystems. Yet, there are opportunities for improvement through regulatory measures, technological advancements, public awareness initiatives, and global collaboration. Efforts to combat white pollution involve implementing plastic bans and regulations, enhancing recycling systems, and promoting sustainable consumption habits. Concurrently, addressing air quality concerns includes monitoring and regulating pollutant emissions, transitioning to renewable energy sources, and fostering innovation in clean technologies. Public education and engagement also play pivotal roles in raising awareness and fostering a collective commitment to environmental protection.

Although addressing white pollution and air quality challenges requires sustained and collaborative efforts, there are promising pathways forward. Embracing sustainable practices, investing in eco-friendly technologies, and strengthening international partnerships can drive progress toward mitigating plastic pollution, preserving air quality, and nurturing a healthier environment for generations to come.

References

Agency for Toxic Substances and Disease Registry (ATSDR) (2021) Toxicological profile for benzene. Agency for Toxic Substances and Disease Registry (ATSDR)

Andrady AL (2011) Microplastics in the marine environment. Mar Pollut Bull 62(8):1596–1605

Andrady AL (2015) The plastic in microplastics: a review. Mar Pollut Bull 97(1–2):290–297

Aransiola SA, Ikhumetse AA, Babaniyi BR, Abioye OP, Oyedele OJ, Falade NO (2023a) Phytoaccumulation of micro and nanoplastics: root uptake. In: Maddela NR, Reddy KV, Ranjit P (eds) Micro and nanoplastics in soil. Springer, Cham. https://doi.org/10.10007/978-3-031-21195-9_8

Aransiola SA, Victor-Ekwebelem MO, Ajiboye AE, Leh-Togi Zobeashia SS, Ijah UJJ, Oyedele OJ (2023b) Ecological impacts and toxicity of micro and nanoplastics in agroecosystem. In: Maddela NR, Reddy KV, Ranjit P (eds) Micro and nanoplastics in soil. Springer, Cham. https://doi.org/10.10007/978-3-031-21195-9_10

Auta SH, Abioye OP, Aransiola SA, Bala JD, Chukwuemeka VI, Hassan A, Aziz A, Fauziah SH (2022) Enhanced microbial degradation of PET and PS microplastics under natural conditions in mangrove environment. J Environ Manag 304(2022):114273. https://doi.org/10.1016/j.jenvman.2021.114273

Baldé CP, Forti V, Gray V, Kuehr R, Stegmann P (2017) The global E-waste monitor 2017: quantities, flows, and resources. United Nations University (UNU)/United Nations Institute for Training and Research (UNITAR) co-hosted SCYCLE Programme, International Telecommunication Union (ITU), & International Solid Waste Association (ISWA)

Bell ML, Goldberg R, Hogrefe C, Kinney PL, Knowlton K, Lynn B, Patz JA (2011) Climate change, ambient ozone, and health in 50 US cities. Clim Chang 102(1–2):161–176

Block ML, Calderón Garcidueñas L (2009) Air pollution: mechanisms of neuroinflammation and CNS disease. Trends Neurosci 32(9):506–516

Brook RD, Rajagopalan S, Pope CA, Brook JR, Bhatnagar A, Diez-Roux AV, Kaufman JD (2010) Particulate matter air pollution and cardiovascular disease: an update to the scientific statement from the American Heart Association. Circulation 121(21):2331–2378

Cooper T, Mok C (2021) Planned obsolescence: a sustainable urban future perspective. In: Sustainable urban future. Springer, Cham, pp 15–28

Costa LG, Giordano G (2007) Developmental neurotoxicity of polybrominated diphenyl ether (PBDE) flame retardants. Neurotoxicology 28(6):1047–1067

De Wit CA (2002) An overview of brominated flame retardants in the environment. Chemosphere 46(5):583–624

Derraik JG (2002) The pollution of the marine environment by plastic debris: a review. Mar Pollut Bull 44(9):842–852

Dietert RR, Dietert JM (2008) Potential for early-life immune insult including developmental immunotoxicity in autism and autism spectrum disorders: focus on critical windows of immune vulnerability. J Toxic Environ Health, Part B 11(8):660–680

Donahue NM, Robinson AL, Stanier CO, Pandis SN (2012) Coupled partitioning, dilution, and chemical aging of semivolatile organics. Environ Sci Technol 46(7):3763–3770

Ellen MacArthur Foundation (2016) The new plastics economy: rethinking the future of plastics. Retrieved from https://www.ellenmacarthurfoundation.org/publications/the-new-plastics-economy-rethinking-the-future-of-plastics

Ellen MacArthur Foundation (2019) The new plastics economy: catalyzing action. Retrieved from https://www.ellenmacarthurfoundation.org/assets/downloads/EllenMacArthurFoundation_TheNewPlasticsEconomy_CatalysingAction_13-1-17.pdf

Ellen MacArthur Foundation (2020) Circular economy. Retrieved from https://www.ellenmacarthurfoundation.org/circular-economy

Environmental Protection Agency (EPA) (2021) Ground-level ozone pollution. Environmental Protection Agency (EPA)

Eriksen M, Lebreton LCM, Carson HS, Thiel M, Moore CJ, Borerro JC, Reisser J (2014) Plastic pollution in the world's oceans: more than 5 trillion plastic pieces weighing over 250,000 tons afloat at sea. PLoS One 9(12):e111913

European Commission (2021) Extended producer responsibility (EPR). Retrieved from https://ec.europa.eu/environment/waste/eu-legislation/extended_producer_en

Grandjean P, Landrigan PJ (2014) Neurobehavioural effects of developmental toxicity. Lancet Neurol 13(3):330–338

Halden RU (2016) Plastics and health risks. Annu Rev Public Health 37:81–82

Hawken P, Lovins A, Lovins LH (2013) Natural capitalism: creating the next industrial revolution. Routledge

Hoornweg D, Bhada-Tata P (2012) What a waste: a global review of solid waste management. World Bank

International Energy Agency (IEA) (2021) Renewables. Retrieved from https://www.iea.org/topics/renewables

Jambeck JR, Geyer R, Wilcox C, Siegler TR, Perryman M, Andrady A, Law KL (2015) Plastic waste inputs from land into the ocean. Science 347(6223):768–771

Jiang S, Bo L, Gong C, Du J, Liu W (2016) Cardiovascular toxicity of ambient polycyclic aromatic hydrocarbons and dioxins: roles of vascular inflammation. Chemosphere 148:438–447

Kelly FJ, Fussell JC (2015) Air pollution and public health: emerging hazards and improved understanding of risk. Environ Geochem Health 37(4):631–649

Kroll JH, Seinfeld JH (2008) Chemistry of secondary organic aerosol: formation and evolution of low-volatility organics in the atmosphere. Atmos Environ 42(16):3593–3624

La Merrill MA, Vandenberg LN, Smith MT, Goodson W, Browne P, Patisaul HB, Zoeller RT (2013) Consensus on the key characteristics of endocrine-disrupting chemicals as a basis for hazard identification. Nat Rev Endocrinol 11(3):133–141

Lebreton LCM, van der Zwet J, Damsteeg JW, Slat B, Andrady A, Reisser J (2017) River plastic emissions to the world's oceans. Nat Commun 8(1):15611

Lelieveld J, Evans JS, Fnais M, Giannadaki D, Pozzer A (2015) The contribution of outdoor air pollution sources to premature mortality on a global scale. Nature 525(7569):367–371

Li WC, Tse HF, Fok L (2018) Plastic waste in the marine environment: recommendations for research and policy in China. Mar Policy 88(110):118. https://www.sciencedirect.com/science/article/pii/S0048969716310154

Masiol M, Squizzato S, Pavoni B (2015) PM10 and PM2.5 concentrations in Milan: seasonal trends and daily variability. Atmos Pollut Res 6(2):270–277

Musa IO, Auta HS, Ilyasu US, Aransiola SA, Makun HA, Adabara NU, Abioye OP, Aziz A, Jayanthi B, Maddela NR, Prasad R (2024) Micro and nanoplastics in environment—degradation, detection, ecological impact. Int J Environ Res 18:1. https://doi.org/10.1007/s41742-023-00551-9

Nixon JD, Evans DE, Furneaux BD (2020) Circular economy for plastics: an analysis across the supply chain. Springer, Cham

Parker L (2018) Consuming plastic. Nat Mater

Peden DB (2019) The epidemiology and genetics of asthma risk associated with air pollution. J Allergy Clin Immunol 143(4):1134–1144

Plastics Europe (2020a) Mechanical recycling: overview of existing technologies. Retrieved from https://www.plasticseurope.org/en/resources/publications/334-mechanical-recycling-overview-existing-technologies

Plastics Europe (2020b) Plastics—the facts 2020: an analysis of European plastics production, demand and waste data. Plastics Europe

Raga GB, Baumgardner D, Castro T, Peralta O, García-García F, Salcido A, Sosa G (2014) Anthropogenic and natural influences on the variability of continental surface aerosol. Atmos Environ 89:634–642

Rochester JR (2013) Bisphenol a and human health: a review of the literature. Reprod Toxicol 42:132–155

Rochman CM, Hoh E, Hentschel BT, Kaye S (2013) Long-term field measurement of sorption of organic contaminants to five types of plastic pellets: implications for plastic marine debris. Environ Sci Technol 47(3):1646–1654

Rusyn I, Corton JC, Lash LH (2014) Chapter 6—dose-response assessment. In: Safe SS, Mather AE, Moore MD (eds) Comprehensive toxicology, 2nd edn. Academic, pp 123–142

Santos L, de Oliveira ML, Farina M, da Cunha DL, de Queiroz GA, Ferraudo AS (2019) Solid waste management in developing countries: perspectives from Brazil. Resour Conserv Recycl 141:211–221

Sathyanarayana S, Karr CJ, Lozano P, Brown E, Calafat AM, Liu F, Swan SH (2008) Baby care products: possible sources of infant phthalate exposure. Pediatrics 121(2):e260–e268

Scheringer M, Trier X, Cousins IT, de Voogt P, Fletcher T, Wang Z, Hungerbühler K (2014) Helsingør statement on poly- and perfluorinated alkyl substances (PFASs). Chemosphere 114:337–339

Schmidt C, Krauth T, Wagner S (2017) Export of plastic debris by rivers into the sea. Environ Sci Technol 51(21):12246–12253

Shah AS, Langrish JP, Nair H, McAllister DA, Hunter AL, Donaldson K et al (2015) Global association of air pollution and heart failure: a systematic review and meta-analysis. Lancet 386(10052):1631–1639

Sillman S (1999) The relation between ozone, NOx and hydrocarbons in urban and polluted rural environments. Atmos Environ 33(12):1821–1845

Sunderland EM, Hu X, Dassuncao C (2019) Microplastics in aquatic environments: implications for sources, sinks, and solutions. Environ Sci Technol 53(17):9899–9907

Thompson RC, Moore CJ, vom Saal FS, Swan SH (2009) Plastics, the environment and human health: current consensus and future trends. Philos Trans R Soc Lond Ser B Biol Sci 364(1526):2153–2166

U.S. Environmental Protection Agency (EPA) (2020) Advancing sustainable materials management: 2018 fact sheet. Retrieved from https://www.epa.gov/facts-and-figures-about-materials-waste-and-recycling/national-overview-facts-and-figures-materials

U.S. Environmental Protection Agency (USEPA) (2021) Landfill methane outreach program. USEPA

UN Global Compact (2020) 10 principles and global goals. Retrieved from https://www.unglobalcompact.org/what-is-gc/mission/principles

UNEP (United Nations Environment Programme) (2016) Marine plastic debris and microplastics—global lessons and research to inspire action and guide policy change. Retrieved from https://www.unep.org/resources/report/marine-plastic-debris-and-microplastics-global-lessons-and-research-inspire-action

UNESCO (United Nations Educational, Scientific and Cultural Organization) (2020) Education for sustainable development. Retrieved from https://en.unesco.org/themes/education-sustainable-development

United Nations Environment Programme (UNEP) (2021) Addressing marine plastic pollution for a healthy planet. https://www.unep.org/topics/oceanseasandcoasts/ecosystemdegradationpollution/plastic-pollution-and-marine-litter-0

WHO (World Health Organization) (2021) Ambient air pollution: a global assessment of exposure and burden of disease. Retrieved from https://www.who.int/data/gho/data/themes/ambient-air-pollution

Wilcox C, Van Sebille E, Hardesty BD (2015) Threat of plastic pollution to seabirds is global, pervasive, and increasing. Proc Natl Acad Sci 112(38):11899–11904

Wong CS, Green DR, Cretney WJ (2019) Quantitative analysis of plastic debris on recreational beaches in Mumbai, India and its implications for mitigation measures. Mar Pollut Bull 142:522–529

Wright SL, Thompson RC, Galloway TS (2013) The physical impacts of microplastics on marine organisms: a review. Environ Pollut 178:483–492

Wright SL, Rowland HR, Thompson RC, Galloway TS (2015) Production of microplastics in the fragmentation of severe acute respiratory syndrome coronavirus 2 (SARS-CoV 2) masks and potential for air pollution. Environ Sci Technol Lett 4(7):365–370. https://www.sciencedirect.com/science/article/pii/S0048969721046295

WWF (World Wide Fund for Nature) (2019) No plastic in nature: a practical guide for business engagement. Retrieved from https://wwfint.awsassets.panda.org/downloads/plastic_engagement_guide.pdf

Chapter 12
Health Cries Through White Pollution Exposure

Srinivasan Kameswaran and Bellamkonda Ramesh

Abstract Plastic garbage accumulation worldwide has reached catastrophic proportions. Given the varied and multifaceted effects of plastic on biological health, it is necessary to assess these effects from the standpoint of One Health to integrate and better understand the complexity of these processes. All ecosystems on Earth have plastic particles, from nanometers to meters in size, from the highest mountains to the deepest sea trenches. The molecular and cellular levels as well as the organismal, community, and ecosystem levels are all impacted by plastic trash. Together, the absorbed substances and microbes carried by plastic waste, the physical properties of plastics, the molecular properties of plastic polymers, and the numerous ingredients added to the plastics in manufacture all mediate these effects. The following concepts are summarised using a One Health framework: (1) dimensions of biological organization in which plastic affects global health, (2) interactions between the many biological layers caused by plastic, and (3) What are the gaps in our knowledge about how plastic affects biological scales and how those scales interact? Along with possible remedies, we also highlight One Health viewpoints that take into account the interconnectedness of humans, animals, and the environment in order to confront this escalating catastrophe.

Keywords Microplastic toxicity · Marine plastic · Ecosystem health and pollution · Human health · Nanoplastic

S. Kameswaran (✉)
Department of Botany, Vikrama Simhapuri University College, Kavali, Andhra Pradesh, India

B. Ramesh
Biochemistry and Molecular Biology, University of Nebraska Medical Center, Omaha, NE, USA

S. A. Aransiola et al. (eds.), *White Pollution: Biodiversity and hazards in marine plastisphere*, Environmental Contamination Remediation and Management,
https://doi.org/10.1007/978-3-031-95547-1_12

12.1 Plastic Is Widely Used and Has an Impact on Every Area of the Environment

Polymers are widely used in our daily lives. Since the 1950s, plastic's popularity has grown dramatically because of its low cost, strength, durability, and light weight (Thompson et al. 2009). Consequently, plastic pollution has spread to every part of the world, including the deepest ocean (Bergmann et al. 2017; Barrett et al. 2020), the soil (Fuller and Gautam 2016), the open ocean (Eriksen et al. 2013; Cózar et al. 2014), coastlines (Kwon et al. 2014; Courtene-Jones et al. 2021), and the atmosphere (González-Pleiter et al. 2021). According to estimations made by Eriksen et al. (2014), there are already at least 5.25 trillion plastics in the world's oceans, and this amount is predicted to rise. Despite quick efforts to minimize waste, Lau et al. (2020) project that from 2016 to 2040, the amount of plastic pollution entering aquatic and terrestrial ecosystems will increase by "710 million metric" tonnes. Bacterial action, UV radiation, temperatures, and abrasion are some of the biotic and abiotic environmental activities that break down these plastics into smaller fragments with different surface characteristics. Particularly in the aquatic environment, these smaller plastics—which are categorised as nanoplastics (<1 μm) and microplastics (<5 mm)—are the most common kind of solid trash (Jambeck et al. 2015; Gigault et al. 2018; Musa et al. 2024). Furthermore, commercial and industrial products contain both kinds of small plastics, also known as micro- and nanoplastics.

Interactions between humans and animals are inevitable due to the pervasiveness of plastics in the environment. Plastics affect a huge variety of marine organisms (Gall and Thompson 2015). Clams, fish, oysters, mussels, seaweed, and crabs Fish, clams, mussels, oysters, crabs, and seaweed that are intended for "human consumption" contain microplastics (Van Cauwenberghe and Janssen 2014; Li et al. 2015; Rochman et al. 2015; Karami et al. 2017; Su et al. 2018; Waite et al. 2018); in addition, seaweed (Baini et al. 2017), table and sea salt (Yang et al. 2015; Zarus et al. 2021), tea (Hernandez et al. 2019), the beer (Kosuth et al. 2018), Water from the tap and bottles (Kosuth et al. 2018; Zuccarello et al. 2019; Kankanige and Babel 2020). The human body has also been shown to contain microplastics, as evidenced by studies on the placenta (Ragusa et al. 2021), the blood (Leslie et al. 2022), feces (Schwabl et al. 2019; Ibrahim et al. 2021), and the "lung tissues" (Amato-Lourenço et al. 2021).

This kind of broad spectrum of direct exposure across all levels of biology may not have been seen in any other single anthropogenic pollutant. According to Gall and Thompson (2015), plastics cause disruptions to homeostasis at the individual organismal level when they are consumed. Plastic pollution can also affect how well ecosystems function by upsetting and disrupting environments (Aloy et al. 2011; Carson et al. 2011; Richards and Beger 2011) and altering the species balancing throughout environments (Barnes and Milner 2005; Goldstein et al. 2012).

These alterations consequently always have unidentified consequences on health. Reducing the negative effects of plastic on ecosystems, wildlife, and human health calls for a different strategy than conventional species-level risk assessments. The idea of One Health is one such framework. The concept of One Health acknowledges the interdependence of people, animals, and plants as well as the fact that each species' health depends on the state of the entire ecosystem (One Health 2021).

Solving health problems at the local, national, and international levels requires a multi-sectoral, multidisciplinary, and cooperative strategy, according to the One Health viewpoint (One Health 2022). While the study of zoonotic diseases is where One Health research first began, this framework offers a transdisciplinary perspective to (i) evaluate the immediate risk that plastic pollution poses to the health of people, animals, and ecosystems, (ii) explain the socioeconomic effects of the plastic pollution and (iii) Put into practice mitigating measures that collaborate with the public and commercial sectors.

To demonstrate the necessity of an integrated assessment, we here highlight the effects of plastic pollution on various levels of the biological organization by concentrating on pathways of exposure and health concerns at the cellular, particular organismal, population, and ecosystem levels. The objectives of this analysis are to (i) considering our knowledge of the ways in which plastic influences the layers of biological organisation, (ii) give justification for the application of the One Health paradigm for understanding and investigating the health effects of plastic, and (iii) highlight the gaps in the body of information regarding the effects of plastics within the context of health.

12.2 Pathways for Exposure

Plastics can be ingested, inhaled, or come into physical contact with plastics and the chemicals used as plastic additives by humans and other species in a number of ways (Cox et al. 2020; World Health Organisation 2022). According to estimates, Americans ingest anywhere from 39,000 to 52,000 microplastic fragments annually through "food and drink alone" (Cox et al. 2020), or "an average of 0.1–5 g of microplastics" every week (Senathirajah et al. 2021). Apart from dietary consumption, plastics can also be inhaled by people and other land species (Wright and Kelly 2017; Karbalaei et al. 2018; World Health Organisation 2022). Micro- and nanoplastics (MNPs) and plastic fibers are released into the atmosphere by a variety of processes, including the cleaning of synthetic textiles, rubbery tires, dried sewage, farming, and household and urban dust. Even relatively easy operations, such cutting and opening plastic containers and packing, can produce MNPs (Sobhani et al. 2020). Even though the long-term effects of inhaled "MNPs" as well as their subsequent absorption in the lungs are still unknown (Amato-Lourenço et al. 2021), floating contact may occur both indoors from household items and clothing and outdoors from particulate matter (Kasirajan and Ngouajio 2012; Wright and Kelly 2017; Catarino et al. 2018).

Human exposures also result from interaction with objects including plastic toys and personal care products, medical device exposure, and occupational exposure

(Karbalaei et al. 2018; Zarus et al. 2021). Plastic additives, which are substances added to plastic to enhance the functionality of the polymers, are inextricably linked to exposure to plastic (Hahladakis et al. 2018; Wiesinger et al. 2021). Light and heat stabilizers, flame retardants, plasticizers, pigments, antistatic agents, lubricants, biocides, slide agents, and thermal stabilizers are examples of additives (Groh et al. 2019). Although these chemicals make plastics operate better, there is a significant chance that they will contaminate food, water, air, and soil (Hahladakis et al. 2018), with unknown health and environmental effects.

Apart from the purposeful inclusion of chemicals, plastics can transmit germs and environmental toxins. It has been suggested that plastic may contribute to antibiotic resistance and spread infectious diseases by being linked to a variety of microbial species, particularly human pathogens, in collected plastic trash (Rasool et al. 2021). More research is required to determine whether the "Trojan horse" or "vector" effects of adsorption that plastics acquire are physiologically significant, as noted by Thompson et al. (2009) and Rochman et al. (2014). Plastics also retain heavy metals and persistent organic contaminants. When analyzing the effects of plastics on health, we need to take into account not just the plastic polymers themselves but also the possibility of being exposed to a range of chemicals and bacteria.

12.3 Comparisons of the Impact of Plastics on Cells and Organisms Across Species

12.3.1 Plastics' Effects on Animals' Health

Plastic's numerous impacts have been evaluated. Numerous organ impacts in various taxa, such as fish, bivalves, phytoplankton, and zooplankton, have been evaluated. MNPs can influence organisms at the cellular level because of their small size (Prinz and Korez 2020; Banerjee and Shelver 2021). Fish, for instance, are more susceptible to cellular damage in the liver, blood cells, gills, digestive tract, and brain due to MNP-induced reactive oxygen species (ROS) and increased oxidative stress (Pitt et al. 2018b; Hu and Palić 2020; Buwono et al. 2022; Capó et al. 2022; Hoyo-Alvarez et al. 2022; Rangasamy et al. 2022). "MNPs" caused catabolic and anabolic processes, mitochondrial stress, and changes in the hormonal regulation of energy metabolism in fish. These effects may have limited the organism's capacity to adapt "to future" stressors (Brun et al. 2019; Trevisan et al. 2019, 2020; Wang et al. 2022). Success of animals, "reproduction", and fitness are all impacted by these changes (Dreier et al. 2019).

MNPs have the potential to adhere to the cell walls of algae (Nam et al. 2022), thereafter they can augment the production of reactive oxygen species (ROS), diminish the viability of the cells, alter the operations of antioxidant enzymes, and stimulate lipid peroxidation (Das et al. 2022). Plastic implications on an organism's development rate and energy requirements are linked to these cellular alterations.

As per Huang et al. (2019), nanoplastics have the potential to influence various aspects of freshwater algae growth, including metabolism, greatest quantum yield in photosynthetic system II, chlorophyll-a concentrations, and development rates (Das et al. 2022; Aransiola et al. 2023a). However, it is important to note that the impact of plastics is not always immediately apparent.

Many invertebrates are likewise impacted by MNP exposure in terms of development, reproduction, and fitness. Studies on MNP exposure, for instance, have been conducted on copepods (Cole et al. 2019), "clams" ("Corbicula fluminea" (Oliveira et al. 2018)), "mussels" ("Mytilus galloprovincialis" (Avio et al. 2015; Abidli et al. 2021)), and artemia parthenogenetica (Wang et al. 2019a, b, c). The effects of MNP exposure on "prey preference", "lipid content", molting, "feeding rates", "filter-feeding", "development", "survival", and "reproduction" have also been investigated in this research. High-density polyethylene exposure caused inflammatory responses, lysosomal membrane damage, and the buildup of microplastics within the digestive gland of mussels (Mytilus edulis) (von Moos et al. 2012). However, according to Sussarellu et al. (2016), mussels polluted with "polystyrene microplastics" increased their intake of algae by three percent and their ability to absorb organic materials in the diet by eleven percent. This may have been induced as a means of compensating for the increased energy demand brought on by the microplastics' detected digestive interference. A 23% decrease in spermatozoa mobility rate and a 38% decrease in oocyte counts were seen in gametes sampled from oysters (Crassostrea gigas) exposed to "2 mm and 6 mm" polyester fragments. In the case of artificial fertilization using gametes from exposed parents, the offspring exhibited "a 20%" decline in "D-larval" produce, an "18% reduction" in larval dimension at seventeen days after fertilization, and a 6-day delay in the time needed to finish "metamorphosis" (Sussarellu et al. 2016). According to these findings, bivalves' subcellular and physiological health may be impacted by tiny plastic particles.

Few research have examined the effects that plastic exposure on the gut microbiome, even though the gut bacteria of vertebrates are essential to health, and disruption of the microbiota increases the risk of illness. Male zebrafish exposed to polystyrene microplastics showed increased mucus production and notable alterations in the microbiota's species richness and variety, according to a study (Jin et al. 2018). The intestinal microbial population of adult "zebrafish" ("Danio rerio") treated with both "titanium dioxide" nanoparticles and the "plasticizer bisphenol A (BPA)" changed in a different study (Chen et al. 2018).

The immunological system, antioxidant system, neurological system, and reproductive system can all be disrupted by exposure to MNPs. According to Paul-Pont et al. (2016), mussels exposed to a mixture "of 2 and 6 mm polystyrene microplastics" showed changes in the expression of genes encoding antioxidant enzymes in the gills and digestive glands, as well as disruption of cellular equilibrium in "hemocytes" and "infiltration of these cells into tissues" of the gastrointestinal tract. B cell development-related genes Cxcr5 and Tnfsf13b were upregulated in "sheepshead minnows" ("Cyprinodon variegatus") exposed to irregularly shaped microplastics (Choi et al. 2018). Studies conducted by Jiang et al. (2016) and Tao et al. (2016)

have shown that feeding fish with irregularly shaped microplastics causes an increase in inflammation. On the other hand, "gilthead seabream" ("Sparus aurata" L.) exposed to PVC microplastics experienced cellular and oxidative stress instead of any change in humoral or cellular immunity. These findings are consistent with changes in "albumin", "total proteins", and "globulin levels" seen in the serum of juvenile "*Clarias gariepinus*" following consumption of "virgin microplastic" (Karami et al. 2016).

MNP exposure has also been shown to impact fish behavior at the nervous system level. Reduced "locomotor activity" (Chen et al. 2017); decreased predatory "avoidance behavior" as well as dysfunctional biological clock movement (Sarasamma et al. 2020); "increased shoal formation" and "feeding time and less exploration" (Mattsson et al. 2017); decreased "swim speed and erratic swimming" (Barboza et al. 2018); and decreased "predatory performance" (Carlos de Sá et al. 2015; Wen et al. 2018; Auta et al. 2022). After being exposed to nanoplastics, brine shrimp larvae and polychaetes have shown changes in their "locomotory behavior" (Bergami et al. 2016) and "burrowing kinetics" (Silva et al. 2020).

MNPs smaller than 5 μm that are ingested can penetrate the gastrointestinal barrier, enter the bloodstream, and maybe travel to different parts of the body (Roch et al. 2020). Particles as small as a few hundred nm are typically able to pass through the egg chorion, and particles as small as a few dozen μm can pass through the blood-brain barriers (Guerrera et al. 2021). The possibility of plastic particles translocating across tissue barriers presents a further risk to multicellular creatures, as they may impact several physiological systems. Because most of these biological barriers are still developing and therefore make it easier for plastic to spread to many organs, early embryonic stages can be especially vulnerable to the translocation of plastics to other organs (Pitt et al. 2018a).

Plastics that pass through the food chain and between generations are another cause of concern, especially when it comes to fish and other marine life. Zooplankton and phytoplankton, which are then eaten by small fish and moved up the food chain, can interact with MNPs (Benson et al. 2022; Aransiola et al. 2023b). According to da Costa Araújo et al. (2020) and Kim et al. (2022), fish exposed to dietary MNPs may experience reduced growth rate, liver damage, impaired swimming performance, and behavioral abnormalities. According to recent studies, plastics can cause developmental and physiological harm to fish as well as cross-generational transfer of the particles to the progeny through the gonads (Pitt et al. 2018b; Zhao et al. 2021). Given that the majority of plastic particles are hydrophobic, female fish may play a significant role as vectors for the cross-generational transmission of plastics since oocytes, with their bigger size and higher lipid content, may be crucial targets for the bioaccumulation of MNPs (Pitt et al. 2018b). More research is necessary on this subject because it has already been found that plastics can be transported to the "female gonads" and "oocytes" more easily when they connect with blood proteins like vitellogenin, which was previously detected in conjunction with "polystyrene nanoplastics".

These studies make it abundantly evident that plastics have a significant impact on these keystone species' biology and fitness. Crustacean zooplankton, including copepods, daphnids, and brine prawns, are essential to the functioning of communities because they serve as a vital link in the food chain among primary producers as well as secondary consumers. In addition to providing "habitat" and acting as the largest consumers at the base of the food chain, bivalves can also enhance the diversity and richness of coastal regions, establish a connection between the marine and bottom "systems through filter-feeding", and provide an essential source of nutrients for other species. Fish are a staple food on a worldwide scale and are vital to the ocean food chain. In addition to being essential to community structure, microalgae are also essential to aquatic habitats' production. For this vitally important group of animals, exposure to plastic may change their abundance and health. The volume and ubiquity of plastics, as well as their multiple detrimental effects on all species, make it imperative that we better comprehend the ecological effects of such alterations to coastal ecosystems and develop strategies to mitigate them. The effects of these changes are still unknown.

12.3.2 Effects of Plastics on Humans

Human beings are exposed to plastics and their metabolites regularly, but the effects of these exposures on human wellness are still poorly understood. According to existing research, breathing in plastics may cause "respiratory irritation", "dyspnea", reduced "lung capacity", "coughing", "obesity", raised synthesis of phlegm, asthma, heart disease, and "cancer" (Wright and Kelly 2017; Karbalaei et al. 2018; World Health Organisation 2022). Additionally, immunological dysfunction, inflammatory responses, neurotoxicity, neoplasia, and metabolic abnormalities have all been linked to MNPs (Wang et al. 2020; Banerjee and Shelver 2021; Coffin et al. 2022; World Health Organisation 2022). Moreover, consuming microplastics by humans may harm gut health, as seen in fish. The gastrointestinal tract can become inflamed and the gut epithelium can be destroyed by microplastic exposure (Fig. 12.1). This can result in intestinal leakage and pose a serious health risk (Huang et al. 2021). Increased oxidative stress in intestinal epithelial cells is assumed to be the cause of this inflammation. Furthermore, the mucous membrane in the intestines, which acts as a vital chemical barrier in the gut, might be diminished by microplastics (Huang et al. 2021). Furthermore, research has demonstrated that microplastics alter gut microbiota, which may cause the intestinal milieu to become unstable (Yong et al. 2020; Huang et al. 2021).

The adverse impacts of BPA and phthalates have dominated a large portion of research on the health effects of plastics and plastic additives in humans. Since phthalates and BPA are recognized endocrine disruptors, they have an impact on reproduction and development. Men may experience decreased fertility or a higher risk of testicular and prostate cancer as a result, while women may experience endometriosis, cancers related to reproduction, impaired ovarian function, and irregular menstruation (Meeker et al. 2009; Kim and Kim 2020).

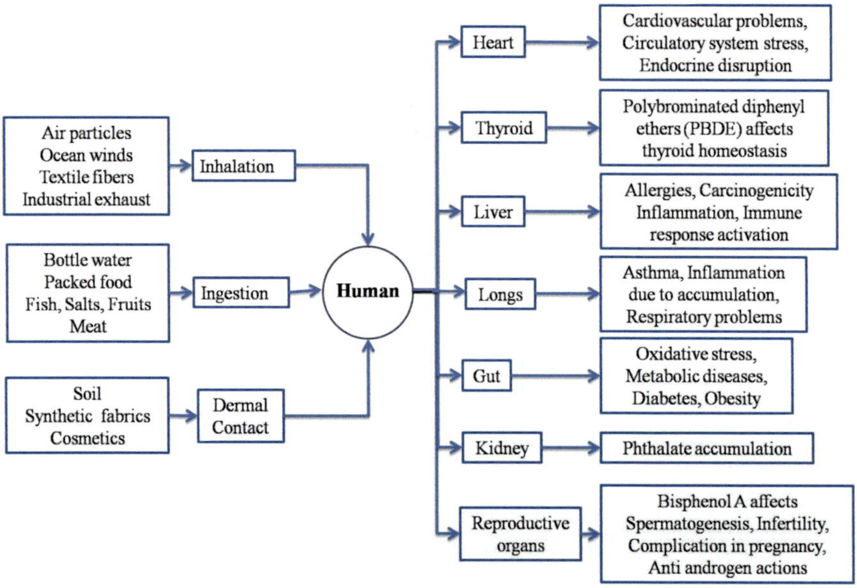

Fig. 12.1 Possible physiological implications of exposure to white pollution in humans

Tumors and disorders of the testicles, prostate, kidney, and immune system may result from exposure to endocrine-disrupting substances during pregnancy (Basak et al. 2020). Furthermore, smaller gestational ages at birth and poorer outcomes from in vitro fertilization are positively connected with phthalate exposure (Latini et al. 2003; Machtinger and Orvieto 2014; Basak et al. 2020). According to Kwon et al. (2020), there is also evidence linking blood BPA levels to compromised thyroid function.

Due to its effects on the "neuroendocrine system" and "inflammatory signaling", "BPA and phthalates" can additionally have neurological effects (Solleiro-Villavicencio et al. 2020; Nadeem et al. 2021). In instances where BPA is exposed, it has been associated with cognitive impairment, "neurobehavioral problems", and "neurodegenerative illness" (Wang et al. 2019a, b, c). BPA can also cross the blood-brain barrier. Studies have indicated a positive correlation between "BPA and phthalates" and "cardiovascular disease", "type 2 diabetes", and elevated blood pressure (Fig. 12.1). Exposure to "BPA and phthalates" is also linked to modifications in cardiovascular health and metabolism (Lang et al. 2008; Gong et al. 2013; Haq et al. 2020; Mariana and Cairrao 2020). Additional "epigenetic effects" of BPA have been demonstrated to include changes in "DNA methylation" in sperm cells, "neuroblastoma cells", "prostate carcinoma cells", and first-trimester trophoblast cells (Manikkam et al. 2013; Senyildiz et al. 2017; Basak et al. 2018; Fatma Karaman et al. 2019). Furthermore, studies have demonstrated "that BPA" can alter "epigenetic levels", which may have an impact on heart development and metabolic dysfunction (Lombo et al. 2015; Junge et al. 2018). A number of the aforementioned research looked at associations between human “BPA and phthalate” concentrations and elevated risks of specific health outcomes, highlighting the possible negative health effects of being exposed to environmentally relevant quantities of these chemicals.

Through current investigations, almost ten thousand chemicals associated with plastic manufacture have been found, approximately 2400 of which are categorized as compounds of potential concern. While BPA and phthalates have received the majority of attention in research on the health impacts of plastics, other substances have also been identified (Hahladakis et al. 2018; Groh et al. 2019; Wiesinger et al. 2021). There is a lack of knowledge regarding the entire range of health concerns provided by plastics, as the present study has concentrated on only a tiny portion of the compounds to which we are probably often exposed. Unambiguous exposures are found when these investigations cover more than a handful of substances. For example, nine out of eighteen toys showed “estrogenic effects”, according to a study by Kirchnawy et al. (2020) that looked at the “estrogenic and androgenic” activities within the salivary of children exposed to the toys. “Seven of the nine toys” that caused an estrogen-like reaction were found to have unidentified plastic ingredients that may have posed a health risk to people, as determined by an examination of 41 proven endocrine disruptors.

Moreover, there is still disagreement over whether plastic additives can leak out of the material. There has been conflicting research on the leachability of specific chemicals from products including kitchen utensils, plastic water pipes, and bottles. Certain studies indicate that the estrogenic activity of water consumed through pipes and bottles of plastic (Wagner and Oehlmann 2011; Liu et al. 2017), but other studies (Corea-Téllez et al. 2008; Aneck-Hahn et al. 2018; Wang et al. 2019a, b, c) have found that the amounts of leached additives are not dangerous to human health. However, the total amount of exposure that a person may have over time and from many sources is not taken into account in these researches.

While current research has not yet quantified such cumulative doses, overall exposure to these substances combined may very well surpass the permitted thresholds. Though human exposures to complex combinations of substances are well recorded, the impact of simultaneous co-exposures to these chemicals on human health is likewise little known (Meeker et al. 2009). Additionally, since plastics degrade and release chemicals due to ultraviolet (UV) light, “mechanical abrasion”, “hydrolysis”, and “oxidation”, these studies do not take into consideration the risk of increased leaching over time (Walker et al. 2021). Human exposure to plastic will only rise as long as plastic manufacturing and use remain constantly elevated. Muncke et al. (2020) have observed that initiatives aimed at reducing trash are also contributing to an increase in the reuse of plastic products, which could potentially raise health hazards because of expanded chemical leaching.

Evidence suggests that the negative effects that plastic has on people’s health are not easily reversed, as being exposed to plastic components may persist even after plastics have been eliminated from one’s surroundings. For instance, despite the introduction of BPA restrictions after a specific amount of time, BPA was found in 23 out of 29 urine samples from employees at a dangerous waste incinerator

(González et al. 2019). Moreover, a 2-month treatment trial conducted on a family that eliminated all plastic sources from their home was unable to significantly lower the levels of "phthalate metabolites" in the urine of any family members (Hutter et al. 2016). As demonstrated by Hutter et al. (2006), phthalates were found in dust in non-negligible amounts in office areas even after phthalate-containing items or sources were eliminated. This underscores the widespread presence of "plastics and plastic additives" in our surroundings and the difficulty of preventing exposure to them, even through tight avoidance measures or local mitigation efforts.

The WHO has produced a study that emphasizes the critical need for more research on the impact of MNPs on health because the available data is "incomplete and inadequate for the evaluation of human risk" (World Health Organisation 2022). Plastic materials have a possibility of having a wide range of detrimental consequences on human health, even if research on their health effects is regrettably lagging following consumer use of plastic goods.

Plastics can harm organs, including the respiratory and digestive systems, due to physical characteristics. Chemicals released from these polymers can also have systemic impacts, impacting development, fertility, metabolic processes, and even generational repercussions by "epigenetic alterations", in addition to cells through "oxidative stress and apoptosis". The World Health Organisation (2022) states that there is an "overwhelming consensus" that action should be taken to reduce exposure to MNPs.

12.4 Diverse Factors Contribute to and Result from Exposure to Plastics

The effects of plastic waste are not felt equally by all people, as is the case with many socioeconomic issues. Since the late 1980s, 87% of all exports of plastic pollution have originated from high-income nations (Brooks et al. 2018). These nations are the main exporters of plastic pollution. High-income nations comprise the United States, Japan, Kuwait, Oman, Argentina, and Italy, six of the top 20 plastic polluters (Law et al. 2020). The majority of these exports go to "Asia and the Pacific's" lower-income nations (Brooks et al. 2018). The surplus load of the exports is too much for the waste-management infrastructure of the nations that receive them, which exacerbates the disproportionate effects of plastic pollution in those nations (Ncube et al. 2021). In certain neighborhoods, the problem of plastic trash is exacerbated by housing scarcity and joblessness, which can create situations where people are incentivized to remain in these regions to better cope with the pressing issues of poverty. For example, before it was closed in the 1990s, the Philippines' Smokey Mountain, an uncontrolled dumpsite at least "20 meters high, housed 30,000 homeless" or scavenger "Filipino families for 40 years" (Galarpe 2015).

The decline in the fishing and agricultural sectors, which are the main sources of income for some societies, brought on by climate change is made worse by plastic pollution. For instance, almost 10% of people on Earth primarily rely on marine habitats for their food and means of subsistence, with the bulk of these people (95%) coming from developing countries (Food and Agriculture Organisation 2014; Taylor et al. 2019). The plastic threat primarily affects shallow Pacific islands with sparse arable land. To keep its economy and people afloat, Tuvalu, for example, adheres to "blue economy" policies that depend on the usage of marine resources (International Organisation for Migration and International Labour Organisation 2021). The impacts of plastic manufacture and incineration on the climate, as well as the accumulation of plastic waste in marine habitats, have made the island nation particularly vulnerable. It is important to investigate the "geo-economic", "environmental", "structural", and "socio-political" factors that contribute to the disproportionate effects of plastic garbage on particular populations.

12.5 Plastic's Effects on Entire Ecosystems

Humans' physical wellness along with "societal", "cultural", and "financial stability" is closely correlated with the "health", "function", and "services" provided by ecosystems (Summers et al. 2012). Sentinel species illustrate the myriad impacts plastics have on every level of biological organization, "from the cells to populations", portending a dismal future for the elements that make up the ecosystem, including people.

Numerous marine apex predators, including marine mammals, are among these sentinel species. Due to their lengthy lifespans, capacity to magnify "trophic information" along several "spatiotemporal scales", and sharing of food supplies crucial to human survival and commerce, they are effective indicators of potential harm to the both individual and population-level animal and human well-being (Bossart 2011; Hazen et al. 2019).

The main method of exposure to microplastic for both filter and raptorial predators is assumed to be the trophic transmission of microplastic particles to marine animals from contaminated prey that have consumed microplastics (Zantis et al. 2022). Humans and marine mammals are inextricably linked; as apex predators share resources, human exposure to microplastics through food is worrisome. Still, a more significant concern of collateral effects remains: could plastic pollution pose a threat to the collapse of entire ecosystems?

It is simple to envision the potential long-term impact of plastics on sea creatures and the environments they inhabit, particularly when considered in the context of the social implications of perturbation (Ocean Studies Board et al. 2017; Bucci et al. 2020). The question of whether plastic threatens the functionality of entire ecosystems is not well-studied (Bucci et al. 2020). Ingestion or entanglement with macroplastics might cause physiological and behavioral alterations that have the potential to be fatal or have acute effects on vital rates and population dynamics (Ocean

Studies Board et al. 2017). In a similar vein, macro- and microplastics may both have long-term, sublethal effects on human health that could change vital rates (Ocean Studies Board et al. 2017). Decreased whale numbers, for instance, might result in a disastrous loss of energy at trophic levels below that rely on whale feces and carcasses since whales play a significant role in the nutrient cycle (Roman et al. 2016). A complete ecosystem's remodeling or collapse could follow from this disturbance in energy availability at the lower trophic levels, which may have an impact on all trophic levels, including those that are important to human culture, subsistence, and commerce. Indeed, indigenous populations depend heavily on marine animals for both food and cultural purposes (Huntington et al. 2016). Marine mammals act as obvious sentinels for a range of environmental and ecological risks to the majority of modern humans worldwide (Bossart 2011; Hazen et al. 2019). However, the flesh from polluted marine mammals may directly affect the health of certain indigenous people who eat them. It is commonly known that polar bears, whales, seals, and sea lions swallow plastic, which can either move to or contaminate edible tissues with harmful compounds (Law 2017; Zantis et al. 2021). As a result, marine mammals' consumption of plastic endangers many indigenous populations' way of life and vital life support systems across the globe.

The correct operation of an ecosystem is, of course, influenced by a multitude of factors. The final impact of disruptions caused by plastic exposure on an ecosystem can be determined by a variety of processes, including emigration and immigration, prey-switching, changes in species assemblages, and niche partitioning. Aside from trash, ecosystems also face other human-caused stresses. Protecting ecosystems from plastic pollution is essential to ensuring the best possible health for humans, other living things, and the environment. As a result, research on the relationships between contact with "plastic and climate change", "habitat loss/degradation", "exploitation", and more is necessary.

12.6 Remedies, Modifications, and Upcoming Study Initiatives

As long as there is a desire for plastic among humans, innovative technological solutions that recycle or decompose plastic will be required. These solutions should include consumer behavior campaigns, the adoption of bold laws at all levels of government, and the development of new technologies. From lowering the quantity of fresh plastic entering the environment to eliminating current plastic pollution, these solutions must be applied throughout the whole lifecycle of plastic. A range of plastic capture techniques are among the technological advancements being used in cleanup and remediation operations. These instruments are condensed in "The Inventory," a compilation of 52 innovations aimed at stopping plastic leaks or gathering marine plastics, such as beach cleaning robots, river and ocean debris filters, and ocean plastic skimmers (Schmaltz et al. 2020). The scale and efficacy of these technologies to date do not match the magnitude of the plastic pollution problem, although they are an essential part of our efforts to reduce plastic pollution.

Using plastics-degrading microorganisms to establish a "circular economy of plastic" is another cutting-edge strategy to stop plastic pollution. Microbes have produced catalysts that break down plastic as a result of the century-long rise of plastic in the environment (Sheth et al. 2019). A growing number of highly effective microbially mediated plastic bioremediation systems have been produced by further refining these naturally occurring enzymes. While several of microbial isolates may be capable of decomposing plastic, none have demonstrated this ability rapidly enough (Tournier et al. 2020; Lu et al. 2022). Apart from the significant advancements in the bacterially-mediated destruction of plastic, it is imperative to convert plastic waste into biodegradable forms, including amorphization or micronization.

As new technologies emerge, governments everywhere are turning to policies, laws, and ordinances to address the problem of plastic pollution. Policies can employ economic, educational, and regulatory tools to address plastic waste in several ways. According to a recent analysis of global plastic policies, national and subnational regulations typically employ plastic bans to reduce plastic pollution, but international policies mostly concentrate on plans and future actions (Diana et al. 2022a, b). Significant gaps in the policy space still exist, notably concerning the kinds of plastic that these policies are intended to address, notwithstanding this growing trend. For instance, just three of the 147 national policies to date exclusively target microplastics; instead, macroplastics are the most often targeted plastic-type in national policies worldwide, followed by plastic bags (Diana et al. 2022a, b). Additionally, just 5% of national policies have efficacy studies published in peer-reviewed journals, underscoring the necessity of developing more evidence-based policies going forward (Diana et al. 2022a, b). A legally enforceable worldwide convention aimed at reducing plastic pollution is the remaining area where global policy is noticeably deficient (Karasik et al. 2020). Although there has been a growing trend of implementing legislation to prevent plastic waste, the COVID-19 pandemic has halted progress in many countries due to safety concerns about reusable substances (Karasik et al. 2020). The effects of COVID-19 reinforce the constraints of current policies, necessitating better and more coordinated global policy actions.

The first step in defining the boundaries of waste generation that guarantee Earth's continued status as a "safe operating space" for humanity is to establish planetary boundaries, which could serve as a useful guide for global policy efforts (Folke et al. 2021). According to Steffen et al. (2015), planetary boundaries have been established thus far for the following: genetic diversity, freshwater consumption, ocean acidification, phosphorus and nitrogen biogeochemical fluxes, "genetic diversity", "land-system change", and "climatic change". But planetary bounds for plastics or other new entities have not yet been established by specialists. Determining the planetary boundary for plastic pollution can aid in determining whether or not plastic pollution is causing significant and irreparable damage to the environment, as well as helping society come up with preventative measures. It's not known if plastic pollution might get to the point where it makes the earth uninhabitable due to its effects on ecosystems, greenhouse gas emissions, and human and animal health. To better comprehend the cumulative and global effects of plastic pollution, recent initiatives have attempted to characterize the hazardous routes that plastic could lead to such permanent damage (Diana et al. 2022a, b). These initiatives represent the first phase in establishing a limit on plastic pollution, which will subsequently make it easier to create international legislation to maintain society within the established bounds.

Ultimately, more research on the effects of plastic is required in addition to better technologies and regulations aimed at combating plastic pollution. Numerous significant gaps in the literature have been identified by a recent assessment of studies looking at the effects of plastic pollution (Bucci et al. 2020). The majority of manipulative laboratory tests have concentrated on microplastics, 96%, while observational or manipulative field investigations have mostly focused on macroplastics (97%). Fewer studies examined other polymer kinds, including "PVC", "PET", "polypropylene", and others, while the majority of microplastic research projects employed polyethylene and polystyrene. In summary, 76% of all studies only looked at the marine environment, whereas freshwater and terrestrial ecosystems have received comparatively little attention. Understanding the effects of diverse plastic materials, sizes, and forms as well as the impact on distinct ecosystems is necessary to obtain a thorough understanding of the global health implications of plastic pollution.

12.7 Conclusions

There is increasing evidence that plastic can impact several levels of the living organization, including the "molecular and cellular" through "organismal" and population scales. Wide-ranging changes are caused by these effects, which include changes to behavior, metabolism, neurology, reproduction and development, microbiota, inflammation, and oxidative stress. Plastics contain chemicals and microorganisms that function as mediators for these effects in addition to the physical consequences of swallowed or absorbed plastic particles.

Even with the increasing amount of study on the effects of plastics on the health of humans, animals, plants, and entire ecosystems, there are still a lot of unanswered issues. To begin with, additional thorough and methodical research is required to explain the large variations in polymer type, plastic particle size, and additive combinations. Research integrating the effects of plastic pollution at the cell, organismal, population, and ecosystem levels is conspicuously lacking. In addition, little is known "about the cumulative exposure to plastics and additives over time across" a range of nature. In addition, the rate of plastic use and manufacturing is significantly higher than the rate at which global policy responses and the adoption of technology that reduces plastic are happening. For the sake of saving all life on Earth, to lessen and stop the growing worldwide plastic waste issue, a One Health approach can help close these knowledge gaps by offering a framework for integrating across biological scales, encouraging transdisciplinary partnerships, and involving stakeholders from a variety of viewpoints.

References

Abidli S, Pinheiro M, Lahbib Y, Neuparth T, Santos MM, Trigui El Menif N (2021) Effects of environmentally relevant levels of polyethylene microplastic on Mytilus galloprovincialis (Mollusca: Bivalvia): filtration rate and oxidative stress. Environ Sci Pollut Res 28:26643–26652

Aloy AB, Vallejo BM, Juinio-Meñez MA (2011) Increased plastic litter cover affects the foraging activity of the sandy intertidal gastropod Nassarius pullus. Mar Pollut Bull 62:1772–1779

Amato-Lourenço LF, Carvalho-Oliveira R, Júnior GR, Dos Santos Galvão L, Ando RA, Mauad T (2021) Presence of airborne microplastics in human lung tissue. J Hazard Mater 416:126124

Aneck-Hahn NH, Van Zijl MC, Swart P, Truebody B, Genthe B, Charmier J et al (2018) Estrogenic activity, selected plasticizers and potential health risks associated with bottled water in South Africa. J Water Health 16:253–262

Aransiola SA, Ikhumetse AA, Babaniyi BR, Abioye OP, Oyedele OJ, Falade NO (2023a) Phytoaccumulation of micro and nanoplastics: root uptake. In: Maddela NR, Reddy KV, Ranjit P (eds) Micro and nanoplastics in soil. Springer, Cham. https://doi.org/10.10007/978-3-031-21195-9_8

Aransiola SA, Victor-Ekwebelem MO, Ajiboye AE, Leh-Togi Zobeashia SS, Ijah UJJ, Oyedele OJ (2023b) Ecological impacts and toxicity of micro and nanoplastics in agroecosystem. In: Maddela NR, Reddy KV, Ranjit P (eds) Micro and nanoplastics in soil. Springer, Cham. https://doi.org/10.10007/978-3-031-21195-9_10

Auta SH, Abioye OP, Aransiola SA, Bala JD, Chukwuemeka VI, Hassan A, Aziz A, Fauziah SH (2022) Enhanced microbial degradation of PET and PS microplastics under natural conditions in mangrove environment. J Environ Manag 304(2022):114273. https://doi.org/10.1016/j.jenvman.2021.114273

Avio CG, Gorbi S, Milan M, Benedetti M, Fattorini D, d'Errico G et al (2015) Pollutants bioavailability and toxicological risk from microplastics to marine mussels. Environ Pollut 198:211–222

Baini M, Martellini T, Cincinelli A, Campani T, Minutoli R, Panti C et al (2017) First detection of seven phthalate esters (PAEs) as plastic tracers in superficial neustonic/planktonic samples and cetacean blubber. Anal Methods 9:1512–1520

Banerjee A, Shelver WL (2021) Micro- and nanoplastic induced cellular toxicity in mammals: a review. Sci Total Environ 755:142518

Barboza LGA, Vieira LR, Guilhermino L (2018) Single and combined effects of microplastics and mercury on juveniles of the European seabass (Dicentrarchus labrax): changes in behavioural responses and reduction of swimming velocity and resistance time. Environ Pollut 236:1014–1019

Barnes DKA, Milner P (2005) Drifting plastic and its consequences for sessile organism dispersal in the Atlantic ocean. Mar Biol 146:815–825

Barrett J, Chase Z, Zhang J, Holl MMB, Willis K, Williams A et al (2020) Microplastic pollution in deep-sea sediments from the great Australian bight. Front Mar Sci 7:576170

Basak S, Srinivas V, Duttaroy AK (2018) Bisphenol-a impairs cellular function and alters DNA methylation of stress pathway genes in first trimester trophoblast cells. Reprod Toxicol 82:72–79

Basak S, Das MK, Duttaroy AK (2020) Plastics derived endocrine disrupting compounds and their effects on early development. Birth Defects Res 112:1308–1325

Benson NU, Agboola OD, Fred-Ahmadu OH, De-la-Torre GE, Oluwalana A, Williams AB (2022) Micro(nano)plastics prevalence, food web interactions, and toxicity assessment in aquatic organisms: a review. Front Mar Sci 9:851281

Bergami E, Bocci E, Vannuccini ML, Monopoli M, Salvati A, Dawson KA et al (2016) Nano-sized polystyrene affects feeding, behavior and physiology of brine shrimp Artemia franciscana larvae. Ecotoxicol Environ Saf 123:18–25

Bergmann M, Wirzberger V, Krumpen T, Lorenz C, Primpke S, Tekman MB et al (2017) High quantities of microplastic in Arctic deep-sea sediments from the HAUSGARTEN observatory. Environ Sci Technol 51:11000–11010

Bossart GD (2011) Marine mammals as sentinel species for oceans and human health. Vet Pathol 48:676–690

Brooks AL, Wang S, Jambeck JR (2018) The Chinese import ban and its impact on global plastic waste trade. Sci Adv 4:eaat0131

Brun NR, van Hage P, Hunting ER, Haramis A-PG, Vink SC, Vijver MG et al (2019) Polystyrene nanoplastics disrupt glucose metabolism and cortisol levels with a possible link to behavioural changes in larval zebrafish. Commun Biol 2:382

Bucci K, Tulio M, Rochman CM (2020) What is known and unknown about the effects of plastic pollution: a meta-analysis and systematic review. Ecol Appl 30:e02044

Buwono NR, Risjani Y, Soegianto A (2022) Oxidative stress responses of microplastic-contaminated Gambusia affinis obtained from the Brantas river in East Java, Indonesia. Chemosphere 293:133543

Capó X, Alomar C, Compa M, Sole M, Sanahuja I, Soliz Rojas DL et al (2022) Quantification of differential tissue biomarker responses to microplastic ingestion and plasticizer bioaccumulation in aquaculture reared sea bream Sparus aurata. Environ Res 211:113063

Carlos de Sá L, Luıs LG, Guilhermino L (2015) Effects of microplastics′ on juveniles of the common goby (Pomatoschistus microps): confusion with prey, reduction of the predatory performance and efficiency, and possible influence of developmental conditions. Environ Pollut 196:359–362

Carson HS, Colbert SL, Kaylor MJ, McDermid KJ (2011) Small plastic debris changes water movement and heat transfer through beach sediments. Mar Pollut Bull 62:1708–1713

Catarino AI, Macchia V, Sanderson WG, Thompson RC, Henry TB (2018) Low levels of microplastics (MP) in wild mussels indicate that MP ingestion by humans is minimal compared to exposure via household fibres fallout during a meal. Environ Pollut 237:675–684

Chen Q, Gundlach M, Yang S, Jiang J, Velki M, Yin D et al (2017) Quantitative investigation of the mechanisms of microplastics and nanoplastics toward zebrafish larvae locomotor activity. Sci Total Environ 584–585:1022–1031

Chen L, Guo Y, Hu C, Lam PKS, Lam JCW, Zhou B (2018) Dysbiosis of gut microbiota by chronic coexposure to titanium dioxide nanoparticles and bisphenol a: implications for host health in zebrafish. Environ Pollut 234:307–317

Choi JS, Jung Y-J, Hong N-H, Hong SH, Park J-W (2018) Toxicological effects of irregularly shaped and spherical microplastics in a marine teleost, the sheepshead minnow (Cyprinodon variegatus). Mar Pollut Bull 129:231–240

Coffin S, Bouwmeester H, Brander S, Damdimopoulou P, Gouin T, Hermabessiere L et al (2022) Development and application of a health-based framework for informing regulatory action in relation to exposure of microplastic particles in California drinking water. Microplast Nanoplast 2:12

Cole M, Coppock R, Lindeque PK, Altin D, Reed S, Pond DW et al (2019) Effects of nylon microplastic on feeding, lipid accumulation, and moulting in a coldwater copepod. Environ Sci Technol 53:7075–7082

Corea-Téllez KS, Bustamante-Montes P, Garcıa-Fabila M, HernándezValero MA, Vázquez-Moreno F (2008) Estimated risks of water and saliva contamination by phthalate diffusion from plasticized polyvinyl chloride. J Environ Health 71(34–9):45

Courtene-Jones W, Maddalene T, James MK, Smith NS, Youngblood K, Jambeck JR et al (2021) Source, sea and sink—a holistic approach to understanding plastic pollution in the southern Caribbean. Sci Total Environ 797:149098
Cox KD, Covernton GA, Davies HL, Dower JF, Juanes F, Dudas SE (2020) Correction to human consumption of microplastics. Environ Sci Technol 54:10974
Cózar A, Echevarrıa F, Gonzalez-Gordillo JI, Irigoien X, Ubeda B, Hernández-León S et al (2014) Plastic debris in the open ocean. Proc Natl Acad Sci USA 111:10239–10244
da Costa Araújo AP, de Andrade Vieira JE, Malafaia G (2020) Toxicity and trophic transfer of polyethylene microplastics from Poecilia reticulata to danio rerio. Sci Total Environ 742:140217
Das S, Thiagarajan V, Chandrasekaran N, Ravindran B, Mukherjee A (2022) Nanoplastics enhance the toxic effects of titanium dioxide nanoparticle in freshwater algae Scenedesmus obliquus. Comp Biochem Physiol C Toxicol Pharmacol 256:109305
Diana Z, Vegh T, Karasik R, Bering J, Llano Caldas JD, Pickle A et al (2022a) The evolving global plastics policy landscape: an inventory and effectiveness review. Environ Sci Pol 134:34–45
Diana ZT, Karasik R, Merril GB, Morrison M, Corcoran KA, Vermeer D et al (2022b) A transdisciplinary approach to reducing global plastic pollution. Front Mar Sci 9:1032381
Dreier DA, Mello DF, Meyer JN, Martyniuk CJ (2019) Linking mitochondrial dysfunction to organismal and population health in the context of environmental pollutants: progress and considerations for mitochondrial adverse outcome pathways. Environ Toxicol Chem 38:1625–1634
Eriksen M, Maximenko N, Thiel M, Cummins A, Lattin G, Wilson S et al (2013) Plastic pollution in the south pacific subtropical gyre. Mar Pollut Bull 68:71–76
Eriksen M, Lebreton LCM, Carson HS, Thiel M, Moore CJ, Borerro JC et al (2014) Plastic pollution in the world's oceans: more than 5 trillion plastic pieces weighing over 250,000 tons afloat at sea. PLoS One 9:e111913
Fatma Karaman E, Caglayan M, Sancar-Bas S, Ozal-Coskun C, ArdaPirincci P, Ozden S (2019) Global and region-specific post-transcriptional and post-translational modifications of bisphenol a in human prostate cancer cells. Environ Pollut 255:113318
Folke C, Polasky S, Rockström J, Galaz V, Westley F, Lamont M et al (2021) Our future in the Anthropocene biosphere. Ambio 50:834–869
Food and Agriculture Organization (2014) The state of world fisheries and aquaculture 2014: opportunities and challenges. Food & Agriculture Organization of the United Nations (FAO), Rome
Fuller S, Gautam A (2016) A procedure for measuring microplastics using pressurized fluid extraction. Environ Sci Technol 50:5774–5780
Galarpe VRK (2015) Review on the impacts of waste disposal sites in the Philippines. Sci Int 29(2):379–385
Gall SC, Thompson RC (2015) The impact of debris on marine life. Mar Pollut Bull 92:170–179
Gigault J, Halle AT, Baudrimont M, Pascal P-Y, Gauffre F, Phi T-L et al (2018) Current opinion: what is a nanoplastic? Environ Pollut 235:1030–1034
Goldstein MC, Rosenberg M, Cheng L (2012) Increased oceanic microplastic debris enhances oviposition in an endemic pelagic insect. Biol Lett 8:817–820
Gong H, Zhang X, Cheng B, Sun Y, Li C, Li T et al (2013) Bisphenol a accelerates toxic amyloid formation of human islet amyloid polypeptide: a possible link between bisphenol a exposure and type 2 diabetes. PLoS One 8:e54198
González N, Cunha SC, Monteiro C, Fernandes JO, Marquès M, Domingo JL et al (2019) Quantification of eight bisphenol analogues in blood and urine samples of workers in a hazardous waste incinerator. Environ Res 176:108576
González-Pleiter M, Edo C, Aguilera Á, Viúdez-Moreiras D, Pulido-Reyes G, González-Toril E et al (2021) Occurrence and transport of microplastics sampled within and above the planetary boundary layer. Sci Total Environ 761:143213
Groh KJ, Backhaus T, Carney-Almroth B, Geueke B, Inostroza PA, Lennquist A et al (2019) Overview of known plastic packaging-associated chemicals and their hazards. Sci Total Environ 651:3253–3268

Guerrera MC, Aragona M, Porcino C, Fazio F, Laurà R, Levanti M et al (2021) Micro and nano plastics distribution in fish as model organisms: histopathology, blood response and bioaccumulation in different organs. Appl Sci 11:5768

Hahladakis JN, Velis CA, Weber R, Iacovidou E, Purnell P (2018) An overview of chemical additives present in plastics: migration, release, fate and environmental impact during their use, disposal and recycling. J Hazard Mater 344:179–199

Haq MEU, Akash MSH, Sabir S, Mahmood MH, Rehman K (2020) Human exposure to bisphenol a through dietary sources and development of diabetes mellitus: a cross-sectional study in Pakistani population. Environ Sci Pollut Res Int 27:26262–26275

Hazen EL, Abrahms B, Brodie S, Carroll G, Jacox MG, Savoca MS et al (2019) Marine top predators as climate and ecosystem sentinels. Front Ecol Environ 17:565–574

Hernandez LM, Xu EG, Larsson HCE, Tahara R, Maisuria VB, Tufenkji N (2019) Plastic teabags release billions of microparticles and nanoparticles into tea. Environ Sci Technol 53:12300–12310

Hoyo-Alvarez E, Arechavala-Lopez P, Jiménez-Garcıa M, Solomando A, Alomar C, Sureda A et al (2022) Effects of pollutants and microplastics ingestion on oxidative stress and monoaminergic activity of seabream brains. Aquat Toxicol 242:106048

Hu M, Palić D (2020) Micro- and nano-plastics activation of oxidative and inflammatory adverse outcome pathways. Redox Biol 37:101620

Huang B, Wei Z-B, Yang L-Y, Pan K, Miao A-J (2019) Combined toxicity of silver nanoparticles with hematite or plastic nanoparticles toward two freshwater algae. Environ Sci Technol 53:3871–3879

Huang Z, Weng Y, Shen Q, Zhao Y, Jin Y (2021) Microplastic: a potential threat to human and animal health by interfering with the intestinal barrier function and changing the intestinal microenvironment. Sci Total Environ 785:147365

Huntington HP, Quakenbush LT, Nelson M (2016) Effects of changing sea ice on marine mammals and subsistence hunters in northern Alaska from traditional knowledge interviews. Biol Lett 12:20160198

Hutter H-P, Moshammer H, Wallner P, Damberger B, Tappler P, Kundi M (2006) Health complaints and annoyances after moving into a new office building: a multidisciplinary approach including analysis of questionnaires, air and house dust samples. Int J Hyg Environ Health 209:65–68

Hutter H-P, Kundi M, Hohenblum P, Scharf S, Shelton JF, Piegler K et al (2016) Life without plastic: a family experiment and biomonitoring study. Environ Res 150:639–644

Ibrahim YS, Tuan Anuar S, Azmi AA, Wan Mohd Khalik WMA, Lehata S, Hamzah SR et al (2021) Detection of microplastics in human colectomy specimens. JGH Open 5:116–121

International Organization for Migration and International Labour Organization (2021) Powering past the pandemic: bolstering Tuvalu's socioeconomic resilience in a covid-19 world. Available at https://publications.iom.int/books/powering-past-pandemic-bolstering-tuvalus-socioeconomicresilience-covid-19-world

Jambeck JR, Geyer R, Wilcox C, Siegler TR, Perryman M, Andrady A et al (2015) Marine pollution. Plastic waste inputs from land into the ocean. Science 347:768–771

Jiang J, Chen Y, Yu R, Zhao X, Wang Q, Cai L (2016) Pretilachlor has the potential to induce endocrine disruption, oxidative stress, apoptosis and immunotoxicity during zebrafish embryo development. Environ Toxicol Pharmacol 42:125–134

Jin Y, Xia J, Pan Z, Yang J, Wang W, Fu Z (2018) Polystyrene microplastics induce microbiota dysbiosis and inflammation in the gut of adult zebrafish. Environ Pollut 235:322–329

Junge KM, Leppert B, Jahreis S, Wissenbach DK, Feltens R, Grützmann K et al (2018) Mediates the impact of prenatal bisphenol a exposure on long-term body weight development. Clin Epigenetics 10:58

Kankanige D, Babel S (2020) Smaller-sized micro-plastics (MPs) contamination in single-use PET-bottled water in Thailand. Sci Total Environ 717:137232

Karami A, Romano N, Galloway T, Hamzah H (2016) Virgin microplastics cause toxicity and modulate the impacts of phenanthrene on biomarker responses in African catfish (Clarias gariepinus). Environ Res 151:58–70

Karami A, Golieskardi A, Ho YB, Larat V, Salamatinia B (2017) Microplastics in eviscerated flesh and excised organs of dried fish. Sci Rep 7:1–9

Karasik R, Vegh T, Diana Z, Bering J, Caldas J, Pickle A, Rittschof D, Virdin J (2020) Years of government responses to the global plastic pollution problem: the plastics policy inventory, NI X, 20(05). Duke University, Durham

Karbalaei S, Hanachi P, Walker TR, Cole M (2018) Occurrence, sources, human health impacts and mitigation of microplastic pollution. Environ Sci Pollut Res Int 25:36046–36063

Kasirajan S, Ngouajio M (2012) Polyethylene and biodegradable mulches for agricultural applications: a review. Agron Sustain Dev 32:501–529

Kim JH, Kim SH (2020) Exposure to phthalate esters and the risk of endometriosis. Dev Reprod 24:71–78

Kim L, Cui R, Kwak JI, An Y-J (2022) Sub-Acute exposure to nanoplastics via two-chain trophic transfer: from brine shrimp Artemia franciscana to small yellow croaker Larimichthys polyactis. Mar Pollut Bull 175:113314

Kirchnawy C, Hager F, Osorio Piniella V, Jeschko M, Washüttl M, Mertl J et al (2020) Potential endocrine disrupting properties of toys for babies and infants. PLoS One 15:e0231171

Kosuth M, Mason SA, Wattenberg EV (2018) Anthropogenic contamination of tap water, beer, and sea salt. PLoS One 13:e0194970

Kwon BG, Saido K, Koizumi K, Sato H, Ogawa N, Chung S-Y et al (2014) Regional distribution of styrene analogues generated from polystyrene degradation along the coastlines of the North-East Pacific Ocean and Hawaii. Environ Pollut 188:45–49

Kwon JA, Shin B, Kim B (2020) Urinary bisphenol a and thyroid function by BMI in the Korean national environmental health survey (KoNEHS) 2012-2014. Chemosphere 240:124918

Lang IA, Galloway TS, Scarlett A, Henley WE, Depledge M, Wallace RB et al (2008) Association of urinary bisphenol a concentration with medical disorders and laboratory abnormalities in adults. JAMA 300:1303–1310

Latini G, De Felice C, Presta G, Del Vecchio A, Paris I, Ruggieri F et al (2003) In utero exposure to di-(2-ethylhexyl)phthalate and duration of human pregnancy. Environ Health Perspect 111:1783–1785

Lau WWY, Shiran Y, Bailey RM, Cook E, Stuchtey MR, Koskella J et al (2020) Evaluating scenarios toward zero plastic pollution. Science 369:1455–1461

Law KL (2017) Plastics in the marine environment. Annu Rev Mar Sci 9:205–229

Law KL, Starr N, Siegler TR, Jambeck JR, Mallos NJ, Leonard GH (2020) The United States' contribution of plastic waste to land and ocean. Sci Adv 6:eabd0288

Leslie HA, van Velzen MJM, Brandsma SH, Vethaak AD, GarciaVallejo JJ, Lamoree MH (2022) Discovery and quantification of plastic particle pollution in human blood. Environ Int 163:107199

Li J, Yang D, Li L, Jabeen K, Shi H (2015) Microplastics in commercial bivalves from China. Environ Pollut 207:190–195

Liu Z-H, Yin H, Dang Z (2017) Do estrogenic compounds in drinking water migrating from plastic pipe distribution system pose adverse effects to human? An analysis of scientific literature. Environ Sci Pollut Res Int 24:2126–2134

Lombó M, Fernández-Díez C, Gonzalez-Rojo S, Navarro C, Robles V, Herráez MP (2015) Transgenerational inheritance of heart disorders caused by paternal bisphenol a exposure. Environ Pollut 206:667–678

Lu H, Diaz DJ, Czarnecki NJ, Zhu C, Kim W, Shroff R et al (2022) Machine learning-aided engineering of hydrolases for PET depolymerization. Nature 604:662–667

Machtinger R, Orvieto R (2014) Bisphenol a, oocyte maturation, implantation, and IVF outcome: review of animal and human data. Reprod Biomed Online 29:404–410

Manikkam M, Tracey R, Guerrero-Bosagna C, Skinner MK (2013) Plastics derived endocrine disruptors (BPA, DEHP and DBP) induce epigenetic transgenerational inheritance of obesity, reproductive disease and sperm epimutations. PLoS One 8:e55387

Mariana M, Cairrao E (2020) Phthalates implications in the cardiovascular system. J Cardiovasc Dev Dis 7(3):26

Mattsson K, Johnson EV, Malmendal A, Linse S, Hansson L-A, Cedervall T (2017) Brain damage and behavioural disorders in fish induced by plastic nanoparticles delivered through the food chain. Sci Rep 7:11452

Meeker JD, Sathyanarayana S, Swan SH (2009) Phthalates and other additives in plastics: human exposure and associated health outcomes. Philos Trans R Soc Lond Ser B Biol Sci 364:2097–2113

Muncke J, Andersson A-M, Backhaus T, Boucher JM, Carney Almroth B, Castillo A et al (2020) Impacts of food contact chemicals on human health: a consensus statement. Environ Health 19:25

Musa IO, Auta HS, Ilyasu US, Aransiola SA, Makun HA, Adabara NU, Abioye OP, Aziz A, Jayanthi B, Maddela NR, Prasad R (2024) Micro and nanoplastics in environment—degradation, detection, ecological impact. Int J Environ Res 18:1. https://doi.org/10.1007/s41742-023-00551-9

Nadeem A, Ahmad SF, Al-Harbi NO, Attia SM, Bakheet SA, Alsanea S et al (2021) Aggravation of autism-like behavior in BTBR T+tf/J mice by environmental pollutant, di-(2-ethylhexyl) phthalate: role of nuclear factor erythroid 2-related factor 2 and oxidative enzymes in innate immune cells and cerebellum. Int Immunopharmacol 91:107323

Nam S-H, Lee J, An Y-J (2022) Towards understanding the impact of plastics on freshwater and marine microalgae: a review of the mechanisms and toxicity endpoints. J Hazard Mater 423:127174

Ncube LK, Ude AU, Ogunmuyiwa EN, Zulkifli R, Beas IN (2021) An overview of plastic waste generation and management in food packaging industries. Recycle Today 6:12

Ocean Studies Board, Division on Earth and Life Studies, National Academies of Sciences Engineering and Medicine and Committee On The Assessment Of The Cumulative Effects Of Anthropogenic Stressors On Marine Mammals (2017) Approaches to understanding the cumulative effects of stressors on marine mammals. National Academies Press, Washington, DC

Oliveira P, Barboza LGA, Branco V, Figueiredo N, Carvalho C, Guilhermino L (2018) Effects of microplastics and mercury in the freshwater bivalve Corbicula fluminea (Müller 1774): filtration rate, biochemical biomarkers and mercury bioconcentration. Ecotoxicol Environ Saf 164:155–163

One Health (2021) OIE—world organisation for animal health. Available at https://www.oie.int/en/what-we-do/global-initiatives/one-health/. Accessed 13 May 2022

One Health (2022). Available at https://www.cdc.gov/onehealth/index.html. Accessed 13 May 2022

Paul-Pont I, Lacroix C, González Fernández C, Hégaret H, Lambert C, Le Goïc N et al (2016) Exposure of marine mussels Mytilus spp. to polystyrene microplastics: toxicity and influence on fluoranthene bioaccumulation. Environ Pollut 216:724–737

Pitt JA, Kozal JS, Jayasundara N, Massarsky A, Trevisan R, Geitner N et al (2018a) Uptake, tissue distribution, and toxicity of polystyrene nanoparticles in developing zebrafish (Danio rerio). Aquat Toxicol 194:185–194

Pitt JA, Trevisan R, Massarsky A, Kozal JS, Levin ED, Di Giulio RT (2018b) Maternal transfer of nanoplastics to offspring in zebrafish (Danio rerio): a case study with nanopolystyrene. Sci Total Environ 643:324–334

Prinz N, Korez Š (2020) Understanding how microplastics affect marine biota on the cellular level is important for assessing ecosystem function: a review. In: YOUMARES 9—the oceans: our research, our future. Springer, Cham, pp 101–120

Ragusa A, Svelato A, Santacroce C, Catalano P, Notarstefano V, Carnevali O et al (2021) Plasticenta: first evidence of microplastics in human placenta. Environ Int 146:106274

Rangasamy B, Malafaia G, Maheswaran R (2022) Evaluation of antioxidant response and Na-K-ATPase activity in zebrafish exposed to polyethylene microplastics: shedding light on a physiological adaptation. J Hazard Mater 426:127789
Rasool FN, Saavedra MA, Pamba S, Perold V, Mmochi AJ, Maalim M et al (2021) Isolation and characterization of human pathogenic multidrug resistant bacteria associated with plastic litter collected in Zanzibar. J Hazard Mater 405:124591
Richards ZT, Beger M (2011) A quantification of the standing stock of macro-debris in Majuro lagoon and its effect on hard coral communities. Mar Pollut Bull 62:1693–1701
Roch S, Friedrich C, Brinker A (2020) Uptake routes of microplastics in fishes: practical and theoretical approaches to test existing theories. Sci Rep 10:3896
Rochman CM, Hentschel BT, Teh SJ (2014) Long-term sorption of metals is similar among plastic types: implications for plastic debris in aquatic environments. PLoS One 9:e85433
Rochman CM, Tahir A, Williams SL, Baxa DV, Lam R, Miller JT et al (2015) Anthropogenic debris in seafood: plastic debris and fibers from textiles in fish and bivalves sold for human consumption. Sci Rep 5:14340
Roman J, Nevins J, Altabet M, Koopman H, McCarthy J (2016) Endangered right whales enhance primary productivity in the Bay of Fundy. PLoS One 11:e0156553
Sarasamma S, Audira G, Siregar P, Malhotra N, Lai Y-H, Liang S-T et al (2020) Nanoplastics cause neurobehavioral impairments, reproductive and oxidative damages, and biomarker responses in zebrafish: throwing up alarms of wide spread health risk of exposure. Int J Mol Sci 21:1410
Schmaltz E, Melvin EC, Diana Z, Gunady EF, Rittschof D, Somarelli JA et al (2020) Plastic pollution solutions: emerging technologies to prevent and collectmarineplastic pollution. Environ Int 144:106067
Schwabl P, Köppel S, Königshofer P, Bucsics T, Trauner M, Reiberger T et al (2019) Detection of various microplastics in human stool: a prospective case series. Ann Intern Med 171:453–457
Senathirajah K, Attwood S, Bhagwat G, Carbery M, Wilson S, Palanisami T (2021) Estimation of the mass of microplastics ingested—a pivotal first step towards human health risk assessment. J Hazard Mater 404:124004
Senyildiz M, Karaman EF, Bas SS, Pirincci PA, Ozden S (2017) Effects of BPA on global DNA methylation and global histone 3 lysine modifications in SH-SY5Y cells: an epigenetic mechanism linking the regulation of chromatin modifying genes. Toxicol in Vitro 44:313–321
Sheth MU, Kwartler SK, Schmaltz ER, Hoskinson SM, Martz EJ, Dunphy-Daly MM et al (2019) Bioengineering a future free of marine plastic waste. Front Mar Sci 6:1162680
Silva MSS, Oliveira M, Valente P, Figueira E, Martins M, Pires A (2020) Behavior and biochemical responses of the polychaeta Hediste diversicolor to polystyrene nanoplastics. Sci Total Environ 707:134434
Sobhani Z, Lei Y, Tang Y, Wu L, Zhang X, Naidu R et al (2020) Microplastics generated when opening plastic packaging. Sci Rep 10:1–7
Solleiro-Villavicencio H, Gomez-De León CT, Del Rıo-Araiza VH, Morales-Montor J (2020) The detrimental effect of microplastics on critical periods of development in the neuroendocrine system. Birth Defects Res 112:1326–1340
Steffen W, Richardson K, Rockström J, Cornell SE, Fetzer I, Bennett EM et al (2015) Sustainability. planetary boundaries: guiding human development on a changing planet. Science 347:1259855
Su L, Cai H, Kolandhasamy P, Wu C, Rochman CM, Shi H (2018) Using the Asian clam as an indicator of microplastic pollution in freshwater ecosystems. Environ Pollut 234:347–355
Summers JK, Smith LM, Case JL, Linthurst RA (2012) A review of the elements of human well-being with an emphasis on the contribution of ecosystem services. Ambio 41:327–340
Sussarellu R, Suquet M, Thomas Y, Lambert C, Fabioux C, Pernet MEJ et al (2016) Oyster reproduction is affected by exposure to polystyrene microplastics. Proc Natl Acad Sci USA 113:2430–2435
Tao S, Zhang Y, Yuan C, Gao J, Wu F, Wang Z (2016) Oxidative stress and immunotoxic effects of bisphenol a on the larvae of rare minnow Gobiocypris rarus. Ecotoxicol Environ Saf 124:377–385

Taylor SFW, Roberts MJ, Milligan B, Ncwadi R (2019) Measurement and implications of marine food security in the Western Indian ocean: an impending crisis? Food Secur 11:1395–1415

Thompson RC, Moore CJ, vom Saal FS, Swan SH (2009) Plastics, the environment and human health: current consensus and future trends. Philos Trans R Soc Lond Ser B Biol Sci 364:2153–2166

Tournier V, Topham CM, Gilles A, David B, Folgoas C, Moya-Leclair E et al (2020) An engineered PET depolymerase to break down and recycle plastic bottles. Nature 580:216–219

Trevisan R, Voy C, Chen S, Di Giulio RT (2019) Nanoplastics decrease the toxicity of a complex PAH mixture but impair mitochondrial energy production in developing zebrafish. Environ Sci Technol 53:8405–8415

Trevisan R, Uzochukwu D, Di Giulio RT (2020) PAH sorption to nanoplastics and the trojan horse effect as drivers of mitochondrial toxicity and PAH localization in zebrafish. Front Environ Sci 8:78

Van Cauwenberghe L, Janssen CR (2014) Microplastics in bivalves cultured for human consumption. Environ Pollut 193:65–70

von Moos N, Burkhardt-Holm P, Köhler A (2012) Uptake and effects of microplastics on cells and tissue of the blue mussel Mytilus edulis L. after an experimental exposure. Environ Sci Technol 46:11327–11335

Wagner M, Oehlmann J (2011) Endocrine disruptors in bottled mineral water: estrogenic activity in the e-screen. J Steroid Biochem Mol Biol 127:128–135

Waite HR, Donnelly MJ, Walters LJ (2018) Quantity and types of microplastics in the organic tissues of the eastern oyster Crassostrea virginica and Atlantic mud crab Panopeus herbstii from a Florida estuary. Mar Pollut Bull 129:179–185

Walker I, Montaño MD, Lankone RS, Fairbrother DH, Ferguson PL (2021) Influence of CNT loading and environmental stressors on leaching of polymer-associated chemicals from epoxy and polycarbonate nanocomposites. Environ Chem 18:131

Wang C, Gao W, Liang Y, Jiang Y, Wang Y, Zhang Q et al (2019a) Migration of chlorinated paraffins from plastic food packaging into food simulants: concentrations and differences in congener profiles. Chemosphere 225:557–564

Wang H, Zhao P, Huang Q, Chi Y, Dong S, Fan J (2019b) Bisphenol-a induces neurodegeneration through disturbance of intracellular calcium homeostasis in human embryonic stem cells-derived cortical neurons. Chemosphere 229:618–630

Wang Y, Mao Z, Zhang M, Ding G, Sun J, Du M (2019c) The uptake and elimination of polystyrene microplastics by the brine shrimp, Artemia parthenogenetica, and its impact on its feeding behavior and intestinal histology. Chemosphere 234:123–131

Wang Q, Bai J, Ning B, Fan L, Sun T, Fang Y et al (2020) Effects of bisphenol a and nanoscale and microscale polystyrene plastic exposure on particle uptake and toxicity in human caco-2 cells. Chemosphere 254:126788

Wang C, Hou M, Shang K, Wang H, Wang J (2022) Microplastics (Polystyrene) exposure induces metabolic changes in the liver of rare minnow. Molecules 27:584

Wen B, Zhang N, Jin S-R, Chen Z-Z, Gao J-Z, Liu Y et al (2018) Microplastics have a more profound impact than elevated temperatures on the predatory performance, digestion and energy metabolism of an Amazonian cichlid. Aquat Toxicol 195:67–76

Wiesinger H, Wang Z, Hellweg S (2021) Deep dive into plastic monomers, additives, and processing aids. Environ Sci Technol 55:9339–9351

World Health Organization (2022) Dietary and inhalation exposure to nanoand microplastic particles and potential implications for human health. Available at https://www.who.int/publications/i/item/9789240054608

Wright SL, Kelly FJ (2017) Plastic and human health: a micro issue? Environ Sci Technol 51:6634–6647

Yang D, Shi H, Li L, Li J, Jabeen K, Kolandhasamy P (2015) Microplastic pollution in table salts from China. Environ Sci Technol 49:13622–13627

Yong CQY, Valiyaveettil S, Tang BL (2020) Toxicity of microplastics and nanoplastics in mammalian systems. Int J Environ Res Public Health 17:1509

Zantis LJ, Carroll EL, Nelms SE, Bosker T (2021) Marine mammals and microplastics: a systematic review and call for standardisation. Environ Pollut 269:116142

Zantis LJ, Bosker T, Lawler F, Nelms SE, O'Rorke R, Constantine R et al (2022) Assessing microplastic exposure of large marine filter-feeders. Sci Total Environ 818:151815

Zarus GM, Muianga C, Hunter CM, Pappas RS (2021) A review of data for quantifying human exposures to micro and nanoplastics and potential health risks. Sci Total Environ 756:144010

Zhao X, Liu Z, Ren X, Duan X (2021) Parental transfer of nanopolystyrene-enhanced tris(1,3-dichloro-2-propyl) phosphate induces transgenerational thyroid disruption in zebrafish. Aquat Toxicol 236:105871

Zuccarello P, Ferrante M, Cristaldi A, Copat C, Grasso A, Sangregorio D et al (2019) Exposure to microplastics (<10 mm) associated to plastic bottles mineral water consumption: the first quantitative study. Water Res 157:365–371

Chapter 13
Characterization Technologies of White Pollution in the Atmosphere

Mustapha Abdulsalam, Innocent Ojeba Musa, Miracle Uwa Livinus, Shehu-Alimi Elelu, Ganiyat Omotayo Ibrahim, Salami Olaitan Lateefat, and Auwal Sagir Muhammad

Abstract This study is aimed at addressing the urgent problem of air plastic pollution thus implying the investigation of possible reduction strategies. The study analyzes the microplastics in the air and examines their sources, containment patterns, and environmental impacts. The study attempts at the problematic issue of environmental microplastic contamination, which may threaten both ecological and human health. A multi-disciplinary approach is used in the study where the different characterization approaches such as aerosol sampling via high-resolution microscopy, remote sensing, and chemical (atmospheric analysis) are evaluated. Evidence shows the behavior of particles driven by various mechanisms, urban to non-urban atmospheres among them. In the study, the issues of particle size distributions, sampling site selection, and methodological challenges are covered as these issues become more important to be considered in the future. The key recommendations include the development of advanced characterization methods, the establishment of partnerships of interdisciplinary groups, the execution of longitudinal studies to detect temporal trends, and the translation of the research knowledge into applicable policies. Local authorities, and initiatives both have to participate in appearing eco-

M. Abdulsalam (✉) · I. O. Musa
Department of Microbiology, Skyline University Nigeria, Kano, Nigeria
e-mail: mustapha.abdulsalam@sun.edu.ng

M. U. Livinus
Department of Biochemistry, Skyline University Nigeria, Kano, Nigeria

S.-A. Elelu
Department of Chemistry, Howard University, Washington, DC, USA

G. O. Ibrahim
Department of Chemistry, Nottingham Trent University, Nottingham, UK

S. O. Lateefat
Department of Chemistry and Molecular Biology, University of Gothenburg, Gothenburg, Sweden

A. S. Muhammad
School of Informatics, Xiamen University, Xiamen, China

S. A. Aransiola et al. (eds.), *White Pollution: Biodiversity and hazards in marine plastisphere*, Environmental Contamination Remediation and Management,
https://doi.org/10.1007/978-3-031-95547-1_13

friendly and waste treatment to reduce the amount of microplastics out of the atmosphere. However, it is stated that it is teamwork that is mandatory to avoid microplastics threatening ecosystems. The study demonstrates how everyone's duty as stewards of the planet is to respect nature. The study asserts that their actions should be ethical and long-lasting, requiring a great deal of effort to establish the framework for a world with clean air and a healthy atmosphere.

Keywords Characterization technologies · Atmospheric plastic pollution · Microplastics · Sustainability · Multidisciplinary collaboration

13.1 Introduction

The modern world has made science and society as interconnected as ever before raising some poignant ecological problems, causing global concern. This study aims to show how characterization technologies have become paramount and how strategies for the future are going to be effective in the fight against environmental degradation. In particular, its focus is on explaining the typical use of persistent plastics, referred to as "white pollution", as the primary factor of pollution in the global oceans and ecosystems (Shen et al. 2020; Aransiola et al. 2023b). These plastics even if not completely biodegradable tend to pollute the air, water, and land. This study stands out by investigating the pollution of the white hole of the atmosphere and it provides sources from the monitoring of the air concentration (Musa et al. 2024; Auta et al. 2022). Additionally, it focuses on the importance of a model change in the experimental method of identifying bio-plastics or not. Microplastic levels, air quality indexes, and sources of plastic particles are some parameters micro-managed to come up with a clear distinction of the levels of atmospheric whitening. The interdisciplinary nature of environmental science, chemistry, and technology is a clear demonstration of the strategic approach to combat the multi-faced nature of this problem (Khoaele et al. 2023).

Many researchers, like Williams et al. (2019), indicate the global importance of the plastic pollution crisis by depicting its ever-increasing levels, contamination of aquatic habitats, and resulting human health hazards due to seafood consumption. These studies broaden the view of plastics on a life cycle basis, from manufacturing to disposal, showing that measures to prevent the proliferation of plastics are highly necessary. The regional differences in consumption habits, waste management systems, and air pollution responses compound the challenge, thus demanding localized solutions adopted in different locations (Miller et al. 2021). Literature on the complexity of plastic pollution and its complex impacts provides the critical base for forming efficient solutions to combat the soaring effect on a global level (Aransiola et al. 2023a).

The line of historical development in the literature represents a gradual but effective process of improvement in the concept of plastic pollution. Rachel Carson was a pioneer whose path-breaking article paved the way for many research focuses on plastic waste. The quest for microplastic research development and characterization technology improvement yields great achievements that are pacing us toward current understanding. The usefulness of this study is due to the preventive point of view approaching solving key gaps in researching atmospheric white pollution. Plastic pollution was already well-studied and supplied many people with more information about it, but its atmospheric part has not been primarily discovered yet. These caterézed ways are the blind spots of the environmental study which act as the unique special data processing techniques for the pollution of air by the composition, sources, and properties of the plastic particles. This study takes over this position by clearing this out and presenting a critical argument for solid and sustainable decisions, policy development, and white pollution management.

13.2 White Pollution in the Atmosphere: Overview

13.2.1 Definition and Types of White Pollution

"White pollution" to some extent reflects the extensive problem of plastic waste, which, by evoking environmental anxieties, merges with other environmental problems. The white pollution process is driven by human activities and manifests in the appearance of small plastic particles scattered throughout the atmosphere (Hameed et al. 2020). It is fueled by the exploitation of inappropriate disposal methods, the absence of efficient and effective waste management systems, and the entry of other pollutants into our atmosphere. Hence, the identification of the presence of macro- and microplastics helps in analyzing the depth of air pollution with white plastic trash.

13.2.2 Sources and Pathways to the Atmosphere

Three distinct sources contribute to white pollution in the atmosphere: surface waters, lakes, groundwater, and marine sources in addition to human systems. Although land plastics can break down when they're large, marine plastics can be taken offshore by the wind and wave action. In addition to natural sources of plastic emissions (e.g., spatial activities like the pyrolysis of plastics, industrial releases, and air-blown plastic falls play a significant role in the emission of plastic in the air. The geometry, weather states, and human activities affect the intricate ways by which plastics enter the atmosphere. Establishing the origins and pathways is critical in the right direction for the development of appropriate strategies to clear up atmospheric white pollution (Hader et al. 2020). Figure 13.1 demonstrates step-by-step ways in which microplastics weave their way around the marine ecosystem that commences from their manufacturing to their effects on marine life (evolutionary).

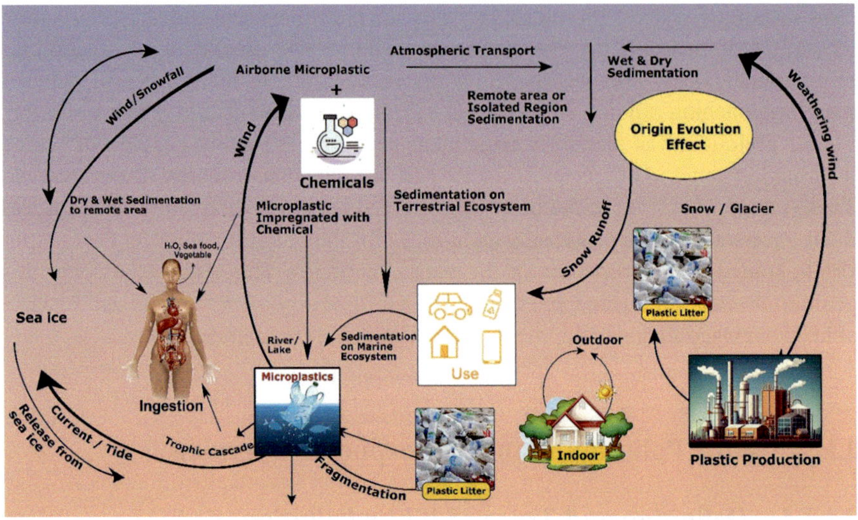

Fig. 13.1 Interconnected pathways of microplastics in marine ecosystem: From production to evolutionary impacts

Coming from manufacturing of plastic or waste disposal is the primary source of microplastics that find their way to the ocean via airborne or runoff, or through direct dumping. These contaminants can be found in various shapes and forms once in the ocean. For instance, they may enter the food chain through marine organisms or they can ride with the ocean currents and deposit on the seafloor. Such microplastics may incorporate owed chemicals and pollutants acting as a transmitting form of threat to marine life and the ecosystem. Ingestion of microplastics and exposure to associated industrial chemicals over a long period can lead to evolutionary impacts on marine organisms which may change their behavior as well as physiology. Such evolutionary effects might also influence their reproductive success. Knowing and sorting out the relationships of these linked pathways are major for the reduction of microplastics and the better preservation of healthy ecosystems in the oceans.

13.2.3 Environmental Impacts

Besides the clear white plastic waste, the bright air pollution also brings great trouble. Airborne particles irrespective of their sizes may be Macros or micros, carry similar impacts on the climate, the health of humans, and the eco-systems (Aves 2021). This particulate matter could penetrate the lungs and introduce added toxins into the body, and after that, it could end up in water bodies and dryland areas. Furthermore, they can also contain toxins that they may consume. It is to be noted that plastics can carry pathogens, which are probably carried through the air and can disrupt the nutrient cycles and biodiversity (Angnunavuri et al. 2020). This ability of them to travel to far-off places and then make their settling remotely increases the issue of global health security. Hence, the phenomenon of ecological consequences cannot be underestimated, since due awareness is the key to successfully fighting atmospheric white pollution (Winter and Koger 2004).

13.3 Air Sampling Techniques

The initial investigations of the accurate sampling techniques have highlighted the significance of sampling for the correct pollution characterization and extent.

13.3.1 High-Volume Air Samplers

The prevalent way of determining microplastics in the air transcribes in the use of high-capacity air samplers that are fitted with filters that are specially designed to capture dust efficiently irrespective of their size. Subsequent sampling further assesses the airborne plastic pollution in the atmosphere (Nguyen et al. 2019). Therefore, this method gives high efficiency in area profiles with higher pollution expected for industrial and urban areas (Mohtar et al. 2018).

13.3.2 Passive Samplers

Innovative methodologies and passive measurement approaches give the possibility to the evaluation of microparticle concentrations in the atmosphere. Part of these devices include aerosol position-point deposition inertial samplers, designed to collect aerosols as they settle and before they reach any surface in various environments (Dong et al. 2023). The elemental composition of the particle surfaces can give useful information that is relevant in determining the geographical distribution and the accumulation rates of microplastics in the atmosphere. It is stated by Chen et al. (2021) that such samplers help in the identification of periodic or seasonal fluctuations and overall plastic presence in the atmosphere by a long-term monitoring approach.

13.3.3 Impingers

Despite their complexity, impingers are still one of the air sampling methods that have been developed to provide a broad range of analytical options. During the process practically, air is pushed through a liquefied component that holds the particles on the way. Impingers have the potential to grab micro-particles as well as catch microplastics, thus, ensuring the details of the smaller components of air pollution. More accurate information about the amount and composition of particles, as well as the liquid medium, is retrieved after the investigation (Wright et al. 2021).

13.3.4 Sampling Protocols and Considerations

It is crucial to develop and create SOPs to attain the optimal performance of air-sampling methods. Since the data on air sampling, filtration processing, and contamination control from different studies are the main resources used, the uniformity of these procedures is of great importance. The proper data collection will depend on the correctness of fields which may include height, length, and sample location (Prata et al. 2020; Wright et al. 2021). These sampling techniques will be chosen according to the characteristics that the study site has and the goals of the study and the results will be assessed depending on the procedure which will be done to address all possible sources of pollution during the sampling process.

13.4 Microplastic Identification Methods

Estimation of the quantity of microplastic and its impact on the environment demands precise detection. The focus of this section is on the primary techniques for the detection of microplastics using FTIR Spectroscopy, Raman Spectroscopy, and microscopic techniques, drawing an audit of their present drawbacks and advancements.

13.4.1 Fourier Transform Infrared Spectroscopy (FTIR)

Identification of microplastics is one of the analytical processes that use FTIR spectroscopy, as this technique uses the absorption of infrared radiation by bonds in the sample material as a principle. The spectral pattern obtained by FTIR is used for the identification of specific functional groups present in plastics which then assists in the classification process of various polymers. This method is characterized by high accuracy and, consequently, the togetherness of diverse environmental samples makes it a substantial contribution to the identification of polymer types in microplastics (Veerasingam et al. 2021).

13.4.2 Raman Spectroscopy

Raman spectroscopy is a powerful tool used to identify microplastics as it focuses on the unique polymeric signature inelastic scattering of monochromatic light with molecular vibrations (Levermore et al. 2020). This technique facilitates the detection of microplastics through the FTIR profile, enabling scientists to differentiate polymers whose FTIR spectra are similar.

13.4.3 Microscopy Techniques

The identification of micropollutants using visual tools is mainly dependent on the application of microscopy techniques such as electron, fluorescence, and optics. In optical microscopy, the high-resolution images are precisely analyzed concerning size and shape (Gaston et al. 2020). TEM and SEM are particular examples of structural analysis due to their specializations. Microplastics tagged with fluorescence can be observed successfully using fluorescence microscopy. The microscopy methods, along with spectral methods give the final visual confirmation that the microplastics do indeed exist (Delgado-Gallardo et al. 2021).

13.4.4 Challenges and Advancements in Identification

Despite the high success rate of the current ways of identification, there are a lot of problems in the analysis of microplastics. The detection of small particles and fragments that are too small to be seen or could be misidentified using conventional detection instruments is one of the main problems (Ivleva 2021). More so, an accurate characterization of plastics is hampered by the degradation of microplastics and the existence of organic coatings. With solving these problems, the efficiency of microplastic identification is improved (Samanta et al. 2022). Machine learning algorithms, image recognition (automated), and multi-modal analyses are all progressing features. A microplastic identification suite was developed by coupling Raman spectroscopy, Fourier infrared transform spectroscopy, and other microscopic techniques (Xie et al. 2023). Besides the aforementioned approaches combined with further development of the methods, it not only helps us to determine the presence of microplastics but also opens up the possibility of more advanced measurements in the future that are capable of detecting microplastics in different environments.

13.5 Quantification of Microplastics in the Atmosphere

To evaluate the level of plastic pollution, we need to measure the concentration of microplastics in the air. This includes techniques of quantification like PCR, image analysis, and gravimetric analysis, and interpretation of data together with others as well.

13.5.1 *Gravimetric Analysis*

The gravimetric technique deserves particular attention among techniques applied for measuring microplastics in air samples. This method employs air filtration on a filter, generating a difference in microplastics. The tweezers are used to separate the microplastic on the sampled air filter so that its mass concentration can be calculated by removing the microplastics from the weighed filter and finding out the difference in weight. The gravimetric analysis involves quantitative measurements of the atmospheric plastic micro-particle loadings and thus its strength owes to its accuracy and usability (Peñalver et al. 2021).

13.5.2 *Image Analysis*

Analogously, image analysis quantifies microplastics by visual means. Then pieces of microplastics are looked at with a microscope or an imaging device after collecting a sample from the air. Following, analysis software is running the acquired pictures that, automatize the microplastic particle counting and measurement. The quantification of microplastic pollution increased due to the image analyses especially for large samples with different microplastic sizes and structures (Sridharan et al. 2021).

13.5.3 *Polymerase Chain Reaction (PCR)*

PCR, a molecular biology method, makes use of genetic markers to conduct assessments on microplastics. Based on PCR that calculates and quantifies the extracted DNA or RNA content from microplastic samples, the primary identification of various types of plastic can be done (Arias-Andres et al. 2018). The special nature of PCR to distinguish between various polymers increases its worth even though it is less common in atmospheric research as far as the discrimination of microplastics in air samples is concerned.

13.5.4 Data Interpretation and Limitations

While interpreting data in the case of microplastic quantification it is important to keep in mind the methodology preferred, the probability of error, and the results itself. Hence it is crucial to evaluate several variables including variability in the atmospheric conditions, the efficiency of sampling as well as the possibility of contamination during analysis (Sridharan et al. 2021). Other drawbacks include difficulties in separating microplastics from particles of similar size and difficulties in accurately determining the size of nanoplastics using conventional techniques. With the establishment of standard methods and conscious of the presence of possible bias (Wright et al. 2021), correct data interpretation is possible. Using a gravimetric analysis with both image analysis and molecular techniques such as PCR gives a multipath strategy to microplastics in the air (Clough et al. 2023). Although there are potential pros and cons when used together, these techniques provide us with an understanding of microplastic types, quantities, and the possibility of their effects in the air (Razeghi et al. 2021). Further advances in analytical approaches and standardization efforts will allow accuracy and consistency in microplastic quantification to be increased in atmospheric research studies.

13.6 Aerosol Monitoring

Understanding the dispersion, distribution, logistics, and properties of atmospheric microplastics needs frequent aerosol monitoring. The main techniques used for aerosol monitoring are addressed in this section, such as size distribution analysis, aerosol impactors, particle counters, and also the knowledge gained from tracking the microplastic aerosols.

13.6.1 Aerosol Impactors

These are pieces of equipment developed for the separation of particles from the air according to their size. These usually are made up of a row of plates or surfaces with holes of varying sizes. When these openings are used for air passing, particles of different sizes are collected on the surface (Wright et al. 2021). This method allows the particles of specified size ranges to be selected selectively. This feature of the method can be applied to identify the fraction of aerosols that may contain microplastics. Evaluation of acquired samples allows to reveal the presence of microplastics in various size ranges.

13.6.2 Particle Counters

These are filters that are designed to separate particles into their respective sizes. Generally, they have several plates or surfaces of different sizes of openings. Smaller particles are collected on the surface while air flows through the apertures (Sridharan et al. 2021). This allows investigators to selectively focus on certain particle size groups, which in turn creates the ability to concentrate on the components of aerosols that may contain microplastics. The analysis of samples that came later, however, reveals the quantity and presence of microplastics and their size ranges.

13.6.3 Size Distribution Analysis

It comprises reporting the whole range of particle sizes present in particle size distributions in aerosol samples. This can be achieved through the implementation of several approaches ranging from scanning electron microscopy and laser diffraction (Yao et al. 2022). The aerosolized microplastics need to be understood in terms of size distribution if the scientists are to be able to determine their transport distances, inhalation risks as well as environmental fates. Integrative analysis of monitoring methods with size distribution of microplastics takes the characterization of atmospheric microplastics to the next level (Zhang et al. 2020).

13.6.4 Insights into Aerosolized Microplastics

Data from aerosol monitoring research provide unique information on how microplastics act in the atmosphere. The dynamics of aerosolized microplastic can be better understood through the data obtained using impactors, particle counters, and size distribution analyses (Rao et al. 2024.) These realizations underlie assessments of the potential risk of exposure, tracing the origins and trajectories of atmospheric microplastics, and evaluation of the efficiency of mitigation methods.

In the case of aerosol monitoring of microplastics, the totality of impactors, particle counters, and size distribution analysis brings in a holistic viewpoint. Using them individually gives us data but when combined this is the most beneficial as it provides us with the information about amount, type, and characteristics of microplastics in the atmosphere (Evangeliou et al. 2022). Despite the advancement of monitoring technologies, aerosol monitoring will continue to be important in grasping the complexities of microplastics in the atmosphere and their health effects (Allen et al. 2022).

13.7 Remote Sensing of Atmospheric Plastic Pollution

For a more comprehensive assessment of the air pollution brought by plastic waste, remote sensing techniques provide a unified perspective. The body of the paper deals with some of the most important remote sensing techniques, like satellite images and drone aerial surveys, and discusses the advantages and limitations of these technologies.

13.7.1 Satellite Imagery

Satellite images can watch changes in the environment everywhere in the world. Using Earth's surface images taken by satellites with optical and infrared sensors about the plastic pollution in the atmosphere, we obtain unique data about the scale and distribution of the plastic particles. Satellite imagery by the detection of the large-scale patterns and hotspots associated with plastic pollution enhances, our comprehension, of the spatial dynamics of atmospheric microplastics (Sakti et al. 2023).

13.7.2 Aerial Surveys Using Drones

It provides a flexible and tailored practical way of using remote sensing. Permitting the deployment of drones that carry high-resolution sensors or cameras to acquire images of specific areas at lower altitudes. The technique, in turn, provides for a more in-depth evaluation of the amount of plastic pollution in the air, including the determination of smaller particles and the focus on the concentrations of microplastics in particular regions. The fact that drones can specifically map plastic pollution in remote areas is very helpful (Topouzelis et al. 2019).

13.7.3 Applications and Challenges

Applications

- *Monitoring Spatial Distribution:* Both satellite imagery and drone surveys are among the efficient remote sensing technologies in providing an overview of the location of airborne plastic pollution (Salgado-Hernanz et al. 2021). This is extremely useful not only in the identification of high plastic concentration areas but also in the assessment of global plastic dispersal patterns.
- *Temporal Monitoring:* Such methodologies help in providing an understanding of microplastics' atmospheric dynamics by letting researchers track the variation

with time. The remote sensing application during extended periods lets one be able to spot changes, patterns, and seasonal variations, as well as outcomes of the measures (Veettil et al. 2022).

- *Environmental Impact Assessment:* Risk assessment of the environment is made easier by remote sensing which also highlights the existing health threats to humans and ecosystems (Mukonza and Chiang 2022).

Challenges

- *Resolution Limitations:* Even though remote sensing gives an overview of the situation, the resolution might lead to the difficulty of detecting tiny microplastics. The restrictions of ground-based surveys and the use of high-resolution satellite imagery are eased in part (Topouzelis et al. 2019).
- *Discrimination of Materials:* It is quite challenging for remote sensing techniques to determine various materials with precision, hence, it becomes very tedious to distinguish microplastics from other airborne particles. (Hu 2021).
- *Cost and Technical Expertise:* It is not expensive but it needs to have specific technical knowledge to collect and analyze data from the remote sensor. Small research organizations or initiatives with no adequate funds may prove the challenge (Chi et al. 2016).

Remote sensing atmospheric plastic pollution holds great promise for the fast-growing field of extended monitoring and assessment. Combing drone aerial surveys with satellite imagery provides complementary information and a holistic vision of the mechanism and the pattern of microplastics in the atmosphere (Chellasamy et al. 2023). Using remote sensing is going to be more and more vital to the understanding of atmospheric plastic pollution and its further consequences because the technology is being developed and refined.

13.8 Chemical Analysis of Plastic Particles

Analyzing the chemical composition and ultimately the effects of plastic particles in the atmosphere is impossible without performing highly advanced chemical analysis. This section explains in detail, the most pertinent tools of analytics use like Liquid Chromatography-Mass Spectrometry (LC-MS) and Gas Chromatography-Mass Spectrometry (GC-MS), which help in unveiling the complex chemical structure of the molecules.

13.8.1 Gas Chromatography-Mass Spectrometry (GC-MS)

GC-MS is a flexible tool that can be applied for both qualitative and quantitative analysis of organic compounds even plastics' associated ones. By mass spectrometry, the GCMS exactly determines the concentration and the molecular mass of the elements by measuring their mass-to-charge ratio, and the gas chromatography part separates compounds by their volatility. This is a very useful way for realizing and measuring the presence of monomers, degradation products, and additives to plastics. The micro-plastic constituent chemical components can be detected through the GC-MS analysis of the different matrix environments (Chen et al. 2020).

13.8.2 Liquid Chromatography-Mass Spectrometry (LC-MS)

Contrarily, LC-MS achieves the goal of GC-MS by detecting less volatile compounds including other plastic particles that are related. After the molecules are saturated with a mobile phase and separated via liquid chromatography, the task of identifying and quantifying compounds belongs the mass spectrometry. LC-MS outperforms all other analytical technologies when it comes to elucidating details of plasticizers, antioxidants, and other non-volatile components of microplastics (Luo et al. 2024). Consequently, GC-MS and LC-MS, together, provide a synergetic effect, which in turn helps in providing more enhanced analytical possibilities accounting for a more thorough look at the chemical make-up of plastic particles.

13.8.3 Chemical Composition Insights

Even understanding the plastic particle composition is crucial for understanding the sources, processes of degradation, and environmental implications of them. This is possible through the elucidation of the chemical composition of the particles. Additives (e.g., phthalates and antioxidants), monomers (e.g., bisphenol A), and specific markers that belong to a specific polymer type are regularly identified compounds (Chun et al. 2022). Researchers can trace the origins of microplastics, analyze the mechanisms of their deterioration, and evaluate the risks to humans and ecosystems thanks to the identification of the chemical properties of the plastic particles (Ivleva 2021).

GC-MS and LC-MS chemical analysis methods are the main tools for establishing the various attributes of plastic particles dispersed in the air. The pinpointing of specific chemicals that relate to plastic encourages the development of detailed knowledge about the end-of-life and transport of microplastics that help to design constructive policies and practical solutions for mitigating the problem (Ahmed et al. 2024). In light of the disclosure and advancement of analytical techniques, chemical analysis will still be a fundamental part of multidisciplinary research and paving the way for the amelioration of the issue of atmospheric plastics pollution (Saliu et al. 2022). The calibration curve technique is usually applied to the proper determination of a particular hazardous plastic-associated compounds identified through gas chromatography-mass spectrometry (GC-MS) or liquid chromatography-mass spectrometry (LC-MS). This is a general quantification formula:

$$Concentration\left(C\right)=\frac{PeakArea}{Calibration\ Curve\ Slope\left(S\right)\times Injection\ Volume\left(IV\right)}$$

Where

Concentration (C) is the concentration of the specific plastic-associated compound.
Peak Area (PA) is the area under the compound's chromatographic peak.
The Calibration Curve Slope (S) is the slope of the calibration curve generated using known concentrations of the compound.
Injection Volume (IV) is the volume of the sample injected into the chromatograph.

The calibration curve is integral to the purpose of converting the measured peak area into corresponding concentrations of the substance. The gradient of the calibration curve corresponds to the response factor, which is used for peak area to concentration conversion.

13.9 Challenges and Considerations in Characterization

The characterization of atmospheric plastic pollution goes hand in hand with certain problems and factors to reckon with. The next part deals with the main challenges, namely particle size distribution problems, sampling site options, uncertainty of standardization and methodology, and data interpretation challenges.

13.9.1 Particle Size Distribution Challenges

They arise because of the range of sizes of plastic particles attached to the atmosphere. The traditional sampling methods might preferably collect the larger particle size and, in this way, the smaller microplastics might be neglected. On the other hand, delicate techniques may have a difficult time grasping the large end of the scale. It is crucial to take on this challenge to truly grasp the big picture of atmospheric microplastics (Caputo et al. 2021). Statistical techniques or algorithms can be used as a tool to solve a problem with particle size distribution. The curve fitting of the observed data is an approach that is preferred by applying a statistical distribution model. The normal distribution (Gaussian distribution) is applied in such cases (Elperin et al. 2016). The formula for a normal distribution probability density function (PDF) is as follows:

$$f\left(x*\mu,\sigma\right)=\frac{1}{\sigma\sqrt{2\pi}}e^{-\frac{(x-\mu)^2}{2\sigma^2}} \tag{13.1}$$

Where:

$f(x|\mu,\sigma)$ is the probability density function.
x represents the particle size.
μ is the mean of the distribution.
σ is the standard deviation of the distribution.

13.9.2 Sample Collection Site Considerations

These characterize plastic pollution in the atmosphere and are a must if we are to do so with some accuracy. In an attempt to minimize the time and risk for sample collection, location sampling should be carried out based on different environmental conditions, possible point sources of pollution, and regional variations in plastic distribution (Miller et al. 2021). The site selection process must be done with meticulousness because obviously, divergent streams of plastic will be seen either under urban, industrial, or remote conditions.

13.9.3 Standardization and Methodology Issues

Technique and standardization problems make it complex to perform the studies in a similar fashion which leads to the lack of consistency and comparability between studies. The absence of standardized sampling, analysis, and reporting protocols makes establishing the universal standards quite an arduous task. Contrasting methods may result in disparities in measured levels and make the amalgamation of the collected data more challenging. The scientific world is required to develop multi-step efforts to standardize protocols, which is necessary to achieve the goal of consistent and reliable results (Lusher and Primpke 2023).

13.9.4 Data Interpretation Challenges

These data provide complex results on man-made atmospheric pollution. A multi-disciplinary approach is required to understand how microplastics are detected in an area, their origin, and what the possible effect on the environment is. Moreover, the understanding of the ecological and human health implications of atmospheric microplastics faces lots of complications, as the complete toxicity data, on microplastics, is not available yet (Rahman et al. 2021). The field of atmospheric plastic pollution characterization must overcome these hurdles to be able to catch up. Informed data interpretation, standard research protocols, rigorous methodologies, and meticulous site selection all give strength to the validity and usefulness of research findings. This development process will be a collaborative effort that takes into consideration interdisciplinary approaches to beat these obstacles and demonstrate a better understanding of the complex dynamics involved.

13.10 Future Directions in Characterization Technologies

A new venue for learning with application is provided by the advancement of characterization technologies for airborne plastic contamination. Further directions are discussed in this section concerning technological progress, the integration of data from multiple data sources, and the envisaged changes in environmental law and regulations.

13.10.1 Advancements in Technology

They will be a turning point in enhancing the sensitivity, specificity, and range of instruments utilized to determine plastic pollution in the atmosphere. In the future, real-time monitoring is probably going to be simplified by advancements in automation, miniaturization, and sensor technologies (Chapman et al. 2020). That in turn will facilitate a dynamic perception of the role of microplastics in the atmosphere's dynamics. Also, adoptions of artificial intelligence and machine learning could help in the automation of the examination of huge datasets; hence, microplastics can be easily detected and measured (Fang et al. 2023).

13.10.2 Integration of Data from Various Methods

An additional area of research that is quite worthwhile is the incorporation of diverse data from different methods. Inter-linkage of the diverse methods of characterization which include chemical analysis, aerosol monitoring, and remote sensing will lead to a detailed map of the fate of atmospheric plastic pollutants (Forbes 2019). Integration of data being communicated from different sources enhances the reliability of results, makes cross-validation simpler, and helps in creating an all-inclusive assessment of the dispersal, origins, and impact of microplastics in the air (Chapman et al. 2020).

13.10.3 Implications for Environmental Policies and Regulations

The developments in characterization technologies lead directly to the changes in environmental policies and regulations. We can expect a change in environmental policies and regulations as we come to grips with the amount of atmospheric plastic pollution we are experiencing (Mitrano et al. 2021). Policymakers have the opportunity to design specific actions aimed at avoiding and reducing pollution through the use of microplastic sources data, concentrations, and environmental effects (Kumar et al. 2021). That is absolutely nothing short of banning plastic production, perfecting the waste disposal procedures, and creating laws focused on specific issues regarding plastic pollution in the atmosphere.

Longitudinal studies are critical for the assessment of temporal dynamics of atmospheric microplastic concentrations since their measurements are repeated over a long term (Zhang et al. 2020). These investigations provide a unique temporal aspect that allows scientists to assess the developing patterns of ongoing microplastic presence. Identifying these time complexities heavily depends on the use of mathematical models, basically time-series analysis or regression analysis (Peng et al. 2006). Trends, fluctuations, and cyclic patterns can be seen in longitudinal studies, but time series analysis breaks down time-ordered data to detect underlying time structures (Rubio-Herrero et al. 2022). Regression models turn out to be very useful instruments for estimating and measuring time trends, for they show the correlation between atmospheric microplastic counts and time. This comprehensive approach can be depicted in the equation for indicating temporal trends, which represents a significant milestone in revealing the story of atmospheric microplastics and has king bearings for environmental comprehension and management (Li et al. 2022). In the real world, it might be other models, if the type of data is more complex and the way it is structured over time.

Linear Regression Model:

$$Y_t = \beta_0 + \beta_1 . X_t + \in_t \tag{13.2}$$

Where

Y_t is the atmospheric microplastic concentration at time t.
X_t is the time variable.
β_0 is the intercept, representing the baseline concentration.
β_1 is the slope, indicating the rate of change in concentration over time.
ϵ_t is the error term.

Estimating the relationship between time and microplastic concentration is made easier by this model, whose significant coefficient β_1 indicates a noteworthy temporal trend.

13.10.4 Time-Series Analysis

Seasonal decomposition of time series (STL), autoregressive integrated moving average (ARIMA) models, and other forecasting techniques can be used for more complex time-series analysis (Günsel 2016). This is a basic formula for an ARIMA (p, d, q) model:

$$Y_t = \phi_1 Y_{t-1} + \phi_2 Y_{t-2} + \phi_p Y_{t-p} + \in_t + \phi_1 \in_{t-1} + \phi_2 \in_{t-2} + \ldots + \phi_q \in_{t-q} \tag{13.3}$$

Where:

$\phi_1,\phi_2,\ldots,\phi_p$ are autoregressive parameters.
$Y_{t-1},Y_{t-2},\ldots,Y_{t-p}$ are residuals.
$\theta_1,\theta_2,\ldots,\theta_q$ are moving average parameters.

The characterization of plastic pollution in the atmosphere will be influenced by data integration, policy implications, and technological innovation. Technology will move the development in the field of technique, giving researchers new weapons for further investigation of microplastics in the atmosphere (Chen et al. 2021; Mallah et al. 2024). The use of various data aggregation methods will lead to a more sophisticated and holistic portrait, which will better equip us to tackle the issues emanating from the atmospheric pollution of plastic. All in all, the gathered knowledge drives the development of environmental laws and policies meant for the effective reduction of the adverse effects of plastics on the environment.

13.11 Conclusion

The pollution of the atmosphere by plastics is considered to be an interdisciplinary problem and hence requires the involvement of researchers and researchers spanning different fields, including natural and social sciences, as well as technical studies. Different researches confirm that microparticles of plastic demonstrate properties that remind us of a musical creation, and take us to interpretations about the delicate secrets of falling plastic. This is the most decisive step in the way of the accounting of moving threads and myriads of agents causing air pollution. In their microplastic shape, they can get to cities, industrial zones, and even remote areas for various objectives stressing the need for the development of sensitive sensor technologies that are now an urgent requirement. Characterization is a vehicle that provides the possibility of implementation of different solutions against plastic-caused air pollution. Both visualization techniques imagery to reveal the unseen and microscope analysis to search for the examined elements- are necessary for understanding the role of atmosphere microplastics. The technique of field sampling combines and utilizes imagery, optical sensor readings, aerosol analysis, and microscopy data, opening the complexity of monitoring aerosol pollution from plastic. Critical strategies including details on particle size distribution analyses, site selection for sampling, materials standardization, and data interpretation are the solid foundation for understanding the problem. Although the opinions about atmospheric plastics pollution are growing, more analyses are relevant and should be done as soon as possible. A capacity for communication between researchers, policymakers, and environmentalists is indispensable in forging new ways and developing practical solutions. Therefore, the evolution issues necessitate the development of technologies for characterization and the collaboration between sectors in carrying out longitudinal studies, converting findings into policies, and spreading educational information. In this collaborative project, the aim is to bring out the cyclone boundary, drawing across the disciplinary boundaries.

References

Ahmed SF, Islam N, Tasannum N, Mehjabin A, Momtahin A, Chowdhury AA, Mofijur M (2024) Microplastic removal and management strategies for wastewater treatment plants. Chemosphere 347:140648

Allen D, Allen S, Abbasi S, Baker A, Bergmann M, Brahney J, Wright S (2022) Microplastics and nanoplastics in the marine-atmosphere environment. Nat Rev Earth Environ 3(6):393–405

Angnunavuri PN, Attiogbe F, Mensah B (2020) Consideration of emerging environmental contaminants in Africa: review of occurrence, formation, fate, and toxicity of plastic particles. Sci Afr 9:e00546

Aransiola SA, Ikhumetse AA, Babaniyi BR, Abioye OP, Oyedele OJ, Falade NO (2023a) Phytoaccumulation of micro and nanoplastics: root uptake. In: Maddela NR, Reddy KV, Ranjit P (eds) Micro and nanoplastics in soil. Springer, Cham. https://doi.org/10.10007/978-3-031-21195-9_8

Aransiola SA, Victor-Ekwebelem MO, Ajiboye AE, Leh-Togi Zobeashia SS, Ijah UJJ, Oyedele OJ (2023b) Ecological impacts and toxicity of micro and nanoplastics in agroecosystem. In: Maddela NR, Reddy KV, Ranjit P (eds) Micro and nanoplastics in soil. Springer, Cham. https://doi.org/10.10007/978-3-031-21195-9_10

Arias-Andres M, Klümper U, Rojas-Jimenez K, Grossart HP (2018) Microplastic pollution increases gene exchange in aquatic ecosystems. Environ Pollut 237:253–261

Auta SH, Abioye OP, Aransiola SA, Bala JD, Chukwuemeka VI, Hassan A, Aziz A, Fauziah SH (2022) Enhanced microbial degradation of PET and PS microplastics under natural conditions in mangrove environment. J Environ Manag 304(2022):114273. https://doi.org/10.1016/j.jenvman.2021.114273

Aves AR (2021) Airborne microplastics in Antarctica and New Zealand. Master of Antarctic Studies, University of Canterbury

Caputo F, Vogel R, Savage J, Vella G, Law A, Della Camera G, Calzolai L (2021) Measuring particle size distribution and mass concentration of nanoplastics and microplastics: addressing some analytical challenges in the sub-micron size range. J Colloid Interface Sci 588:401–417

Chapman J, Truong VK, Elbourne A, Gangadoo S, Cheeseman S, Rajapaksha P, Cozzolino D (2020) Combining chemometrics and sensors: toward new applications in monitoring and environmental analysis. Chem Rev 120(13):6048–6069

Chellasamy G, Ramasundaram S, Veerapandian M, Chandran M, Dhanasekaran B, Oh TH, Yun K (2023) A systematic review on fate and behavior of microplastics towards the environment. TrAC Trends Anal Chem 169:117390

Chen G, Fu Z, Yang H, Wang J (2020) An overview of analytical methods for detecting microplastics in the atmosphere. TrAC Trends Anal Chem 130:115981
Chen J, Wang W, Liu H, Xu X, Xia J (2021) A review of the occurrence, distribution, characteristics, and analysis methods of microplastic pollution in ecosystems. Environ Pollut Bioavailab 33(1):227–246
Chi M, Plaza A, Benediktsson JA, Sun Z, Shen J, Zhu Y (2016) Big data for remote sensing: challenges and opportunities. Proc IEEE 104(11):2207–2219
Chun S, Muthu M, Gopal J (2022) Mass spectrometry as an analytical tool for detection of microplastics in the environment. Chemosensors 10(12):530
Clough R, Fisher A, Gibson B, Russell B (2023) Atomic spectrometry update: review of advances in the analysis of metals, chemicals, and materials. J Anal At Spectrom 37:2207–2281
Delgado-Gallardo J, Sullivan GL, Esteban P, Wang Z, Arar O, Li Z, Sarp S (2021) From sampling to analysis: a critical review of techniques used in the detection of micro-and nanoplastics in aquatic environments. ACS ES&T Water 1(4):748–764
Dong H, Wang X, Xu L, Ding J, Wania F (2023) A flow-through passive sampler for microplastics in air. Environ Sci Technol 57(6):2362–2370
Elperin T, Fominykh A, Krasovitov B (2016) Effect of raindrop size distribution on scavenging of aerosol particles from Gaussian air pollution plumes and puffs in turbulent atmosphere. Process Saf Environ Prot 102:303–315
Evangeliou N, Tichý O, Eckhardt S, Zwaaftink CG, Brahney J (2022) Sources and the fate of atmospheric microplastics revealed from inverse and dispersion modeling: from global emissions to deposition. J Hazard Mater 432:128585
Fang C, Luo Y, Naidu R (2023) Microplastics and nanoplastics analysis: options, imaging, advancements and challenges. TrAC Trends Anal Chem 166:117158
Forbes P (2019) Atmospheric chemistry analysis: a review. Anal Chem 92(1):455–472
Gaston E, Woo M, Steele C, Sukumaran S, Anderson S (2020) Microplastics differ between indoor and outdoor air masses: insights from multiple microscopy methodologies. Appl Spectrosc 74(9):1079–1098
Günsel CB (2016) Environmental problems in Northern Cyprus: a descriptive statistical analysis and forecasting of air pollution and solid waste. Master's thesis, Middle East Technical University
Häder DP, Banaszak AT, Villafañe VE, Narvarte MA, González RA, Helbling EW (2020) Anthropogenic pollution of aquatic ecosystems: emerging problems with global implications. Sci Total Environ 713:136586
Hameed M, Bhat RA, Singh DV, Mehmood MA (2020) White pollution: a hazard to the environment and sustainable approach to its management. In: Innovative waste management technologies for sustainable development. IGI Global, pp 52–81
Hu C (2021) Remote detection of marine debris using satellite observations in the visible and near-infrared spectral range: challenges and potentials. Remote Sens Environ 259:112414
Ivleva NP (2021) Chemical analysis of microplastics and nanoplastics: challenges, advanced methods, and perspectives. Chem Rev 121(19):11886–11936
Khoaele KK, Gbadeyan OJ, Chunilall V, Sithole B (2023) The devastation of waste plastic on the environment and remediation processes: a critical review. Sustain For 15(6):5233
Kumar R, Verma A, Shome A, Sinha R, Sinha S, Jha PK et al (2021) Impacts of plastic pollution on ecosystem services, sustainable development goals, and the need to focus on circular economy and policy interventions. Sustain For 13(17):9963
Levermore JM, Smith TE, Kelly FJ, Wright SL (2020) Detection of microplastics in ambient particulate matter using Raman spectral imaging and chemometric analysis. Anal Chem 92(13):8732–8740
Li M, Wang Y, Xue H, Wu L, Wang Y, Wang C, Han N (2022) Scientometric analysis and scientific trends on microplastic research. Chemosphere 304:135337
Luo D, Chu X, Wu Y, Wang Z, Liao Z, Ji X, Shang X (2024) Micro-and nano-plastics in the atmosphere: a review of occurrence, properties, and human health risks. J Hazard Mater 465:133412

Lusher AL, Primpke S (2023) Finding the balance between research and monitoring: when are methods good enough to understand plastic pollution? Environ Sci Technol 57(15):6033–6039

Mallah MA, Yu Z, Qu Z, Ali M, Liu N (2024) Recognition and detection technology for microplastic, its source, and health effects. Environ Sci Pollut Res 31(8):1–25

Miller E, Sedlak M, Lin D, Box C, Holleman C, Rochman CM, Sutton R (2021) Recommended best practices for collecting, analyzing, and reporting microplastics in environmental media: lessons learned from comprehensive monitoring of San Francisco Bay. J Hazard Mater 409:124770

Mitrano DM, Wick P, Nowack B (2021) Placing nanoplastics in the context of global plastic pollution. Nat Nanotechnol 16(5):491–500

Mohtar AAA, Latif MT, Baharudin NH, Ahamad F, Chung JX, Othman M, Juneng L (2018) Variation of major air pollutants in different seasonal conditions in an urban environment in Malaysia. Geosci Lett 5(1):1–13

Mukonza SS, Chiang JL (2022) Satellite sensors as an emerging technique for monitoring macro- and microplastics in aquatic ecosystems. Water Emerg Contam Nanoplastics 1(4):17

Musa IO, Auta HS, Ilyasu US, Aransiola SA, Makun HA, Adabara NU, Abioye OP, Aziz A, Jayanthi B, Maddela NR, Prasad R (2024) Micro and nanoplastics in environment—degradation, detection, ecological impact. Int J Environ Res 18:1. https://doi.org/10.1007/s41742-023-00551-9

Nguyen B, Claveau-Mallet D, Hernandez LM, Xu EG, Farner JM, Tufenkji N (2019) Separation and analysis of microplastics and nanoplastics in complex environmental samples. Acc Chem Res 52(4):858–866

Peñalver R, Costa-Gómez I, Arroyo-Manzanares N, Moreno JM, López-García I, Moreno-Grau S, Córdoba MH (2021) Assessing the level of airborne polystyrene microplastics using thermogravimetry-mass spectrometry: results for an agricultural area. Sci Total Environ 787:147656

Peng RD, Dominici F, Louis TA (2006) Model choice in time series studies of air pollution and mortality. J R Stat Soc Ser A Stat Soc 169(2):179–203

Prata JC, Castro JL, da Costa JP, Duarte AC, Rocha-Santos T, Cerqueira M (2020) The importance of contamination control in airborne fibers and microplastic sampling: experiences from indoor and outdoor air sampling in Aveiro, Portugal. Mar Pollut Bull 159:111522

Rahman A, Sarkar A, Yadav OP, Achari G, Slobodnik J (2021) Potential human health risks due to environmental exposure to nano-and microplastics and knowledge gaps: a scoping review. Sci Total Environ 757:143872

Rao W, Fan Y, Li H, Qian X, Liu T (2024) New insights into the long-term dynamics and deposition-suspension distribution of atmospheric microplastics in an urban area. J Hazard Mater 463:132860

Razeghi N, Hamidian AH, Wu C, Zhang Y, Yang M (2021) Scientific studies on microplastic pollution in Iran: an in-depth review of the published articles. Mar Pollut Bull 162:111901

Rubio-Herrero J, Marrero CO, Fan WTL (2022) Modeling atmospheric data and identifying dynamics temporal data-driven modeling of air pollutants. J Clean Prod 333:129863

Sakti AD, Sembiring E, Rohayani P, Fauzan KN, Anggraini TS, Santoso C et al (2023) Identification of illegally dumped plastic waste in a highly polluted river in Indonesia using Sentinel-2 satellite imagery. Sci Rep 13(1):5039

Salgado-Hernanz PM, Bauzà J, Alomar C, Compa M, Romero L, Deudero S (2021) Assessment of marine litter through remote sensing: recent approaches and future goals. Mar Pollut Bull 168:112347

Saliu F, Biale G, Raguso C, La Nasa J, Degano I, Seveso D, Modugno F (2022) Detection of plastic particles in marine sponges by a combined infrared micro-spectroscopy and pyrolysis-gas chromatography-mass spectrometry approach. Sci Total Environ 819:152965

Samanta P, Dey S, Kundu D, Dutta D, Jambulkar R, Mishra R et al (2022) An insight on sampling, identification, quantification, and characteristics of microplastics in solid wastes. Trends Environ Anal Chem 36:e00181

Shen M, Song B, Zeng G, Zhang Y, Huang W, Wen X, Tang W (2020) Are biodegradable plastics a promising solution to solve global plastic pollution? Environ Pollut 263:114469

Sridharan S, Kumar M, Singh L, Bolan NS, Saha M (2021) Microplastics as an emerging source of particulate air pollution: a critical review. J Hazard Mater 418:126245

Topouzelis K, Papakonstantinou A, Garaba SP (2019) Detection of floating plastics from satellite and unmanned aerial systems (Plastic Litter Project 2018). Int J Appl Earth Obs Geoinf 79:175–183

Veerasingam S, Ranjani M, Venkatachalapathy R, Bagaev A, Mukhanov V, Litvinyuk D et al (2021) Contributions of Fourier transform infrared spectroscopy in microplastic pollution research: a review. Crit Rev Environ Sci Technol 51(22):2681–2743

Veettil BK, Quan NH, Hauser LT, Van DD, Quang NX (2022) Coastal and marine plastic litter monitoring using remote sensing: a review. Estuar Coast Shelf Sci 279:108160

Williams M, Gower R, Green J, Whitebread E, Lenkiewicz Z, Schröder P (2019) No time to waste: tackling the plastic pollution crisis before it's too late. Tearfund, London

Winter DDN, Koger SM (2004) The psychology of environmental problems. Psychology Press

Wright SL, Gouin T, Koelmans AA, Scheuermann L (2021) Development of screening criteria for microplastic particles in air and atmospheric deposition: critical review and applicability towards assessing human exposure. Microplast Nanoplast 1(1):1–18

Xie L, Luo S, Liu Y, Ruan X, Gong K, Ge Q et al (2023) Automatic identification of individual nanoplastics by Raman spectroscopy based on machine learning. Environ Sci Technol 57(46):18203–18214

Yao Y, Glamoclija M, Murphy A, Gao Y (2022) Characterization of microplastics in indoor and ambient air in northern New Jersey. Environ Res 207:112142

Zhang Y, Kang S, Allen S, Allen D, Gao T, Sillanpää M (2020) Atmospheric microplastics: a review on current status and perspectives. Earth Sci Rev 203:103118

Chapter 14
Policies and Regulations of White Pollution in the Environment

Gabriel Gbenga Babaniyi, Olusegun Opeyemi Oni, Ulelu Jessica Akor, and Oluwatosin Emmanuel Daramola

Abstract The level of consciousness among Nigerians concerning the environmental repercussions of their actions remains deficient, whether intentional or unintentional. The constitution asserts its supremacy, rendering inconsistent laws null and void. In spite of the National Environmental Standards and Regulations Enforcement Agency (NESREA) having comprehensive regulations across various sectors, their effectiveness in environmental management is impeded by inadequate awareness and enforcement. Developing nations grapple with mounting pollution challenges, yet technological advancements, such as innovative recycling methods, provide optimism for a sustainable future. In the midst of all this ecological dilemma, a collective endeavor is essential to navigate regulatory complexities and pave a sustainable path forward.

Keywords Single-use plastics (SUPs) · Pollution policies · Pollution

14.1 Introduction

White pollution, often disguised as disposable plastics and styrofoam, poses a significant threat to ecological balance. Its diverse range, from single-use containers to lightweight packaging materials, makes creating a unified regulatory framework an immense challenge. Knoblauch and Mederake (2021) highlight the inherent complexity of white pollution, with sudden spikes in environmental impact

G. G. Babaniyi (✉) · O. O. Oni
Department of Agricultural Development and Management, Agricultural and Rural Management Training Institute (ARMTI), Ilorin, Nigeria

U. J. Akor
Department of Animal Production, University of Ilorin, Ilorin, Nigeria

O. E. Daramola
Department of Chemistry and Biochemistry, Texas Tech University, Lubbock, TX, USA

S. A. Aransiola et al. (eds.), *White Pollution: Biodiversity and hazards in marine plastisphere*, Environmental Contamination Remediation and Management,
https://doi.org/10.1007/978-3-031-95547-1_14

complicating the development of policies to address its multifaceted dimensions. Navigating the global maze of white pollution regulations adds another layer of complexity, as the absence of an international framework leaves nations entangled in a web of disparate policies, hindering collaborative efforts to combat this pervasive environmental threat (Adam et al. 2020). The unpredictable nature of white pollution, marked by discrete upsurges in various areas, necessitates a coordinated global response. However, the complex nature of international relations hinders the formation of a cohesive front against this elusive foe (Vesilind et al. 2013).

Nevertheless, single-use plastics (SUPs) and other plastic pollution have infiltrated every society worldwide. The majority of plastics are SUPs, designed for one-time use before disposal (Gong and Xie 2020). Examples of SUPs include straws, cutlery, plastic bags, microbeads, polystyrene food containers, and sachet water wrappers. While the impact of SUPs is evident on land, the maritime environment is a significant repository. SUPs constitute approximately 80% of the plastic debris reaching the sea floor or shorelines, posing a threat to marine life (Ye et al. 2020). An estimated 4.8–12.7 million metric tons of plastic enter the oceans annually, equivalent to a truckload of plastic entering the oceans every minute. Most marine plastic pollution originates from poorly managed land-based waste systems, accounting for about 80% of plastic debris entering seas (Naidu et al. 2021). In addition to their detrimental aesthetic impact, SUPs pose significant health risks. They obstruct stormwater drains, leading to flooding, property damage, and even fatalities. Fisheries and coastal tourism suffer adverse effects from SUP plastic marine contamination (Gong and Xie 2020).

Yuan and Zhang (2020) note that governments worldwide have adopted various policy approaches to mitigate SUP pollution. While the global north initiated these legislative actions, African nations are now enacting laws to address SUP pollution more comprehensively than ever before. African nations are recognized for having the world's strongest and most stringent anti-plastic laws, creating complexity when enforcing white pollution restrictions (Yin and Yang 2020). The sporadic spikes in pollution events further strain the already overburdened resources of environmental agencies, creating a dilemma for enforcement. The absence of regulated surveillance systems and the covert nature of illicit disposal activities compound the difficulties authorities face in containing the white pollution crisis (Yuan and Zhang 2020).

14.1.1 Control Measure for Environmental Pollution

Developing nations grapple with complex, severe, and swiftly expanding pollution challenges, while contemporary societies express growing global apprehension about environmental issues. The detrimental impact of foreign businesses operating with scant regard for the local environment compounds the potent amalgamation of industrialization, urban expansion, and mass consumer patterns (Wang et al. 2019). Environmental pollution, having the potential to devastate homes and communities, transcends mere health concerns, becoming a more profound societal preoccupation. The developmental trajectory of developing nations is closely entwined with the complexities surrounding pollution issues. Despite this interconnection, numerous developing nations lack the requisite implementation mechanisms to ensure the effectiveness of regulations or have not put in place measures for controlling environmental pollution (Shen et al. 2020).

Following World War II, Japan underwent a rapid surge in economic growth, marked by unprecedented environmental challenges unparalleled elsewhere. The emphasis on economic expansion and profits at the expense of public health during that era can be linked to these environmental issues, prompting the government to hesitate in embracing environmental policies (Mia et al. 2019). As environmental problems escalated, various advocacy groups emerged, altering public opinion and compelling prefectural and national governments to take decisive action. Through extensive trial and error, practical approaches to tackling environmental contamination were eventually implemented, leading to an amelioration in the environmental condition. By delving into Japan's experiences with the challenges arising from its initial reluctance to address environmental issues and the accomplishments of subsequent environmental initiatives, the goal is to contribute to preventing the deterioration of health problems in developing nations and fostering resilient and health-conscious social development.

The central focus in the post-World War II period was the reconstruction of the nation, with a shift towards economic expansion gaining prominence from the late 1950s onward. Notably, efforts were concentrated on advancing heavy industries and constructing petrochemical facilities (Mia et al. 2019). The substantial pollution emanating from the steel, oil, aluminum, and power industries gave rise to unparalleled health challenges, including ailments like "Itai-Itai" disease, Minamata disease, and pollution-induced asthma. These challenges were exacerbated by the absence of effective pollution mitigation measures. In 1960, Osaka witnessed 156 days marked by haze, and its rivers resembled open sewers. According to da Silva Rocha and Salomão (2019), municipal governments in Japan initiated environmental pollution control measures in the 1950s. The Tokyo Metropolitan Government took the inaugural official step in 1959 with the Factory Pollution Control Ordinance (da Silva Rocha and Salomão 2019). Unfortunately, due to industry resistance and unclear criteria, these restrictions proved largely ineffective. Local governments responded by adopting more stringent policies, establishing emission limits, setting up treatment and processing facilities, monitoring pollution emissions, and providing administrative guidance when necessary (Aragaw 2020). Incentives, such as taxes, were also introduced to motivate industries to embrace pollution control measures for environmental protection. However, Aragaw (2020) highlighted a significant surge in pollutant emissions resulting from a combination of factors, including robust economic growth, rapid industrial development (especially in petrochemical and heavy sectors), and unprecedented urban expansion. The issues escalated as new and distinct contaminants continued to emerge.

14.1.2 Emerging Solutions Amidst Chaos

Amidst the chaotic backdrop of white pollution policies, glimmers of hope emerge through innovative solutions. Collaborative efforts between governments, industries, and environmental advocacy groups showcase the potential for holistic approaches. Technological innovations, such as advanced recycling processes and sustainable materials, offer a tantalizing glimpse into a future may be tamed through concerted action.

14.1.3 Definition of Environmental Pollution

In accordance with the Basic Law for Environmental Pollution Control, environmental pollution encompasses any activities, whether conducted by corporations or individuals, that present a danger to the environment or the welfare of others in a specific area, given the existence of a demonstrable causal link (Waldron 2021; Truby et al. 2022). As outlined by Kumar et al. (2021) and Waldron (2021), the environment is susceptible to seven discernible forms of pollution.

***Atmospheric Pollution*:** The pollutants include nitrogen dioxide and sulfur dioxide, in addition to smoke, dust, and emissions from exhaust. Their existence gives rise to respiratory conditions like bronchitis and asthma. "Yokkaichi Asthma" is recognized as a variant of photochemical smog.

***Water pollution*:** Water contaminants originate from polluted wastewater, waste fluids such as petroleum, sludge, household sewage, sewage discharges, general refuse, and agricultural chemicals. The main outcomes involve disagreeable odors and occurrences of poisoning, such as PCB poisoning, Minamata disease, and "Itai-Itai" disease linked to cadmium poisoning.

***Soil pollution*:** Poisoning is induced by substances such as arsenic and heavy metals, with a particular focus on pesticides used in agriculture.

***Noise*:** Contaminants originating from industrial facilities, construction endeavors, vehicular and railway transportation, nocturnal business operations, and promotional activities. The chief symptoms encompass headaches, insomnia, despondency, auditory impairment, and stunted growth.

***Vibration*:** Contaminants emitted by factories, construction sites, cars, trains, and airplanes, with a predominant symptom being dizziness, unease, and structural harm to residences.

***Ground subsidence*:** The generation of these pollutants results from activities like coal extraction, gravel excavation, and the emergence of groundwater. A key symptom associated with them is the deterioration of structural integrity in buildings.

***Noxious odors*:** These encompass pollutants emanating from livestock farms, amassed sewage, sanitation installations, river pollution, vehicular emissions, and various origins. The primary manifestations involve sensations of pain and headaches.

14.2 Environmental Regulations and Policies

Environmental regulation involves imposing restrictions or obligations on individuals, corporations, and other entities to prevent environmental harm or improve degraded ecosystems. Conversely, environmental policy aims to govern the interaction between humans and the natural environment for mutual benefit. Throughout history, it has addressed issues such as pollution control, waste management, and habitat preservation. In Nigeria, the basis of environmental policy can be traced back to the 1999 Constitution, which grants the state the authority, outlined in section 20, to protect and enhance the environment, safeguarding Nigeria's water, air, land, forests, and wildlife (Bueren 2023).

The most impactful environmental legislations in the United States include the Clean Air Act, the Endangered Species Act, the Montreal Protocol, the Clean Water Act, and Reformation Plan No. 3 of 1970. These legal frameworks have led to significant improvements in the health of Americans and the overall state of their environment (McKittrick and Kitchin 2009). Environmental policy employs three fundamental strategies: litigation, command and control, and economic policy tools, each of which can be analyzed and differentiated. The regulations, particularly the Clean Air Act, the Endangered Species Act, the Montreal Protocol, the Clean Water Act, and Reformation Plan No. 3 of 1970, are widely recognized as highly effective environmental mandates in the United States. They have played a pivotal role in substantially enhancing the well-being of Americans and the environmental quality of their surroundings. The domain of environmental policy encompasses three crucial approaches: legal actions, command and control mechanisms, and economic policy tools, all of which warrant examination and comparison (Vesilind et al. 2013).

In the lawsuits approach, individuals affected by external costs from pollution can initiate legal proceedings against polluters, addressing individual cases. A benefit is the accountability it imposes on polluters for specific damages, but a drawback is the potential for a slow and costly process. Command and control involve regulatory measures where specific limits and standards are imposed on polluters (Abdullah 1995). The advantage lies in the clarity of regulatory expectations, but the disadvantage includes potential inflexibility and high compliance costs for businesses. Economic policy tools leverage economic incentives, such as taxes or tradable permits, to influence environmental behavior (Peirce et al. 1998). The advantage is the potential for cost-effective solutions, but the downside involves the complexity of implementing and monitoring such systems. The (environmental) policy cycle aligns with W. Deming's PDCA cycle (Plan, Do, Check, Act), reflecting a continuous improvement process. This cycle involves repeated iterations to enhance environmental policies over time.

14.2.1 Production Processes' Impact on the Environment

Throughout their creation, utilization, or disposal, organic compounds possess the potential to permeate the atmosphere, water systems, and soil. When assessing the environmental repercussions of production, two pivotal factors come into play: the pollution stemming from energy generation for production and the intrinsic pollution inherent in the manufacturing process itself. The environmental impact is contingent on variables such as the quantity of the chemical released, the specific type and concentration of the chemical, and its spatial distribution. Some chemicals pose a risk even in the absence of an immediate and observable impact, giving rise to concerns about their potential integration into the food chain, leading to accumulation and persistence in the environment over extended periods, potentially spanning decades (Oláh et al. 2020).

The overarching category of "environmental regulations" can be deconstructed into various specialized regulatory domains. While a universally agreed-upon taxonomy is lacking, foundational environmental regulations primarily address the ecological ramifications of producing and utilizing organic chemicals. Parallel regulatory frameworks, distinct yet influenced by environmental legal principles, are dedicated to overseeing specific natural resources such as forests, minerals, or fisheries (Brough and Jouhara 2020). An alternative approach to mitigate environmental impact involves the adoption of green manufacturing practices during production and operations. Strategies like lean manufacturing, six sigma, and total quality management (TQM) are deployed to eliminate waste, defects, and inefficiencies in the manufacturing process (Norström et al. 2020; Bajpai et al. 2020). Certain elements, such as environmental impact assessment, may not neatly fit into either category but remain integral components of environmental law. Environmental impact assessment (EA) involves scrutinizing the environmental consequences, both positive and negative, linked to a plan, policy, program, or project before deciding to proceed with the proposed action. The term "environmental impact assessment" (EIA) is commonly used for specific projects, while "strategic environmental assessment" is pertinent to policies, plans, and programs. Environmental assessments may adhere to administrative procedure rules concerning public involvement and decision-making documentation, potentially subject to judicial review.

The organic chemicals industry stands as one of the largest sectors in the United States, eliciting heightened concerns in neighboring communities due to potential environmental hazards. Production facilities in this industry are frequently recognized as significant sources of pollutants in their proximity, subject to regulation under environmental laws governing air (the atmosphere), water (the aquasphere), and land (the terrestrial biosphere) (Oláh et al. 2020). Consequently, the organic chemicals industry, whether justly or unjustly without passing judgment, is perceived as a substantial contributor to pollution in its operational areas, falling under the purview of various environmental regulations.

14.2.2 Air Pollution

Air quality laws play a pivotal role in overseeing the release of air pollutants into the atmosphere, with a specialized subset focusing on maintaining air quality within structures. The primary objective of air quality laws often centers on safeguarding human health by imposing restrictions or elimination of concentrations of airborne pollutants. Additionally, certain initiatives address broader ecological concerns, including constraints on substances impacting the ozone layer and the implementation of emissions trading programs to combat acid rain or address climate change. These regulatory measures involve identifying and categorizing air pollutants, establishing permissible emission levels, and prescribing necessary mitigation technologies (Norström et al. 2020).

Organic chemicals, a significant origin of perilous air pollutants, contribute substances such as BTEX compounds (benzene, toluene, ethylbenzene, and xylene isomers). They also play a crucial role in emitting criteria air pollutants like particulate matter (PM), nitrogen oxides (NOx), carbon monoxide (CO), hydrogen sulfide (H_2S), and sulfur oxides (SOx). Furthermore, the regulation of air quality necessitates identifying substances and energies qualifying as "pollution" for effective control (Zhang et al. 2022). While specific terminology may differ across jurisdictions, substantial consensus exists among many governments regarding the definition of air pollution. For instance, the United States Clean Air Act designates ozone, particulate matter, carbon monoxide, nitrogen oxides (NOx), sulfur dioxide (SO_2), and lead (Pb) as "criteria" pollutants requiring nationwide regulation (Oláh et al. 2020). The EPA has identified over 180 compounds classified as "hazardous" pollutants, subject to stringent control (Chacon 2019). Some compounds are acknowledged as air pollutants due to their harmful impact on the environment (e.g., CFCs causing ozone depletion) or human health (e.g., asbestos in indoor air). A more comprehensive perspective on air pollution may encompass factors like noise, light, and radiation (Kumari 2023).

Controversy has recently emerged in the United States regarding the categorization of carbon dioxide (CO_2) and other greenhouse gases as air pollutants. Currently, the production of organic chemicals, including refinery operations, releases fewer toxic hydrocarbons compared to previous decades. Emissions in the industry arise from various sources, including equipment leaks, high-temperature combustion processes for electricity generation, heating of steam and process fluids, and product transfer. These pollutants are typically released into the environment over the course of a year through routine emissions, fugitive releases, accidental releases, or plant upsets. The combination of volatile hydrocarbons and nitrogen oxides significantly contributes to the formation of ozone, presenting one of the most pressing challenges in air pollution (Chacon 2019).

14.2.3 Water Contamination

Water pollution pertains to the introduction of substances into water sources, rendering the water unsuitable for various activities such as drinking, cooking, cleaning, swimming, and more. The spectrum of pollutants encompasses chemicals, debris, bacteria, and parasites, with all forms of pollution ultimately infiltrating water (Ukaogo et al. 2020; Auta et al. 2022). This contamination not only impacts human activities but also disrupts ecosystems, giving rise to phenomena like eutrophication, leading to the decline of fish and other aquatic organisms. The existence of toxic elements dissolved in water poses a hazard to humans when consumed through fish or other aquatic organisms. Water quality laws play a pivotal role in overseeing the release of pollutants into diverse water resources, including surface water, groundwater, and stored drinking water (Walker et al. 2019; Singh et al. 2020).

Certain water quality laws, particularly those governing drinking water, primarily concentrate on preserving human health. Others broaden their scope to include constraints on altering the chemical, physical, radiological, and biological attributes of water resources, signifying a broader commitment to safeguarding aquatic ecosystems (Walker et al. 2019; Babaniyi et al. 2022; Aransiola et al. 2023a; Musa et al. 2024). Regulatory endeavors encompass the identification and categorization of water pollutants, the establishment of acceptable pollutant concentrations in water resources, and the imposition of limits on pollutant discharges from diverse sources, encompassing sewage treatment and disposal, industrial waste, and agricultural wastewater management. Additionally, control measures extend to managing surface runoff from construction sites and urban environments. Production facilities themselves pose potential threats to both groundwater and surface water, as some utilize deep-injection wells to dispose of wastewater, potentially leading to aquifer and groundwater contamination (Shen et al. 2020). These practices fall under the regulatory framework of the Safe Drinking Water Act (SDWA).

Industrial wastewater, originating from diverse processes like desalting, cooling tower water, stormwater, distillation, or cracking, undergoes rigorous treatment processes, including wastewater treatment plants, before being discharged into surface waters under government-issued permits (Ismail et al. 2019; Ogunlade et al. 2021). These discharges into surface waters are subject to state discharge regulations and governed by the Clean Water Act (CWA). The guidelines established under the CWA set limits on sulfides, ammonia, suspended solids, and other compounds present in wastewater (Shen et al. 2020). Despite these guidelines, historical contamination from prior discharges may persist in surface water bodies.

14.2.4 *Wastewater*

The domain linked with organic chemicals, encompassing the fossil fuel sector, encounters numerous challenges related to wastewater, particularly concerning the presence of chemicals in waste process waters. Persistent industry initiatives are dedicated to addressing potential water contamination resulting from inadvertent petroleum leaks or the discharge of polluted water from diverse processes (Babaniyi et al. 2023b). Sustained efforts are made to mitigate such incidents. In addition to monitoring organic compounds in water, there is an ongoing need to assess the concentration of metals. This is crucial as heavy metals tend to accumulate in the tissues of fish and animals, with concentrations increasing up the food chain (Babaniyi et al. 2023a). Municipal sewage treatment facilities, regardless of their size, encounter common sewage challenges. Essential treatment processes involving solid settling and removal (primary treatment) are mandatory, while the adoption of secondary treatment methods, employing bacteria and aeration to enhance organic degradation, is becoming more prevalent (Ogunlade et al. 2021; Babaniyi et al. 2023a). Chemical production companies are progressively integrating tertiary treatment measures, such as filtration through activated carbon, applications of ozone, and chlorination, to further enhance their wastewater treatment processes.

Wastewater pre-treaters, releasing water into sewer systems, confront new regulatory requirements. Stringent pollutant standards have been established for sewage sludge, requiring identification of toxic substances in water and the development of plans to address any associated issues (Obotey Ezugbe and Rathilal 2020; Khan et al. 2021). Regulators consistently establish and refine water quality standards for priority toxic pollutants, reflecting an ongoing commitment to safeguarding water resources and mitigating potential environmental impacts (Jain et al. 2021; Aransiola et al. 2023b). This collective effort by industries and regulators underscores the significance of responsible wastewater management practices to ensure the preservation of water quality and the well-being of ecosystems.

14.3 Environmental Regulations and Policies in Nigeria

The inception of environmental protection legislation in Nigeria can be traced back to the colonial era, marked by the introduction of various ordinances and laws addressing environmental issues. However, it wasn't until the late 1980s and early 1990s that Nigeria significantly escalated efforts to tackle environmental challenges. The creation of the Federal Ministry of Environment in 1988 marked a crucial development, playing a role in coordinating environmental policies and regulating activities contributing to environmental degradation. Subsequently, in 2007, the National Environmental Standards and Regulations Enforcement Agency (NESREA) was established to enforce environmental standards and regulations (Suleiman et al. 2019). The year 1989 witnessed the enactment of the National

Environmental Protection (NEP) Act, a pivotal legislation aiming to protect, preserve, and enhance the environment, as well as prevent and control pollution. This Act laid the foundation for environmental management, including the regulation of hazardous waste and the implementation of environmental impact assessment procedures (Moses et al. 2019). The Oil Pipeline Act of 1991 further emerged to regulate the construction and operation of oil pipelines, addressing issues like oil spillage and pipeline explosions. Specific laws, such as the Forestry Act and Wildlife Act, have also been enacted to safeguard distinct facets of the environment. Despite these legislative measures, Nigeria grapples with persistent environmental challenges, encompassing deforestation, desertification, soil erosion, air and water pollution, and oil spills. Ongoing efforts involve the enforcement of environmental laws, heightened awareness campaigns, and the implementation of sustainable environmental management practices by the government and other stakeholders (Nwankwo et al. 2020).

Environmental regulations and policies in Nigeria face uncertainty and volatility due to various challenges, including a history of environmental degradation involving oil spills, deforestation, desertification, and pollution. These issues pose substantial threats to public health and the environment (Aidonojie et al. 2020). Consequently, the Nigerian government has instituted numerous policies and regulations to tackle these problems. The Federal Ministry of Environment plays a central role in formulating and implementing environmental regulations and policies. Notably, the National Environmental Standards and Regulations Enforcement Agency (NESREA) Act of 2007 holds significance as a key environmental law. It serves to regulate and enforce environmental standards and regulations in the country (Suleiman et al. 2019). NESREA's responsibilities include monitoring and ensuring environmental compliance in industries, particularly in sectors like oil, gas, and mining, to integrate environmentally responsible practices into their operations (Efobi et al. 2019; Aidonojie et al. 2020). The National Environmental Protection (NEP) Act of 1988 is another influential policy, establishing a comprehensive framework for environmental management in Nigeria. It addresses various concerns, including pollution, waste management, and the protection of natural resources. Despite these regulatory efforts, Nigeria continues to grapple with challenges like inadequate waste management, soil erosion, and deforestation. Moreover, the effectiveness of policies is hampered when there is a lack of resources for their enforcement, resulting in volatility (Atanda 2019).

To overcome these challenges, Nigeria must enhance the implementation of environmental policies and regulations, strengthen waste management systems, and elevate public awareness of environmental impact (Amechi and Oni 2019; Egbetokun et al. 2020). Effective enforcement requires meticulous monitoring and substantial allocation of resources to environmental regulation agencies. Fostering environmentally conscious lifestyles and increasing public awareness about the importance of environmental protection and the negative consequences of indiscriminate waste disposal can reduce uncertainty. For Nigeria to effectively manage its environmental challenges, sustained commitment from the government, active participation from citizens and the private sector, and well-coordinated, regularly reviewed, and forward-looking environmental protection strategies are essential.

14.3.1 Nigeria's Environmental Regulations: An Evolution

In response to environmental challenges, the Nigerian government has instituted a series of regulations aimed at addressing and mitigating the impact of these issues. Noteworthy legislations include the Forestry Act of 1958, Antiquities Act of 1958, Territorial Waters Decree of 1967, Oil in Navigable Waters Decree of 1968, Petroleum Decree of 1969, Quarries Decree of 1969, Sea Fisheries Decree of 1971, Land Use Decree of 1978, National Environmental Protection (Management of Solid and Hazardous Waste) Regulations of 1991, Guidelines and Standards for Environmental Pollution Control in Nigeria of 1991, Environmental Impact Assessment Decree of 1992, Water Resources Act 101 of 1993, and Minerals and Mining Act of 1999. Nigeria has also committed to various international conventions related to the environment, including the Convention on Biological Diversity of 1992, the Montreal Protocol on Substances that Deplete the Ozone Layer, the United Nations Framework Convention on Climate Change (1992), and the Stockholm Convention on Persistent Organic Pollutants (POPs), among others (Aina 1989).

The drive for stringent environmental regulations in Nigeria gained momentum following a national outcry in 1987, triggered by the illicit dumping of toxic waste in Koko, Southwest Nigeria, by an Italian vessel. In response to this incident, the Federal Environmental Protection Agency (FEPA) was established in 1988, prompting each of Nigeria's 36 states to establish their environmental protection agencies known as State Environmental Protection Agencies (SEPAs). These state agencies formulated their respective laws and regulations (Adekunle 1998). With the restoration of democracy in Nigeria in 1998, there was a renewed emphasis on environmental management and protection, leading to the establishment of the Federal Ministry of Environment and State Ministries of Environment at the state levels in 1999. The Federal Ministry of Environment acts as the apex policy organ for all environmental policies in Nigeria, overseeing compliance with regulations by local, state, and federal agencies, as well as companies operating in Nigeria. Various agencies under the ministry are tasked with enforcing regulations governing diverse human activities impacting the environment (Agabi 1995).

Key agencies responsible for enforcing environmental regulations include the National Environmental Standards and Regulations Enforcement Agency (NESREA), established in 2007 to replace FEPA, and the National Oil Spill Detection and Response Agency (NOSDRA), created in 2009 to address environmental damage caused by oil production in the Niger Delta (Mabogunje 1999). This comprehensive regulatory framework underscores Nigeria's commitment to addressing environmental issues both at the national and international levels.

14.3.2 Environmental Legislation and Guidelines

Globally, environmental laws and regulations share a common objective of protecting the environment, human health, and natural resources, although the specifics vary across nations. These legal frameworks encompass a wide range of topics, such as waste management, habitat preservation, air and water quality, and emission reduction. In the Nigerian context, there is no singular comprehensive law, regulation, or dedicated entity exclusively focused on environmental laws and sustainable practices. However, several other laws specifically target environmental regulations and sustainability practices. Some of these include:

Environmental Impact Assessment Act: The legislation commonly referred to as the Environmental Impact Assessment Act primarily centers on the consideration of environmental consequences in both public and commercial undertakings. Specific clauses within the EIA act deal with the prevention of environmental emergencies. Notably, Section 2(1) mandates the evaluation of public or private projects anticipated to have substantial adverse environmental impacts. Section 2(4) necessitates applicants to submit a written application to the Agency before commencing projects, initiating an environmental assessment process to ascertain project approval. Additionally, Section 60 establishes legal responsibility for any breach of the Act's provisions.

Harmful Waste (Special Criminal Provision) Act: The Harmful Waste Act and the Harmful Waste (Special Criminal Provision) Act collectively prohibit the unauthorized transportation, disposal, or depositing of hazardous waste in Nigeria's air, land, or water bodies, emphasizing the need for a valid permit. These stringent measures underscore the nation's commitment to combat environmental harm and enforce responsible waste management practices.

The Constitution of the Federal Republic of Nigeria: The revised Chapter II of the Nigerian Constitution mandates that officials in the legislative, executive, and judicial branches must adhere to and implement the ESG directive principles outlined in Chapter II. For instance, Section 20 outlines the objectives of the Nigerian State, emphasizing the improvement and preservation of Nigeria's air, land, water, forests, and wildlife. Additionally, Section 12 specifies that foreign agreements ratified by the National Assembly should be integrated into Nigerian law, covering those related to environmental concerns.

Climate Change Act: The legal structure addressing Climate Change calls for extensive green growth and the promotion of sustainable economic development, aiming primarily to reduce greenhouse gas emissions.

The Companies and Allied Matters Act (CAMA): CAMA requires directors of Nigerian companies to comply with environmental regulations. Under CAMA, directors have an obligation to assess the potential influence of their company's activities on the immediate local environment where operations are conducted.

14.4 White Pollution

White pollution, often denoting the environmental threat posed by plastic waste, has emerged as a critical global issue, casting a shadow over ecosystems, human well-being, and the delicate equilibrium of nature. The widespread use of plastic materials, valued for their adaptability and convenience, has resulted in a concerning surge in plastic waste, creating an environmental crisis that demands immediate attention and unified action (Shen et al. 2020). At the core of the challenge posed by white pollution lies the extensive utilization and inadequate handling of single-use plastics (Brouwers et al. 2022). These disposable items, ranging from plastic bags and bottles to packaging materials, have become symbolic of convenience-driven consumerism but have inflicted severe harm on the environment. The inherent durability of plastics, intended for longevity, transforms them into persistent pollutants, lingering in the environment for hundreds of years (Díaz-Chirón et al. 2021). Landfills and water bodies bear the brunt of white pollution, as discarded plastics infiltrate these ecosystems, causing havoc on biodiversity and ecosystems. The visible blight of plastic waste, often strewn across landscapes or obstructing waterways, underscores the magnitude of the issue. Beyond aesthetics, the environmental consequences are profound, with wildlife succumbing to ingestion or entanglement in plastic debris (Adam et al. 2020; Brouwers et al. 2022).

The ramifications of white pollution extend beyond the visible aspects, delving into the realm of microplastics. These minute plastic particles, resulting from the disintegration of larger plastic items, permeate soils, water sources (Shen et al. 2020), and even the air we breathe. The pervasive presence of microplastics raises concerns about their potential impact on human health, as these particles can enter the food chain, posing risks that are not yet fully understood. Tackling white pollution requires a comprehensive approach involving legislative measures, public awareness campaigns, and innovative waste management strategies. Governments globally are contending with the challenge, implementing regulations to curb single-use plastics and promote recycling initiatives. Simultaneously, the private sector is exploring alternative materials and sustainable packaging solutions to mitigate the environmental footprint of plastic products (Shen et al. 2020; Huuskonen et al. 2021).

Nevertheless, the fight against white pollution necessitates a shift in consumer behavior. Education and awareness initiatives play a crucial role in fostering responsible consumption patterns, urging individuals to reduce, reuse, and recycle. The plastic industry itself is under pressure to embrace circular economy models, where the lifecycle of plastic products is meticulously managed to minimize environmental impact (Díaz-Chirón et al. 2021; Huuskonen et al. 2021). Ultimately, white pollution epitomizes the unintended consequences of humanity's dependence on plastics. Addressing this environmental crisis requires a coordinated global effort, blending regulatory measures, technological innovation, and a fundamental shift in societal attitudes toward sustainable living. The urgency of tackling white pollution is not merely an environmental imperative; it is a collective responsibility to preserve the planet for present and future generations.

14.5 Conclusion

The enigmatic domain of policies and regulations surrounding white pollution in the environment is a testament to the complex relationship between human activities and the delicate balance of nature. As we grapple with the complexities and bursts inherent in this ecological challenge, a clarion call resonates for a collective effort to disentangle the regulatory quagmire and forge a sustainable path forward. Only through a harmonious symphony of international cooperation, technological innovation, and conscientious policy-making can we hope to navigate the labyrinth and emerge into a world free from white pollution.

References

Abdullah AR (1995) Environmental pollution in Malaysia: trends and prospects. TrAC Trends Anal Chem 14(5):191–198

Adam I, Walker TR, Bezerra JC, Clayton A (2020) Policies to reduce single-use plastic marine pollution in West Africa. Mar Policy 116:103928

Adekunle AD (1998) In search of partners in a context of multiple regulators: the perspective of an operator in Nigeria oil industry. A paper presented at the International seminar on the Petroleum Industry and the Nigerian Environment Abuja Sheraton Hotel & Towers, Africa

Agabi JA (1995) Introduction to the Nigeria environment. In: Nigerian environment, national conservation foundation. Macmillan Nigeria Publishers Limited, Lagos

Aidonojie PA, Anani OA, Agbale OP, Olomukoro JO, Adetunji OC (2020) Environmental law in Nigeria: a review on its antecedence, application, judicial unfairness and prospects. Arch Sci Technol 1(2):212–221

Aina EOA (1989) New direction for sustainable development in Nigeria. A paper delivered at the International Workshop on the Environment and Sustainable Development in Nigeria at the NICONNOGA HILTON Hotel, Abuja, Africa

Amechi EP, Oni BA (2019) Import of electronic waste into Nigeria: the imperative of a regulatory policy shift. Chin J Environ Law 3(2):141–166

Aragaw TA (2020) Surgical face masks as a potential source for microplastic pollution in the COVID-19 scenario. Mar Pollut Bull 159:111517

Aransiola SA, Ikhumetse AA, Babaniyi BR, Abioye OP, Oyedele OJ, Falade NO (2023a) Phytoaccumulation of micro and nanoplastics: root uptake. In: Maddela NR, Reddy KV, Ranjit P (eds) Micro and nanoplastics in soil. Springer, Cham. https://doi.org/10.10007/978-3-031-21195-9_8

Aransiola SA, Victor-Ekwebelem MO, Ajiboye AE, Leh-Togi Zobeashia SS, Ijah UJJ, Oyedele OJ (2023b) Ecological impacts and toxicity of micro and nanoplastics in agroecosystem. In: Maddela NR, Reddy KV, Ranjit P (eds) Micro and nanoplastics in soil. Springer, Cham. https://doi.org/10.10007/978-3-031-21195-9_10

Atanda JO (2019) Green building standards: opportunities for Nigeria. J Clean Prod 227:366–377

Auta SH, Abioye OP, Aransiola SA, Bala JD, Chukwuemeka VI, Hassan A, Aziz A, Fauziah SH (2022) Enhanced microbial degradation of PET and PS microplastics under natural conditions

in mangrove environment. J Environ Manag 304(2022):114273. https://doi.org/10.1016/j.jenvman.2021.114273
Babaniyi GG, Oyemade B, Orija D (2022) Effects of aquatic (freshwater and marine) pollution on microbial enzyme activities. In: Maddela NR, Aransiola SA, Prasad R (eds) Ecological interplays in microbial enzymology. Environmental and microbial biotechnology. Springer, Singapore. https://doi.org/10.1007/978-981-19-0155-3_17
Babaniyi GG, Olagoke OV, Babaniyi BR (2023a) Microbial biomass and activity, enzyme activities, and microbial community composition: long-term effects of aided phytostabilization of trace elements. In: Microbiology for cleaner production and environmental sustainability. CRC Press, pp 301–318
Babaniyi GG, Olagoke OV, Aransiola SA (2023b) 12: extracellular enzymatic activity of bacteria in aquatic ecosystems. In: Microbiology for cleaner production and environmental sustainability. CRC Press, p 277
Bajpai R, Choudhary K, Srivastava A, Sangwan KS, Singh M (2020) Environmental impact assessment of fly ash and silica fume based geopolymer concrete. J Clean Prod 254:120147
Brough D, Jouhara H (2020) The aluminium industry: a review on state-of-the-art technologies, environmental impacts and possibilities for waste heat recovery. Int J Thermofluids 1:100007
Brouwers MG, Bonsor A, Malamud U (2022) A road-map to white dwarf pollution: tidal disruption, eccentric grind-down, and dust accretion. Mon Not R Astron Soc 509(2):2404–2422
Chacon M (2019) Albuquerque air care: vehicle emissions testing Information, city of albuquerque, new mexico. United States of America. retrieved from https://coilink.org/20.500.12592/zv5hc5
da Silva Rocha AB, Salomão GM (2019) Environmental policy regulation and corporate compliance in evolutionary game models with well-mixed and structured populations. Eur J Oper Res 279(2):486–501
Díaz-Chirón L, Negral L, Megido L, Suárez-Peña B, Domínguez-Rodríguez A, Rodríguez S et al (2021) Relationship between exposure to sulphur dioxide air pollution, white cell inflammatory biomarkers and enzymatic infarct size in patients with ST-segment elevation acute coronary syndromes. Eur Cardiol 16:e50
Efobi U, Belmondo T, Orkoh E, Atata SN, Akinyemi O, Beecroft I (2019) Environmental pollution policy of small businesses in Nigeria and Ghana: extent and impact. Environ Sci Pollut Res 26:2882–2897
Egbetokun S, Osabuohien E, Akinbobola T, Onanuga OT, Gershon O, Okafor V (2020) Environmental pollution, economic growth and institutional quality: exploring the nexus in Nigeria. Manag Environ Qual 31(1):18–31
Gong J, Xie P (2020) Research progress in sources, analytical methods, eco-environmental effects, and control measures of microplastics. Chemosphere 254:126790
Huuskonen MT, Liu Q, Lamorie-Foote K, Shkirkova K, Connor M, Patel A et al (2021) Air pollution particulate matter amplifies white matter vascular pathology and demyelination caused by hypoperfusion. Front Immunol 12:785519
Ismail HN, Iqbal A, Nasr L (2019) Employee engagement and job performance in Lebanon: the mediating role of creativity. Int J Prod Perform Manag 68(3):506–523
Jain K, Patel AS, Pardhi VP, Flora SJS (2021) Nanotechnology in wastewater management: a new paradigm towards wastewater treatment. Molecules 26(6):1797
Khan S, Naushad M, Lima EC, Zhang S, Shaheen SM, Rinklebe J (2021) Global soil pollution by toxic elements: current status and future perspectives on the risk assessment and remediation strategies—a review. J Hazard Mater 417:126039
Knoblauch D, Mederake L (2021) Government policies combatting plastic pollution. Curr Opin Toxicol 28:87–96
Kumar R, Verma A, Shome A, Sinha R, Sinha S, Jha PK et al (2021) Impacts of plastic pollution on ecosystem services, sustainable development goals, and need to focus on circular economy and policy interventions. Sustainability 13(17):9963
Kumari DS (2023) The impact of environment related laws on the Indian society. Int J Law Manag Hummanit 6:2064

Mabogunje AL (1999) Nigerian environment in the new millennium. Federal Environmental Protection Agency, Abuja
McKittrick K, Kitchin R (2009) Diaspora. In: Nigel thrift, international encyclopedia of human geography. Elsevier
Mia R, Selim MD, Shamim AM, Chowdhury M, Sultana S, Armin M et al (2019) Review on various types of pollution problem in textile dyeing & printing industries of Bangladesh and recommendation for mitigation. J Text Eng Fash Technol 5(4):220–226
Moses O, Michael EE, Dabel-Moses JN (2019) A review of environmental management and reporting regulations in Nigeria. In: Environmental reporting and management in Africa, pp 159–182. https://doi.org/10.1108/S1479-359820190000008007
Musa IO, Auta HS, Ilyasu US, Aransiola SA, Makun HA, Adabara NU, Abioye OP, Aziz A, Jayanthi B, Maddela NR, Prasad R (2024) Micro and nanoplastics in environment—degradation, detection, ecological impact. Int J Environ Res 18:1. https://doi.org/10.1007/s41742-023-00551-9
Naidu R, Biswas B, Willett IR, Cribb J, Singh BK, Nathanail CP et al (2021) Chemical pollution: a growing peril and potential catastrophic risk to humanity. Environ Int 156:106616
Norström AV, Cvitanovic C, Löf MF, West S, Wyborn C, Balvanera P et al (2020) Principles for knowledge co-production in sustainability research. Nat Sustain 3(3):182–190
Nwankwo W, Olayinka SA, Ukhurebor KE (2020) Green computing policies and regulations: a necessity. Int J Sci Technol Res 9(1):4378–4383
Obotey Ezugbe E, Rathilal S (2020) Membrane technologies in wastewater treatment: a review. Membranes 10(5):89
Ogunlade TM, Babaniyi BR, Afolabi FJ, Babaniyi GG (2021) Physicochemical, heavy metals and microbiological assessment of wastewater in selected abattoirs in Ekiti State, Nigeria. J Environ Treat Tech 9(4):788–795
Oláh J, Aburumman N, Popp J, Khan MA, Haddad H, Kitukutha N (2020) Impact of Industry 4.0 on environmental sustainability. Sustainability 12(11):4674
Peirce JJ, Vesilind PA, Weiner R (1998) Environmental pollution and control. Butterworth-Heinemann
Shen M, Song B, Zeng G, Zhang Y, Huang W, Wen X, Tang W (2020) Are biodegradable plastics a promising solution to solve the global plastic pollution? Environ Pollut 263:114469
Suleiman RM, Raimi MO, Sawyerr OH (2019) A deep dive into the review of national environmental standards and regulations enforcement agency (NESREA) act. Int Res J Appl Sci. ISSN: 2663-5577
Singh N, Tang Y, Zhang Z, Zheng C (2020) COVID-19 waste management: effective and successful measures in Wuhan, China. Resour Conserv Recycl 163105071
Truby J, Brown RD, Dahdal A, Ibrahim I (2022) Blockchain, climate damage, and death: policy interventions to reduce the carbon emissions, mortality, and net-zero implications of non-fungible tokens and Bitcoin. Energy Res Soc Sci 88:102499
van Bueren E (2023, November 21) Environmental policy. Encyclopedia Britannica. https://www.britannica.com/topic/environmental-policy
Vesilind PA, Peirce JJ, Weiner RF (2013) Environmental pollution and control. Elsevier
Ukaogo PO, Ewuzie U, Onwuka CV (2020) Environmental pollution: causes, effects, and the remedies. In: Microorganisms for sustainable environment and health, pp 419–429. Elsevier
Walker M, Head MJ, Lowe J, Berkelhammer M, BjÖrck S, Cheng H, Weiss H (2019) Subdividing the Holocene Series/Epoch: formalization of stages/ages and subseries/subepochs, and designation of GSSPs and auxiliary stratotypes. J Quat Sci 34(3):173–186
Waldron IR (2021) There's something in the water: environmental racism in Indigenous & Black communities. Fernwood Publishing
Wang Y, Sun X, Guo X (2019) Environmental regulation and green productivity growth: empirical evidence on the Porter Hypothesis from OECD industrial sectors. Energy Policy 132:611–619
Ye Z, Yang J, Zhong N, Tu X, Jia J, Wang J (2020) Tackling environmental challenges in pollution controls using artificial intelligence: a review. Sci Total Environ 699:134279

Yin GZ, Yang XM (2020) Biodegradable polymers: a cure for the planet, but a long way to go. J Polym Res 27(2):38
Yuan B, Zhang Y (2020) Flexible environmental policy, technological innovation and sustainable development of China's industry: the moderating effect of environment regulatory enforcement. J Clean Prod 243:118543
Zhang X, Han L, Wei H, Tan X, Zhou W, Li W, Qian Y (2022) Linking urbanization and air quality together: a review and a perspective on the future sustainable urban development. J Clean Prod 346:130988

Chapter 15
Climate Change and White Pollution as a Result of Renewable and Non-renewable Energy Consumption in Growing Economies

Hakan Acaroğlu, Atabak Najafi, and Mustafa Güllü

Abstract Climate change has been an important concern for humanity in recent time. Renewable energy consumption (REC) and non-renewable energy consumption (NREC) have an impact on ecosystem in terms of economics and sustainable future. There is a critical trade off between environmentally friendly energy consumption (i.e., solar, wind, hydrogen consumption.) and the opposite one (i.e., fossil fuels consumption). This chapter presents an insight between the renewable and non-renwable energy by considering white pollution (i.e., plastic production and waste management) as a result of consumption of energy, and its impact on climate change in growing economies. The used methodology is the structural development and historical perspective of energy consumption, climate change, gross domestic product (GDP) per capita and their by-product. The REC has positive impact on environment and white pollution (NREC) the opposite. Therefore, decision makers should be careful on energy consumption behaviour in growing economies. However, there should be a positive approach to maintain a balance in managing nature for safer environment.

Keywords Climate change · White pollution · Renewable energy consumption · Non-renewable energy consumption · GDP per capital

---

H. Acaroğlu (✉)
Department of Economics, Faculty of Economics and Administrative Sciences, Eskisehir Osmangazi University, Eskisehir, Turkey
e-mail: hacaroglu@ogu.edu.tr

A. Najafi
Department of Electrical and Electronics Engineering, İstanbul Esenyurt University, İstanbul, Turkey

M. Güllü
Republic of Turkey Ministry of National Education, Ankara, Turkey

S. A. Aransiola et al. (eds.), *White Pollution: Biodiversity and hazards in marine plastisphere*, Environmental Contamination Remediation and Management,
https://doi.org/10.1007/978-3-031-95547-1_15

15.1 Introduction

Energy is the most important requirement for sustaining human life. However, the main aim of energy supply policies should be to eliminate the environmental threat. Today's understanding of development, has been replaced by the understanding of technology and resource use that can provide energy more efficiently and more economically, instead of energy abundance and consumption. This new perspective has brought about "healthy environment", "energy security" and "energy diversity" policies. Reducing the effects of global warming and climate change, forms the basis of these new policies.

Although climate change is one of the most important environmental problems, it also plays a triggering role in other environmental issues. Contrary to this situation, climate change only entered the international agenda in 1980s, except for some initiatives and conferences on the environment during the Cold War, and generally remained the subject of only technical studies and scientific research. However, scientific developments and the increasing frequency of natural disasters have brought the issue to the international agenda and awareness has begun to increase day by day.

As the significant ecological and environmental damages of global warming, and when climate change entered the world agenda, national and international measures began to be considered. However, the conference organized by the United Nations in 1972 can be considered as the first international step. At this conference, sessions on climate change were held and this issue was brought to the world agenda. The first step in starting this process was the establishment of the "Intergovernmental Panel on Climate Change" in 1988 with the support of the United Nations Environment Program and the World Meteorological Organization (Mercan and Karakaya 2015).

In order to understand the relationship between ecology and energy, it is necessary to have knowledge and practical understanding in four areas and their interactions with each other. These 4 areas are grouped under 2 main headings. The first is the human dimension of the subject, that is, basically sociology and economics; The second is the engineering dimension, that is, energy and production technologies. Unfortunately, scientific studies on this subject without being expert these four areas are as non-functional. Regarding sustainable energy policies by considering white pollution (i.e., plastic production and waste management) as a result of consumption of energy, the necessity of establishing energy management and renewable energy policies to reduce the effects of global warming and climate change is among the main measures to be taken (IPCC 2016).

15.2 Literature on Climate Change, and Energy Consumption

15.2.1 Climate Change, Its Components with Economic Indicators

15.2.1.1 Precipitation and Temperature

When climate change and energy consumption are concerned, the subject can be investigated by sub-titles through literature: components of climate change or precipitation and temperature. Sea levels are rising as glaciers melt in our world. NASA predicts that the Arctic Ocean will remain ice-free in summer by 2050, meaning animals such as arctic foxes, polar bears and reindeer will lose their habitat. It is predicted that sea levels will rise by approximately 30–120 cm by 2100 due to melting glaciers, which will cause more floods and tides. As the temperature of the sea increases, the water will rise even more. In addition, bacteria and viruses that have remained inactive in glaciers and permafrost for centuries may pose a threat again in the warming world due to climate change.

The increase in temperature will be even more effective in countries that are already hot. The number and intensity of droughts and heat waves are expected to increase. As a result, while the incidence of hurricanes and floods is increasing in some parts of the world, long-lasting droughts and desertification are effective in some regions. Acaroğlu and Güllü (2022) examined the effects of renewable energy consumption (REC) and non-renewable energy consumption (NREC) on climate change with the ARDL method and Toda-Yamamoto test from 1980 to 2019 in Turkey, used temperature and precipitation as dependent variables as climate change determinants. They found that the correlation between REC and temperature is negative and a 1% increase in REC degreases temperature by 0.031%. In a similar study, Brini (2021), also examined the relationship between REC, NREC, and climate change in 16 African countries, determined that NREC has a negative impact on climate change in the long-run, while REC has a positive effect. The researcher used precipitation and average temperature data to represent climate change as dependent variables in his model, and used renewable and non-renewable electricity consumption and GDP per capita data as explanatory variables and he determined that REC can help to mitigate climate change in African countries. These studies can be followed through Table 15.1 at below.

15.2.1.2 GDP Per Capita

As countries grow, they should be more careful about environment. Economic growth is closely related with energy consumption. Chen et al. (2021) investigated the effects of climate change on clean energy investments from 1996 to 2018 by using long-term global panel data and instrumental variable quantile regression

Table 15.1 Literature studies for the components of climate change: precipitation and temperature

Author	Country	Method	Period	Subject	Findings
Brini (2021)	16 selected African countries	ARDL-PMG, ARDL-MG, Granger Causality	1980–2014	The relationship between NREC and REC and climate change	NREC has a negative impact on climate change, while REC has a positive effect.
Acaroğlu and Güllü (2022)	Turkey	ARDL and Toda-Yamamoto test	1980–2019	The effects of REC and NREC on climate change	A 1% increase in REC reduces the temperature by 0.031%.

model for 44 countries, found that climate change encourages clean energy investments in countries who has more clean investments. The more society becomes wealthier the more precautions for environment are taken. For instance, Campagnolo and De Cian (2022) evaluated the macroeconomic effects of climate change and carbon tax for Italian households and regions with a recursive-dynamic general equilibrium model, found that, as a result of their exposure to higher temperature, relatively wealthier households reduced their gas and oil fuel expenditures, while lower households increased their electricity expenditures more.

In line with growing economies through energy consumption, energy expenditures are better move from fossil fuels to renewable energy for sustainable future. Because renewable energy sources are critical for struggling with climate change and this will be more critical for environment in future. Ji et al. (2022) investigated whether there are barriers in front of China related with climate change, to achieve the goal of being carbon neutral and the world's largest energy consuming economy by 2060, found that climate change could lead less than 5% reduction on solar energy potential. However, they concluded that this reduction will not prevent China from achieving its goals and it is possible to realize a net zero electricity system with the installation of 7.46 TW solar PV panels. In another study, Matthew et al. (2019b) examined the effects of climate change on economic activities with a panel data set covering 174 countries, and found that real output per capita growth was negatively affected by persistent changes in temperature, but had no significant relationship with precipitation. Matthew et al. (2019b) also estimated that, according to their counterfactual analysis by 2100, an increase in average global temperature of 0.04 °C per year would reduce world real GDP per capita by more than 7%, and by taking into account country-specific variability they found that the loss would increase to 13%. Sikder et al. (2022), who examine the relationship between GDP, industrialization, energy and urbanization with CO_2 using panel ARDL, concluded that a 1% increase in energy use increases CO_2 emissions by 2.32% and that CO_2 emissions in developing countries are mostly driven by GDP growth and energy use. Table 15.2 shows the related literature for economic activities (GDP per capita) and climate change.

Table 15.2 Literature for GDP per capita and components of climate change

Author	Country	Method	Period	Subject	Findings
Kahn et al. (2021) and Matthew et al. (2019b)	174 countries	Panel ARDL model	1960–2014	The effects of climate change on economic activities	Real output per capita growth is negatively affected by changes in temperature.
Paglialunga et al. (2022)	More than 150 countries	Unbalanced panel	2003–2017	The relationship between climate change and domestic inequality	Climate change distorts the income distribution.
Chen et al. (2021)	44 countries	Instrumental variable quantile regression model	1996–2018	The effects of climate change on clean energy investments	Climate change encourages investment more in countries who make more clean investments.

15.2.1.3 CO_2 Emissions and Environment

Global CO_2 emissions from industrial processes and energy combustion grew 0.9% or 321 Mt in 2022 to a new all-time high of 36.8 Gt (IEA 2023a). Ding et al. (2022) who examine the effects of carbon, fossil energy, and clean energy markets on climate change using daily carbon and energy market data, stated that the fight against climate change continues, fossil fuel-based energy is the main source of total carbon emissions, and short-term investors in the carbon market should pay more attention to abnormal fluctuations in the fossil energy market. Akan (2023), who investigates the negative and positive complementarity of total or sectoral renewable energy use in decreasing or increasing GHG emissions, discussed 20 OECD countries from 1990 to 2019. According to the results of the study, the author stated that it would be optimistic to claim that the increase in renewable energy consumption will reduce CO_2 or GHG emissions in the long run, and that the renewable energy-climate change relationship will have diverse consequences on the basis of countries, and that policy strategies should be unique to each country.

For instance in India, Das et al. (2022), who examine the relationship between REC, economic development, and environmental-sustainability with dynamic ARDL simulations, stated that there is a unidirectional causality between renewable energy consumption and environmental sustainability and also strict policy decisions should be made to reduce ecological footprints, promote renewable energy, and environmental sustainability. For other country, for examples China, Wang (2022) examined the relationship between REC and CO_2 emissions specifically for China and with the ARDL model, concluded that the effect of China's renewable energy consumption on carbon emissions is significant, and that renewable energy consumption has a negative effect on carbon emissions in the long run. Ali et al. (2022), who examine the relationship between REC and NREC, urban population, R&D expenditures, technological innovation, and carbon emission intensity from

1990 to 2019 in China, concluded that REC has a positive impact on carbon emission intensity by 0.27% and NREC by 0.75% in the long run. Also, they recommended that continued investment in renewable energy sources should include energy efficiency improvement, forest restoration, and clean energy innovation initiatives. Sun et al. (2022b) examined the relationships between REC, green innovation, and carbon neutrality with the NARDL model in China, reveals that positive shocks in green technology reduce carbon emissions, while negative shocks increase it. In addition, they found that while positive shocks in REC cause a decrease in carbon as Urban et al. (2009) examines whether the consumption of renewable and low-carbon energies will be an opportunity for China to combat climate change and replace fossil fuels, concluded that by increasing the share of renewable and nuclear energies, energy-related CO_2 emissions could be avoided from 17% to 57% until 2030. Sun et al. (2022b) also concluded that economic growth and population also contribute to higher CO_2 emissions in China.

Atsu and Adams (2021), examine the relationship of energy consumption and financial development with environmental quality in BRICS countries with CS-ARDL and FMOLS techniques, concluded that fossil fuel consumption contributes to CO_2 emissions, and REC reduces CO_2 emissions. According to these results, they stated that technological factors and government policies should be developed to encourage the innovation needed to develop low-carbon options in climate change mitigation. Kuo et al. (2022), examine the relationship between environmental quality measures and natural resources, REC, NREC, and economic growth for ten Asian nations with the CS-ARDL method, concluded that non-renewable energy sources and economic expansion increase carbon emissions. On the other hand, Kuo et al. (2022) finds that the use of renewable energy reduces carbon emissions, therefore they recommend policy recommendations such as the use of renewable energy sources reduce CO_2 emissions.

In developed countries for example, Emre Caglar (2020) examines the relationship between REC and NREC, foreign direct investment, economic growth and carbon emissions for 9 countries with the highest score in the Climate Change Performance Index (CCPI) 2018 report and named as "green category" and a long-term co-integration relationship has not been determined for these nine countries, that when NREC increases in Denmark and Finland, CO_2 emissions will increase immediately, investments in renewable energy technologies in Denmark and France lead to a decrease in CO_2 emissions in the short run, and they have come to the conclusion that the use of policy tools for these nine countries may have a permanent effect on mitigating CO_2 emissions. Using the ARDL model for Australia, Matthew et al. (2019a), take tourist arrivals, energy consumption, GDP, financial development, fixed capital formation, and population as independent variables; and CO_2 carbon emissions as dependent variable and it is emphasized that tourism, energy consumption, and GDP growth have a positive, significant, and long-term relationship with carbon emissions and are an obstacle to achieving zero carbon. The average global temperature rise is currently approximately 1.2 degrees Celsius (°C) above pre-industrial levels. This year, many parts of the world have experienced extreme weather events, including heat waves and droughts. These put significant pressure on power systems by decreasing supply from sources such as hydroelectric and thermal energy and significantly increasing demand for space cooling. Isiksal and Assi (2022) examined the effects of sustainable energy, financial progress, technological innovations, economic freedom, GDP, and environmental pollution in European Economic Area countries with PMG/ARDL and CS-ARDL methods, stated that environmental pollution is decreasing with the increase in sustainable energy use. They point out that increasing sustainable energy use through financing models and research and development could be beneficial in supporting green development.

For least developed countries Sufyanullah et al. (2022) found that the effects of carbon emissions on urbanization, energy consumption and economic growth in Pakistan using ARDL model and VECM, stated that as urbanization increases, CO_2 emissions increase, and to avoid this, preventive steps such as establishing an energy efficient and environmentally friendly transportation infrastructure and smart cities, and adopting green technology in manufacturing sectors should be taken. Suman (2021), examines the role of renewable energy technologies in reducing the effects of climate change for Nepal, states that renewable energy plays a very important role in climate change mitigation and adaptation in Nepal, and that they found that the gradual increase in REC from 2017 to 2018 reduced GHG emissions to 221.129 tCO_2e.

As shown in Fig. 15.1, in the stated policy scenario (STEPS), emissions remain broadly constant until the late 2020s and then gradually decline to 30 Gt CO_2 in 2050. Emissions would decrease by more than 5% per year in the Net Zero Emissions (NZE) to 24 Gt CO_2 in 2030 and then to net zero in 2050.Under the Advanced Commitment Scenario (APS), emissions would decrease by just over 2% per year to 31 Gt CO_2 in 2030 and then to 12 Gt CO_2 in 2050 (IEA 2023c). Table 15.3 shows the related literature for CO_2 emissions, energy, and environment.

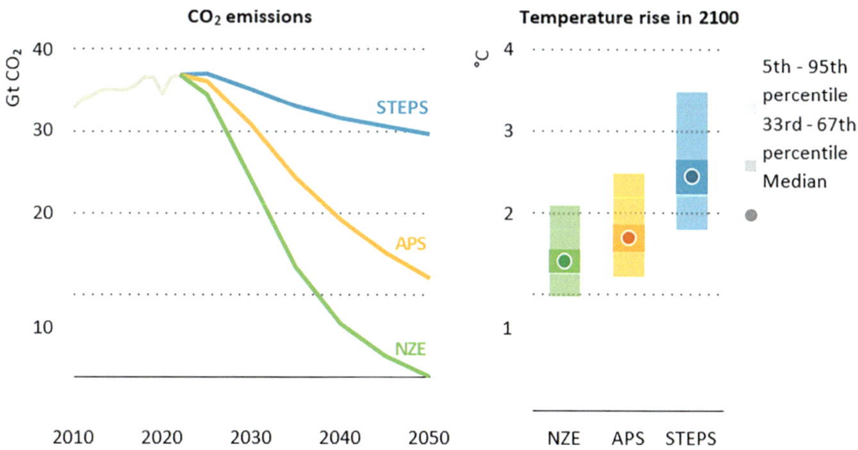

Fig. 15.1 Global CO_2 emissions related to energy and industrial processes and emergence of temperature levels before industrialization. (Source: Authors' own drawn)

Table 15.3 The related literature for CO_2 emissions, energy, and environment

Author	Country	Method	Period	Subject	Findings
Isiksal and Assi (2022)	European Economic Area	PMG/ARDL, CS-ARDL	1996–2018	Sustainable energy, financial progress, technological innovations, economic freedom and the impact of environmental pollution	With the increase in the use of sustainable energy, environmental pollution decreases.
Adedoyin et al. (2021)	Japan	ARDL-KRLS	1970–2018	The effect of the Economic Complexity Index and air transport on CO_2 emissions	An increase in GDP per capita increases CO_2 emissions, while an increase in REC decreases it.
Sufyanullah et al. (2022)	Pakistan	ARDL ve VECM	1975–2018	The relationship between urbanization, energy consumption and economic growth, and CO_2 emissions	As urbanization increases, CO_2 emissions increase.
Sun et al. (2022b)	China	NARDL	1990–2018	Renewable energy, green innovation and carbon neutrality relationships	Positive shocks in green technology and renewable energy reduce carbon emissions.
Akan (2023)	20 OECD countries	Augmented and non-augmented ARDL and vector error correction	1990–2019	The relationship between REC and CO_2 or GHG emissions	REC growth cannot reduce emissions in the long run.
Sikder et al. (2022)	23 developing countries	Panel ARDL	1995–2018	The relationship between GDP, industrialization, energy, and urbanization with CO_2	CO_2 emissions come mostly from GDP growth and energy use.
Wang (2022)	China	ARDL and Granger causality	2007–2019	The relationship between REC and carbon emissions	Renewable energy consumption has a negative impact on carbon emissions in the long run.

(continued)

Table 15.3 (continued)

Author	Country	Method	Period	Subject	Findings
Ali et al. (2022)	China	Dynamic ARDL	1990–2019	REC and NREC, with carbon emissions intensity	REC and NREC have a positive effect on the carbon emission intensity.
Atsu and Adams (2021)	BRICS countries	CS-ARDL and FMOLS	1984–2017	The relationship between energy consumption and financial development and CO_2 emissions	Fossil fuel consumption contributes to CO_2 emissions, renewable energy reduces.
Emre Caglar (2020)	Selected 9 countries	Bootstrap ARDL bound test	1970–2014	The relationship between REC, NREC, foreign direct investment, economic growth, and carbon emissions	There is no long-term cointegration relationship for CO_2 emissions, the use of policy tools has a permeant impact on CO_2 emissions.
Kuo et al. (2022)	10 Asian nations	CS-ARDL	1995–2019	Environmental quality measures, the relationship between REC and NREC and economic growth	NREC increases carbon emissions, while REC reduces it.
Khanal et al. (2022) and Matthew et al. (2019a)	Australia	ARDL	1976–2019	The relationship of tourism, energy consumption, GDP, capital and population's relationship with carbon emissions	Tourism, energy consumption, and GDP have a positive, significant and long-term relationship with carbon emissions.
Das et al. (2022)	India	Dynamic ARDL simulations	2000–2020	The relationship between REC and economic development and environmental sustainability	There is a one-way causality between REC and environmental sustainability.

15.2.2 *The Effects of Renewable and Non-renewable Energy Consumption on Environment*

15.2.2.1 Effects of Renewable and Non-renewable Energy Sources on Climate Change

Since the world's climate has a very complex structure, it is very difficult to make an accurate prediction about how climate changes will affect humanity. However, when mass warming, one of the most glaring issues of climate change, is considered alone, the resulting social dimensions take on a frightening appearance. For instance, the wind and precipitation patterns that have dominated the world climate for thousands of years and affected the lives of tens of millions of people will change significantly. Rising sea levels will threaten islands and low-lying coastal areas. One of the most important causes of climate change is the gases released during the burning of fossil fuels. A significant portion of these gases prevent the escape of solar rays, especially the warming infrared rays, reflected from the earth to the atmosphere, thus causing the areas close to the surface to warm up. Since this physical phenomenon is similar to the plastic or glass covers used in greenhouses causing the interior of the greenhouse to heat up, the gases in question are called "greenhouse gases" (GHG). The types of GHG, their rates of rise in the atmosphere and their sources are given below. Carbon dioxide (CO_2) gas: The concentration of CO_2 gas has increased by a lot (i.e., 31%) since 1750 until today. The amount of CO_2 in the atmosphere today has never been this high in the last 420,000 years or even in the last 20 million years. Over the last 20 years, around 75% of the human-caused CO_2 gas released into the atmosphere has come from the fossil fuels, and the rest has come from deforestation as a result of land use change.

These gases form a layer around our earth that makes it difficult for excess heat to escape. This is one of the reasons behind the gradual increase in global temperatures, the melting of glaciers and the recent devastating fires in Australia. Renewable energy systems, on the other hand, do not need to burn fossil fuels to produce electricity. In most cases, they do not emit harmful gases during use. For example, an electric car powered by electricity obtained from renewable energy sources does not create emissions.

15.2.2.2 The Relationship Between Climate Change, Plastics, and Energy Consumption

Plastic pollution that occurs as a result of the disposal of plastic materials into nature, which are used in all areas of life and pose a great threat to the world (Aransiola et al. 2023a; Musa et al. 2024). Plastic pollution is a big threat to our world, is also a contributing factor to climate change. In this section, the connection between climate change and plastic pollution is discussed.

The world we live in was formed billions of years ago. However, our world today is facing problems such as climate change and global warming. The increase in greenhouse gas emissions as a result of human activities causes climate change. In addition, throwing plastic waste into nature or into the sea is one of the factors that trigger climate change.

In our world, 71% of which is covered by water, plastic pollution is strongly faced in the seas and oceans. Since seas and oceans are home to a wide variety of organisms, marine creatures show the greatest effects of plastic pollution. The reason for this is that plastic scraps are discharged from cities into the seas through rivers and sewers. For this reason, plastic waste recycling is a very important issue.

15.2.2.3 How Does Plastic Pollution Affect Climate Change?

Plastic is the raw material of the products that we use in every aspect of our lives because it is cheap and easy. Today, approximately 16% of plastic waste is recycled (EEA 2020). The remaining plastic scraps are taken to the plastic yard to be burned. Most of the plastic waste that does not go to the recycling facility ends up in rivers and oceans. This situation threatens many aquatic species. At the same time, plastic causes problems such as climate change and global warming as it slowly breaks down and releases greenhouse gases. Plastic waste exposed to sunlight and heat turns into microplastic by releasing gases such as methane and ethylene into the nature and breaking into small pieces. Studies also reveal that microplastics affect the ability of marine microorganisms to absorb carbon dioxide and release oxygen At least half of the world's oxygen comes from oceans, produced mostly by plankton. These tiny organisms also absorb carbon through photosynthesis, making the oceans a major carbon sink. Microplastics negatively affect the growth, reproduction and carbon sequestration abilities of these organisms (Aransiola et al. 2023b; Auta et al. 2022). Research into climate change and plastic pollution means the harmful effects of plastic pollution on the marine environment are alarming. A warming ocean clogged with plastic creates a negative feedback loop in which plant and animal life are harmed, less carbon dioxide is absorbed, and our ability to curb climate change is further impaired. For this reason, plastic pollution can be prevented, at least partially, by plastic recycling. Figure 15.2 shows the Life cycle emissions of fossil fuel-based plastics. When looking at the entire life cycle of fossil fuel-based plastics, research shows that almost two-thirds of greenhouse gas emissions are produced in the early stages, from fossil fuel extraction to resin production (Sun et al. 2022a).

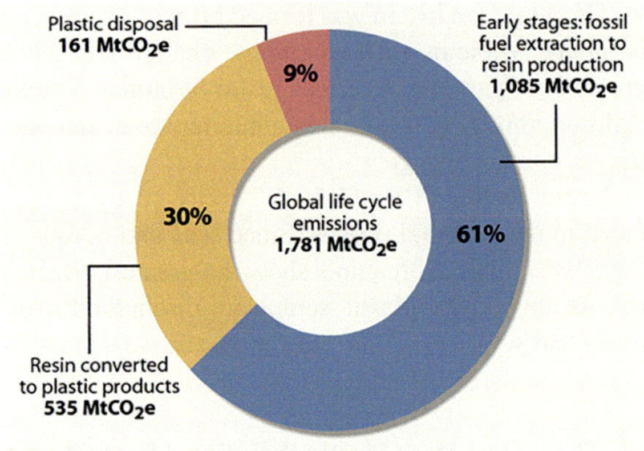

Fig. 15.2 Life cycle emissions of fossil fuel-based plastics. (Source: Authors' own drawn)

15.3 Researches on Climate Change Through the Consumption of Renewable and Non-renewable Energy in Growing Economies

15.3.1 Developed Countries

One of the studies on developed country groups, reference (Isiksal and Assi 2022), concluded that on the European Economic Area, environmental pollution decreases with the increase in sustainable energy use. In their study on Japan, Adedoyin et al. (2021) concluded that the increase in REC reduces emissions. Likewise, Atsu and Adams (2021) examined the relationship between energy consumption and CO_2 emissions for BRICS countries, concluded that renewable energy consumption reduces CO_2 emissions, while fossil fuels increase those. Matthew et al. (2019a), who analyze tourism, energy consumption and the relationship of GDP on carbon emissions for Australia, found a positive, significant, and long-term relationship between the variables.

In some of the studies conducted on developed countries, it was concluded that there was no relationship. Emre Caglar (2020) who works on developed selected 9 countries, concluded that there is no long-run cointegration relationship between REC, NREC, and CO_2 emissions, and Akan (2023) who studies 20 OECD countries, concluded that the increase in REC cannot reduce emissions in the long run.

Global warming can be defined as an extraordinary temperature increase caused by greenhouse gases (carbon dioxide-CO_2, methane-CH4, nitrous oxide-N_2O, etc.) that emerge as a result of ecological-biotic processes and human activities and mix with the air. The deterioration of the composition of these gases, which are present in certain amounts in the atmosphere, causes the long-wave rays coming from the earth to be retained and the average surface temperatures to increase excessively (IEA 2023b).

Figure 15.3 shows the Current and historic contribution to climate change (UNEP 2023). As can be seen in the figure, consumption-based emission inequality exists

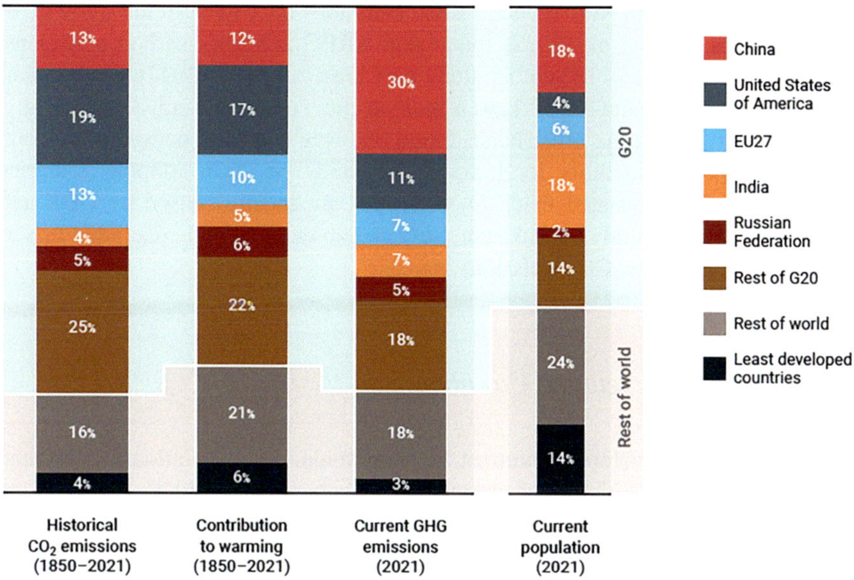

Fig. 15.3 Current and historic contribution to climate change. (Source: Authors' own drawn)

between and within countries. Globally, 10% of the population produces almost half of the greenhouse gas emissions, and two-thirds of this group live in developed countries.

15.3.2 *Developing Countries*

In studies on developing countries, the relations of climate change with REC, NREC, agricultural production, and environmental sustainability have been discussed. Among the studies examining the relationship between climate change and agricultural production in developing countries, Chandio et al. (2020) concluded that temperature and precipitation have a negative impact on agricultural production for China. Likewise, Attiaoui and Boufateh (2019) sates in their study on Tunisia that rainfall scarcity negatively affected grain production. Xiang and Solaymani (2022) concluded in their analysis on Malaysia that climate change has a negative effect on grain yield. Koondhar et al. (2021), who examines the relationship between emissions and grain production for China, concluded that an increase in carbon emissions would decrease grain production. In addition, Pickson et al. (2022), who studied 30 Chinese provinces concluded that temperature has a negative effect on rice production, but the effect of precipitation is positive.

Wang (2022) who investigated the relationship between climate change and REC, NREC, and carbon emissions in developing countries, states that REC has a negative impact on carbon emissions for China. Likewise, Sun et al. (2022b) state that positive shocks in REC reduce carbon emissions for China. In his study of 10 Asian nations, Kuo et al. (2022) found that NREC increases carbon emissions, while REC decreases it. In another study on China, Ali et al. (2022), on the other hand, state that REC and NREC have a positive effect on carbon emission intensity. Das et al. (2022), on the other hand, found one-way causality between REC and environmental sustainability for India. In his study on 23 developing countries Sikder et al. (2022), stated that CO_2 emissions are mostly caused by GDP and energy use. In their study on Pakistan, Sufyanullah et al. (2022) concluded that as urbanization increases, CO_2 emissions increase.

15.3.3 Least Developed Countries

In studies on least developed countries, there are studies on the relationship between climate change and agricultural production. Chhogyel et al. (2020) analyzed the relationship between climate change and agricultural development for Bhutan, stated that climate change causes crop loss. Ceesay and Ben Omar Ndiaye (2022) examined the relationship between food security and climate for Gambia, concluded that the change in precipitation has negative effects on food security.

Warsame et al. (2021), perform their analysis on Somalia, stated that the temperature has a negative impact on crop production, but rainfall has the opposite. Sarker et al. (2012), who analyze the effects of climate change on the production of different rice varieties in Bangladesh, stated that climate has different effects on different types of rice and climate-resistant varieties should be emphasized. On the other hand, Barrios et al. (2008), examine the relationship between agricultural production and climate change on Africa and concluded that the relationship is determinant in some African countries and not in the other.

Reference (Brini 2021), investigate the relationship between NREC, REC and climate change in studies on Least developed countries, conducted his study on 16 selected African countries and found that NREC has a negative impact on climate change, whereas REC has a positive effect.

15.4 Conclusions

When approaching the ecology-energy relationship as a scientist, engineer and economist, the important part of the work is the economic and financial dimension. In this case, if the economic aspect is looked at, the system somebody have designed must have an acceptable unit electricity production cost and be financeable, and only after these two conditions are met can the issue of ecology be discussed. Because, unfortunately, a non-competitive production has no meaning. For instance, the daily and seasonal demand fluctuations of the region where electricity will be used must be met, and while doing this, energy should not be stored because this is the costly part of the job. For this, it is necessary to have up-to-date information about all existing energy production, transmission and distribution technologies and costs. On the other hand, one should also have knowledge about the climate. Because renewable energy technologies can only be used in certain climatic conditions. This is the supply aspect of the business. Let's say the cleanest energy was produced at the cheapest price, in this case also dealing with the demand side of the issue is necessary. After all, there are users of this produced energy. It would be remiss not to address the issue of how these users can use electricity more efficiently. This is where behavioral economics comes into play. For example, a pricing policy that prioritizes energy efficiency can be considered. Because you cannot solve everything with supply-based measures, the demand side of the business must also be utilized to achieve the optimum result.

In order to understand the relationship between ecology and energy, to produce scientific works on this subject and to be beneficial to humanity, it is necessary to have a holistic perspective that includes almost every branch of science. Understanding this is more important than anything right now. Otherwise, it is not possible to get anywhere unless the solutions produced are in demand, that is, unless we can provide maximum benefit to people. A recycling industry designed in accordance with the rules is such a vital issue for a country, for both geopolitics, raw material supply security, therefore sustainable export industry and for many other issues. For instance, accumulation of plastics is a major environmental problem for ecosystem and disposing of waste plastics through reuse and recycle is crucial but not sufficient in today's world and waiting to be solved by conscious utilization and technological developments.

References

Acaroğlu H, Güllü M (2022) Climate change caused by renewable and non-renewable energy consumption and economic growth: a time series ARDL analysis for Turkey. Renew Energy 193:434–447

Adedoyin FF, Ozturk I, Bekun FV, Agboola PO, Agboola MO (2021) Renewable and non-renewable energy policy simulations for abating emissions in a complex economy: evidence from the novel dynamic ARDL. Renew Energy 177:1408–1420

Akan T (2023) Investigating renewable energy–climate change nexus by aggregate or sectoral renewable energy use? Environ Sci Pollut Res 30:2042–2060

Ali U, Guo Q, Kartal MT, Nurgazina Z, Khan ZA, Sharif A (2022) The impact of renewable and non-renewable energy consumption on carbon emission intensity in China: fresh evidence from novel dynamic ARDL simulations. J Environ Manag 320:115782

Aransiola SA, Ikhumetse AA, Babaniyi BR, Abioye OP, Oyedele OJ, Falade NO (2023a) Phytoaccumulation of micro and nanoplastics: root uptake. In: Maddela NR, Reddy KV, Ranjit P (eds) Micro and nanoplastics in soil. Springer, Cham. https://doi.org/10.10007/978-3-031-21195-9_8

Aransiola SA, Victor-Ekwebelem MO, Ajiboye AE, Leh-Togi Zobeashia SS, Ijah UJJ, Oyedele OJ (2023b) Ecological impacts and toxicity of micro and nanoplastics in agroecosystem. In: Maddela NR, Reddy KV, Ranjit P (eds) Micro and nanoplastics in soil. Springer, Cham. https://doi.org/10.10007/978-3-031-21195-9_10

Atsu F, Adams S (2021) Energy consumption, finance, and climate change: does policy uncertainty matter? Econ Anal Policy 70:490–501
Attiaoui I, Boufateh T (2019) Impacts of climate change on cereal farming in Tunisia: a panel ARDL-PMG approach. Environ Sci Pollut Res Int 26:13334–13345
Auta SH, Abioye OP, Aransiola SA, Bala JD, Chukwuemeka VI, Hassan A, Aziz A, Fauziah SH (2022) Enhanced microbial degradation of PET and PS microplastics under natural conditions in mangrove environment. J Environ Manag 304(2022):114273. https://doi.org/10.1016/j.jenvman.2021.114273
Barrios S, Ouattara B, Strobl E (2008) The impact of climatic change on agricultural production: is it different for Africa? Food Policy 33:287–298
Brini R (2021) Renewable and non-renewable electricity consumption, economic growth and climate change: evidence from a panel of selected African countries. Energy 223:120064
Campagnolo L, De Cian E (2022) Distributional consequences of climate change impacts on residential energy demand across Italian households. Energy Econ 110:106020
Ceesay EK, Ben Omar Ndiaye M (2022) Climate change, food security and economic growth nexus in the Gambia: evidence from an econometrics analysis. Res Globalization 5:100089
Chandio AA, Jiang Y, Rehman A, Rauf A (2020) Short and long-run impacts of climate change on agriculture: an empirical evidence from China. Int J Clim Change Strateg Manag 12:201–221
Chen X, Fu Q, Chang C-P (2021) What are the shocks of climate change on clean energy investment: a diversified exploration. Energy Econ 95:105136
Chhogyel N, Kumar L, Bajgai Y, Hasan MK (2020) Perception of farmers on climate change and its impacts on agriculture across various altitudinal zones of Bhutan Himalayas. Int J Environ Sci Technol 17:3607–3620
Das N, Bera P, Panda D (2022) Can economic development & environmental sustainability promote renewable energy consumption in India? Findings from novel dynamic ARDL simulations approach. Renew Energy 189:221–230
Ding Q, Huang J, Zhang H (2022) Time-frequency spillovers among carbon, fossil energy and clean energy markets: the effects of attention to climate change. Int Rev Financ Anal 83:102222
EEA (2020) Air quality in Europe. EEA
Emre Caglar A (2020) The importance of renewable energy consumption and FDI inflows in reducing environmental degradation: bootstrap ARDL bound test in selected 9 countries. J Clean Prod 264:121663
IEA (2023a) CO_2 emissions in 2022. IEA
IEA (2023b) Renewables 2022. IEA
IEA (2023c) Word energy outlook. IEA
IPCC (2016) Renewable energy sources and climate change mitigation, Special report of the intergovernmental panel on climate change. Cambridge University Press
Isiksal AZ, Assi AF (2022) Determinants of sustainable energy demand in the European economic area: evidence from the PMG-ARDL model. Technol Forecast Soc Chang 183:121901
Ji L, Wu Y, Sun L, Zhao X, Wang X, Xie Y, Guo J, Huang G (2022) Solar photovoltaics can help China fulfill a net-zero electricity system by 2050 even facing climate change risks. Resour Conserv Recycl 186:106596
Kahn ME, Mohaddes K, Ng RN, Pesaran MH, Raissi M, Yang JC (2021) Long-term macroeconomic effects of climate change: a cross-country analysis. Energ Econ 104:105624. https://doi.org/10.1016/j.eneco.2021.105624
Koondhar MA, Aziz N, Tan Z, Yang S, Raza Abbasi K, Kong R (2021) Green growth of cereal food production under the constraints of agricultural carbon emissions: a new insights from ARDL and VECM models. Sustain Energy Technol Assess 47:101452
Kuo Y, Maneengam A, Phan The C, Binh An N, Nassani AA, Haffar M, Qadus A (2022) Fresh evidence on environmental quality measures using natural resources, renewable energy, non-renewable energy and economic growth for 10 Asian nations from CS-ARDL technique. Fuel 320:123914

Matthew EK, Kamiar M, Ryan NCN, Hashem P, Mehdi R (2019a) Long-term macroeconomic effects of climate change: a cross-country analysis. Energy Policy Research Group, Cambridge Judge Business School, University of Cambridge

Matthew EK, Kamiar M, Ryan NCN, Pesaran MH, Mehdi R, Jui-Chung Y (2019b) Long-term macroeconomic effects of climate change: a cross-country analysis, IMF Working Papers 19. International Monetary Fund

Mercan M, Karakaya E (2015) Sera gazi saliminin azaltiminda alternatif politikalarin ekonomik maliyetlerinin incelenmesi: Türkiye için genel denge analizi. Erciyes Üniversitesi İktisadi ve İdari Bilimler Fakültesi Dergisi 0:123–159

Musa IO, Auta HS, Ilyasu US, Aransiola SA, Makun HA, Adabara NU, Abioye OP, Aziz A, Jayanthi B, Maddela NR, Prasad R (2024) Micro and nanoplastics in environment—degradation, detection, ecological impact. Int J Environ Res 18:1. https://doi.org/10.1007/s41742-023-00551-9

Paglialunga E, Coveri A, Zanfei A (2022) Climate change and within-country inequality: new evidence from a global perspective. World Dev 159:106030

Pickson RB, He G, Boateng E (2022) Impacts of climate change on rice production: evidence from 30 Chinese provinces. Environ Dev Sustain 24:3907–3925

Sarker MAR, Alam K, Gow J (2012) Exploring the relationship between climate change and rice yield in Bangladesh: an analysis of time series data. Agric Syst 112:11–16

Sikder M, Wang C, Yao X, Huai X, Wu L, Kwame Yeboah F, Wood J, Zhao Y, Dou X (2022) The integrated impact of GDP growth, industrialization, energy use, and urbanization on CO2 emissions in developing countries: evidence from the panel ARDL approach. Sci Total Environ 837:155795

Sufyanullah K, Ahmad KA, Sufyan Ali MA (2022) Does emission of carbon dioxide is impacted by urbanization? An empirical study of urbanization, energy consumption, economic growth and carbon emissions—using ARDL bound testing approach. Energy Policy 164:112908

Suman A (2021) Role of renewable energy technologies in climate change adaptation and mitigation: a brief review from Nepal. Renew Sust Energ Rev 151:111524

Sun X, Xie M, Mai L, Zeng EY (2022a) Biobased plastic: a plausible solution toward carbon neutrality in plastic industry? J Hazard Mater 435:129037

Sun Y, Razzaq A, Sun H, Irfan M (2022b) The asymmetric influence of renewable energy and green innovation on carbon neutrality in China: analysis from non-linear ARDL model. Renew Energy 193:334–343

UNEP (2023) Emission gap record, broken record. UNEP

Urban F, Benders RMJ, Moll HC (2009) Renewable and low-carbon energies as mitigation options of climate change for China. Clim Chang 94:169–188

Wang L (2022) Research on the dynamic relationship between China's renewable energy consumption and carbon emissions based on ARDL model. Res Policy 77:102764

Warsame AA, Sheik-Ali IA, Ali AO, Sarkodie SA (2021) Climate change and crop production nexus in Somalia: an empirical evidence from ARDL technique. Environ Sci Pollut Res 28:19838–19850

Xiang X, Solaymani S (2022) Change in cereal production caused by climate change in Malaysia. Eco Inform 70:101741